Library of
Davidson College

Growth and Reforms in Centrally Planned Economies

George R. Feiwel

The Praeger Special Studies program—utilizing the most modern and efficient book production techniques and a selective worldwide distribution network—makes available to the academic, government, and business communities significant, timely research in U.S. and international economic, social, and political development.

Growth and Reforms in Centrally Planned Economies
The Lessons of the Bulgarian Experience

PRAEGER SPECIAL STUDIES IN INTERNATIONAL ECONOMICS AND DEVELOPMENT

Praeger Publishers New York Washington London

Library of Congress Cataloging in Publication Data

Feiwel, George R
 Growth and reforms in centrally planned economies.

 (Praeger special studies in international economics and development)
 Bibliography: p. 338
 1. Bulgaria—Economic policy—1944-   I. Title.
HC403.F44   330.9'4977'03   76-12849
ISBN 0-275-23330-8

PRAEGER PUBLISHERS
111 Fourth Avenue, New York, N.Y. 10003, U.S.A.

Published in the United States of America in 1977
by Praeger Publishers, Inc.

All rights reserved

© 1977 by Praeger Publishers, Inc.

Printed in the United States of America

for Abram Bergson

PREFACE

Men do not live by economic growth, statistical GNP, or even economics alone. An economy's resources could be used for various ends—good or evil—depending on the decision makers' scales of values. Given ends may be achieved more or less efficiently, at higher or lower human and material costs, and with more or less desirable or adverse nonmaterial consequences. Efficiency is not an end of economic activity; merely a means for accomplishing that end, however determined. Efficiency does matter because a more efficacious husbandry of resources enables the economy to achieve a higher realization of the ends of economic activity or—what is not equivalent (for example, unemployment of labor)—realization of given ends with fewer resources.

The normative question concerns the ends to which resources ought to be deployed and how the product of economic activity ought to be distributed. There is usually a clash among various objectives of economic policy, for instance, growth, consumption, efficiency, full employment, and more egalitarian income distribution. The ultimate aim of economic activity ought to be to deploy resources to enhance continuously living standards and the quality of life of the population as a whole. This is not tantamount to maximization of immediate consumption, mainly because without a prudent accumulation of capital, no sustained improvement of living standards could take place. The intertemporal allocation of resources between present needs and amenities and those of the future is a key problem. A compromise has to be struck to the usual accumulation-consumption dilemma. While the short-run effects of the increase in the accumulation rate are manifestly clear, possibly—but not always—a higher accumulation rate can yield a higher standard of living in the long run. Capital accumulation derives its value from the future living standards it supports, though for reasons of national prestige or imitation, a particular rate of accumulation or specific kinds of investments are often components of the preference function itself.

In the post-World War II period a high premium has been placed on the rate of economic growth. But a high growth rate of output (aggregate or per capita) can hardly be the sole desideratum; nor is maximization of the growth rate a sensible aim of economic policy. The highest attainable growth rate cannot be identified as the touchstone of economic progress. A higher annual growth rate is not necessarily preferable to a lower one, if only because it might have been achieved at the detriment of the long-term growth rate. If a country outperforms another or registers one year a higher growth rate than that recorded in the previous year, these are not sufficient indications of improvement. The growth rate as such is essentially a quantitative measure of economic change that understates the critical and refractory qualitative dimensions.

Conventional GNP (or its variants) is a measure of production and not one of consumption or of economic welfare. Nor is welfare yielded only by produced goods and services. Centering on the growth rate as a criterion of progress detracts attention from the following: composition of production, consumption, the quality of life, income distribution, and costs (including nonmaterial costs) of achieving output. A significant share of production may well be illusory from the standpoint of enhancing welfare; some may even be deleterious or perilous. Nor can GNP be satisfactorily adapted for measuring net economic welfare, notwithstanding recent pioneering attempts by William Nordhaus and James Tobin to do so. Very tentative estimates for the United States (1929-65) have shown that net economic welfare has been growing at a considerably lower rate than net national product. If—in addition to all the other complicated recalculations of Soviet-type economies' (STEs) GNP required for comparisons with the West—one were to recompute it to reflect net economic welfare (a task yet to be undertaken by some courageous souls), one would probably find that the gap between the rates of growth of net national product and net economic welfare would be even larger than that in the West. This would be due, in part, to the following: discriminatory allocation of inputs to consumption in both quantitative and qualitative terms; production of much wasteful, plan-satisfying, low quality output; substitution of unwanted goods for those desired, forced on the consumer by shortages and maldistribution; the very limited availability and poor quality of services; and all sorts of shopping frustrations.

There are different ways of achieving identical growth rates. These ways are important for society not only because they bear on working and living conditions at a given time; but insofar as they affect the will to produce, creativity, and future knowledge and skills (these ways also influence future growth rates and living standards). Certain ways may tend to produce transitory high growth rates, but they may be particularly destructive to the human element—an indictment against them not only on moral but also on economic grounds. There are indeed ways of increasing output that may be economical (from the standpoint of efficiency), yet are reprehensible in terms of widely held human values. It might be advantageous to forfeit higher growth rates for the sake of improving the quality of life, for more welfare-oriented output composition, and for more harmonious and sustained growth.

All this is not an accolade to those who argue that higher growth rates of GNP result in "gross national pollution," that they are achieved at the price of irreparable decay of environment and welfare, and that ipso facto a smaller GNP would result in "greater national pleasure." This is not the place to resolve the controversy about the possibility, desirability, and necessity of economic growth. Whatever the pros and cons are, growth does matter, particularly to those who have not yet experienced its delights or banes. Moreover, in practice it is difficult to make sweeping changes in allocation of resources and distribution of consumption. Such alterations are more feasible from the increment of GNP. Indeed, a smaller GNP will only aggravate the social demerits of a system.

As shown authoritatively by Simon Kuznets, a distinctive characteristic of modern economic growth in industrialized capitalist economies is that the high growth rates of per capita national output were accomplished primarily by improvements in quality—and to a much lesser extent by increased quantity—of inputs; that is, essentially by a rise in productivity traceable to rapid advances in technical, organizational, and managerial know-how. Thus the high rates of technical advance and efficiency have been partly instrumental in maintaining relatively modest shares of capital formation and high shares of consumption in national product. By contrast the very impressive growth rates of STEs have been accomplished primarily by growing quantities of inputs with disappointing increases in productivity.

It is hazardous to make comparisons particularly when a host of various factors is at work. However, some striking similarities between the STEs and Japan may be noted: both succeeded in recording exceptionally high growth rates, both channeled a relatively high share of GNP to capital formation, and both gave priority to growth promoting activities. Differences were many. Japan energetically exploited the advantages of backwardness; its investments primarily embodied creatively adapted superior foreign technology. The Japanese exhibited a strong entrepreneurial spirit. Government policy and flexible working arrangements facilitated adjustment to change and a forceful penetration of foreign markets. The Japanese were phenomenally successful in extracting more marketable output out of every unit of labor and capital combined. The STEs failed miserably in all these areas where the Japanese have been so successful.

Since the early 1960s the STEs' growth momentum and rate of increase of productivity have flagged. Growth can no longer be propelled by relying on increasing commitment of resources due to the relative exhaustion of resources and to the proliferating competitive claims on them. More recently there have been disquieting signs about the continuation of Japan's success story. It appears that to some extent the remarkable Japanese and STEs' growth momentums were facilitated by "transitory" opportunities. In Japan the advantages of backwardness are receding, labor is becoming more demanding, and the need to improve housing and public services can no longer be postponed. Similarly in the STEs, increasing attention has to be paid to the worker-consumer, limiting the share of GNP diverted to capital formation. However, aside from other possible improvements, the STEs still have a vast source of potential growth in better allocation of the investment fund, in improvement of utilization of existing capacity, and in borrowing and adapting foreign know-how—if only they could devise the means of effectively doing so.

In the last decade or so the STEs have sought an additional source of growth in reform of the functioning system. The institutional framework of a social system is a fundamental component of its economic dynamics. Among the multiple determinants of the rate and pattern of economic activity, the economy's functioning system is a weighty but not necessarily overriding factor. The efficacy of the working arrangements for resource allocation and the behavior of microunits are not independent of macroeconomic conditions. Working arrangements are merely means for realizing given ends. The merit of such arrangements can be assessed only in the light of how well it performs this function. The end-means structure requires a congruence among the functioning system and the other components of the economic process. Rationality requires that dissonance among the components of the economic process be removed by modifying the discordant components to satisfy the conditions of consistency and efficiency. Efficiency is a stronger requirement than consistency and coherence. Mobilization and steering of resources into appropriate activities and keeping the production flows and the economy moving are signs of economic vitality; economic efficiency means employing these resources to obtain as much output as possible from them by applying the best available techniques. The choice of an appropriate growth rate, structure, techniques of production, and of better working arrangements should jointly produce better results than those that could be achieved by attacking the economic riddle on only one front.

Theoretically the centrally planned economy provides opportunities for a wide gamut of arrangements for resource allocation, of which the STE is only one alternative. "Planning or the market" as diametrically opposed instruments of resource allocation is largely an anachronism. The formidable problem is the proper blend of plan and market to achieve most effectively the aims of economic activity, however determined. But given institutions are not set up only because they are the best "production techniques," but also because decision makers have usually strong preferences for specific institutional arrangements. Aside from the folklore of these arrangements and specific vested interests for maintaining them, rigidities develop so that even intolerably ineffective institutions tend to perpetuate themselves. Redesign of a system or its components meets with formidable obstacles. On the other hand, when the economy is under stress, there is often a tendency to tinker with separate components of the system, which either could remove some dissonance, or could introduce new inconsistencies. These might later be eliminated by reverting to the old ways of doing things, or by progressive changes in the other components.

In STEs conformism to the notorious "command system" is also an anachronism. At the present juncture the focus is on different ways of organizing the socialist economy. Yet in practice the STEs have evinced remarkable resistance to fundamental organizational innovations. The substance of some of the practical changes is often exaggerated. In fact, notwithstanding the significance of their differences, the various reforms implemented in practice at different times in different STEs have been no more than bungling half-measures. We have yet to see a welfare-oriented, consistent, and coherent reform of the functioning system in a STE, accompanied by a revision of the growth strategy and by a system of checks and balances to curb the omnipotence of the system's directors. Above all, economic reform has to embrace the whole gamut of economic activity and to harmonize all segments of activity. The STE system has been challenged for its failure to live up to the ideals of socialism and for the dehumanizing effect of the industrialization rush. To qualify as reforms, system changes should palpably improve the daily conditions of life and work of the mass of population; they should stimulate the willingness to produce and innovate at all levels; and should profoundly affect the structure of wants and activate an endogenous mechanism for their satisfaction.

This study is a part of a larger project on growth, plan, and market of which several volumes have already been published. Initially the research was centered on the aspirations and performance of economies that adopted the Soviet model of development and working arrangements, with the aim of expanding the project to embrace the experience of countries with other social and economic systems. Every successive stage is a modest attempt at distilling from a country's experience lessons of wider interest and validity and at generalizing on the varieties of experiences studied.

In conformity with common usage, I have employed the term STE to designate the USSR and the East European countries that have followed the same pattern of development. The term is obviously a misnomer, for similarity should not conceal diversity. The so-called Soviet model has, of course, been applied to countries at different stages of development, with different factor endowment, and of different historical and traditional backgrounds. While there is some conspicuous uniformity, these countries' experiences differ and certain distinct characteristics are instructive. Not only were there differences in adaptation of the model in different countries and at different times, but there are varying opportunities in different countries for continuing growth along more or less the same lines, or for changing the course of economic activity. Also, the system's directors' perception of feasibility constraints differ.

Having previously examined the Soviet model and performance, and its application to the industrialized Czechoslovak economy and the "intermediate" Polish economy, an attempt is made in this study to analyze the results of the adaptation of this model to an underdeveloped country—Bulgaria. After the end of World War II, Bulgaria, together with Rumania, was the least developed country to come within the Soviet orbit, but unlike Rumania it was poorly endowed with natural resources. Bulgaria had a relatively ample labor supply, with very low living standards. The Soviet mode of development and working arrangements were vigorously adopted, but in contrast to the other STEs from the outset Bulgaria has paid more attention to the development of agriculture. Bulgaria has one of the highest investment rates in the world, characterized by the abruptness of the take-offs and the velocity of investment expansion, and registers impressive growth rates as measured by the traditional criteria. The rapidity of the transformation and the scale of the structural break were remarkable, but at high material and nonmaterial costs whose legacy haunts the system to this day. The economic costs were mitigated by what appears to have been generous treatment of Bulgaria by the USSR (and under its aegis by some of the other STEs), so that rather than being exploited, Bulgaria was to some extent the "exploiter." This was partly due to the historical umbilical cord between Bulgaria and Russia and to the strategic role of Bulgaria within the Balkans. Also, both in its domestic and foreign policy, Bulgaria has been one of the most subservient and docile satellites of the USSR.

In this study an attempt was made to examine Bulgarian growth performance in comparison with those of other STEs and, to a more limited extent, of STEs in comparison with Western countries at a similar level of development. While it is tempting to compare countries at similar levels of development, one should bear in mind that it is the relative performance of a STE within the bloc that really matters for economic policy decisions.

Different approaches to planning assign different scope to the capacity, limitations, and efficiency of deliberate decision making to transform or control the course of economic life. Planning techniques and arrangements are influenced by planning philosophies, and have to be congruent with growth strategy. The meaning and content of Bulgarian economic reforms has been shifting. It would have been interesting to see what would have happened had more radical market-oriented reforms been implemented. For that purpose a fairly thorough study of the Hungarian reforms was undertaken, which for reasons of space had to be excluded from this study, but which underlies the analysis and conclusions.

ACKNOWLEDGMENTS

It is always a pleasant task to thank those whose help has in some way contributed to improving the content of a study, but I am bankrupt in discharging this task. Obviously, all those who have influenced me during my formative years and whose generous help has been acknowledged in my previous studies have left an imprint on this work as well.

During the inordinately long gestation process of this study I had the good fortune to exchange views with many members of the economic fraternity in the United States and abroad. My thanks go out to them collectively as well as to the participants of various seminars that I had the privilege of conducting and to my students. I would be remiss not to single out Harvard's Russian Research Center, with which I have had friendly ties for many years. I have benefited from its stimulating intellectual atmosphere and working conditions during my numerous visits. I had the pleasure of visiting Bulgaria on several occasions. The hospitality of its people was warm and overwhelming —in inverse relation to the officials' willingness to provide practical information on what is really being done. My discussions with some Bulgarian economists afforded some very helpful insights into the special problems faced by Bulgaria and clarified some fuzzy issues. I was able to fill the information gap at various institutes, organizations, and libraries (including some in Eastern Europe).

From the very outset this study has benefited from the encouragement and sympathetic interest of Abram Bergson, to whom it is gratefully and respectfully dedicated. I have profited from the advice and exchange of ideas with certain people on specific subjects discussed in this book. In particular I wish to thank Joseph Berliner, Alexander Erlich, Simon Kuznets, Harvey Leibenstein, and Joan Robinson. I must reiterate here my intellectual debt to the late Michal Kalecki, with whom I had the privilege of discussing problems of Bulgarian development and growth and planning in general in Cambridge in the spring of 1969.

This study was originally undertaken under a grant from the National Science Foundation. Research abroad was in part supported by the American Council for Learned Societies and the American Philosophical Society. I also benefited from financial support from the College of Business Administration, University of Tennessee, and its Center for Business and Economic Research provided assistance in typing statistical tables. The financial assistance was helpful and is gratefully acknowledged. The real cost of countless hours of labor and far from "welfare-enhancing" scrutiny of the Bulgarian press was undertaken by my wife who sustained me throughout the arduous task of researching and completing this study. Whatever merits this book might have are due to those who have rendered a helping hand; all its faults and fallacies can be attributed directly to me.

CONTENTS

	Page
PREFACE	vi
ACKNOWLEDGMENTS	xiii
LIST OF TABLES	xix
LIST OF FIGURES	xxvi
LIST OF ABBREVIATIONS AND ACRONYMS USED IN TEXT	xxvii
LIST OF BIBLIOGRAPHICAL ABBREVIATIONS	xxviii

Chapter

1	HIGH-PRESSURE ECONOMY	1
	Center-Periphery Interaction	1
	Determining Economic Growth Rate	3
	Soviet Growth Strategy	8
	Labor Productivity	10
	Realistic Planning	13
	Notes	14
2	GROWTH AND INSTABILITY	16
	Comparative Stages of Development	16
	Growth and Composition of Output	19
	Structural Change	20
	Shifts in Bulgarian Industrialization	22
	Industrial Growth Performance	23
	Agricultural Performance	28
	Capital Formation	29
	Consumption and Accumulation	29
	Investment	31
	Distribution of Investment	32
	Investment in Industry	35

Chapter	Page
Fluctuations and Deceleration of Growth Rates	38
Postwar Reconstruction and Industrialization	43
The New Course	44
The Second Industrialization Rush and Downturn	45
Decelerating Growth Rates	48
Human Resources and Employment	49
Growth of the Labor Force	50
Sectoral Distribution of Labor Force	52
Emergent Labor Shortages	54
Notes	56
3 RESOURCE UTILIZATION	**61**
Productivity	61
Labor Productivity	62
Capital Productivity	64
Foreign Trade	67
Growth Rates	69
Structure	71
Intra-CMEA Trade	73
Trade with the West	75
Bulgarian Trade Relations	76
The Traditional Planning System	81
Notes	85
4 ECONOMIC REFORM: TRANSITIONAL PERIOD	**91**
Theses on the New System of Planning and Management	91
Planning	93
The Financial System	94
Prices	95
Experience with Reform Introduction	96
The 1968 Reform in Reverse	100
Notes	103
5 PRICES AND DECISIONS	**105**
The Dual-Price System	108
Turnover Tax	113
On Reform Proposals	115
Retrospect on Price Revisions	120
The 1967-71 Price Revision	122
A Note on Agricultural and Retail Prices	129

Chapter		Page
	Domestic and Foreign Trade Prices	131
	Notes	135
6	REDESIGN OF THE SYSTEM	139
	Administrative Structure	139
	The Association	140
	Planning	143
	Profit Distribution	144
	Financing of Investments	145
	Special Purpose Funds for Financing Technical Progress	148
	Financing of Working Capital	149
	The Counterplan	149
	Wages and Premiums	152
	Foreign Trade Management	155
	Concluding Remarks	158
	Notes	162
7	CONTINUITY AND CHANGE	164
	Sketch of Economic Performance	164
	The 1966-70 Five-Year Plan	166
	The 1971-75 Five-Year Plan	167
	National Income	168
	Industrial Output	168
	Productivity	171
	Investment	172
	Living Standards	175
	Qualitative Indexes	176
	Labor Productivity	178
	Output Quality	179
	Measurement Problems	180
	Labor Problems	181
	Malfunctioning of Supply	183
	Machine Building and Foreign Trade	186
	Technical Progress	192
	Central Planning	196
	Computerization	196
	Plan Construction	199
	The Ministries	200
	Complexes	201
	Postulates for Future Development	203
	Notes	204

Chapter		Page
8	ACCENT ON LIVING STANDARDS?	210
	Consumption Structure	212
	Supply of Consumer Goods	216
	Housing	220
	Collective Consumption	221
	Wage, Price, and Tax Policy	222
	Alienation and Economic Crimes	227
	Notes	230
9	PLANNING PHILOSOPHIES AND REFORMS	233
	Growth and Reform	233
	The Socialist Economy	235
	Varieties of Socialist Economies	236
	Planometrics	240
	Incentives	241
	Prices	243
	Inflation	248
	The Association	250
	Technical Progress	252
	Pressures and Slack	253
	Attitudes to Reform	254
	Welfare and Democracy	257
	Notes	259

Appendix

A	APPENDIX TO CHAPTER 6: REVENUE GENERATION AND DISTRIBUTION	261
B	STATISTICAL APPENDIX	264
	SELECTED BIBLIOGRAPHY	338
	ABOUT THE AUTHOR	347

LIST OF TABLES

Table		Page
5.1	Relative Share of A and B in Industrial Output, Inputs, and Financial Accumulation, 1957-61	110
5.2	Changes in Profitability, as Percentage of Costs, Between 1956 and 1961	121
5.3	Profitability, as Percentage of Costs, of Industrial Branches, 1959-62	123
5.4	Industrial Enterprises According to Profitability, as Percentage of Costs, 1962	124
7.1	Percentage Annual Growth Rates of Machine Building in CMEA Countries, 1966-75	170
8.1	Per Capita Personal Consumption in CMEA Countries in 1970	213
8.2	Changes in the Structure of per Capita Personal Consumption, 1960-70	214
8.3	Estimated Stocks of Selected Consumer Durables in Some CMEA Countries, 1960-70	215
8.4	Growth of the Collective Consumption Fund, 1952-71	222
8.5	Nominal Wages of Blue- and White-Collar Workers, 1948-67	223
B.1	Area and Population of CMEA Countries	264
B.2	GNP per Capita of Selected European Countries in U.S. Dollars, 1955	265
B.3	Estimates of GDP (Gross Domestic Product) per Capita in CMEA and Some Western Countries, 1950-67	266
B.4	Per Capita GDP, Investment and Consumption in Eastern Europe and Some Western Countries in 1965	267

Table		Page
B.5	Comparisons of Estimated per Capita Level of Industrial Output in Eastern Europe, Selected Years, 1955-67	268
B.6	Alton-Project Estimates of Levels of per Capita Agricultural Output: CMEA and United States, 1959, 1966, and 1970	269
B.7	Average Annual Growth Rates of National Income in CMEA Countries, 1951-74	270
B.8	Fluctuating Annual Growth Rates of National Income in CMEA Countries, 1951-74	271
B.9	Growth and Fluctuations of per Capita National Income in CMEA Countries, 1951-68	272
B.10	Alton-Project (Provisional) Estimates of GNP Dynamics in Eastern Europe, 1950-72	273
B.11	Dynamics of National Income in Eastern Europe, 1950-72	274
B.12	Comparison of Official and Alton-Project Revised Estimates of Growth of Bulgarian National Product, 1950, 1955, and 1960-74	275
B.13	Shares of Industry (1), Construction (2), and Agriculture (3) in National Income Produced in CMEA Countries, 1950-72	276
B.14	Average Annual Rates of Growth of Value Added in Key Sectors of CMEA Countries, 1951-65	277
B.15	Dynamics of Value Added in Key Sectors of CMEA Countries, 1950-71	278
B.16	Alton-Project Estimates (Revised) of Dynamics of GNP by Sectors of Origin in Eastern Europe, 1965-74	279
B.17	Alton-Project Estimates (Revised) of Dynamics of GNP by Sectors Origin in Bulgaria, 1950-74	280

Table		Page
B.18	Distribution of National Income into Consumption (1) and Accumulation (2) in Some CMEA Countries, 1950–73	281
B.19	Average Annual Rates of Growth of Consumption and Accumulation in Some CMEA Countries, 1951–67	282
B.20	Rates of Growth of Consumption (1) and Accumulation (2) in Some CMEA Countries, 1955–73	283
B.21	Growth of National Income, Consumption and Accumulation in Bulgaria, 1939–70	284
B.22	Percentage Distribution of Accumulation in Some East European Countries, 1960–64	285
B.23	Growth of Investments in CMEA and Some Western Countries, 1950–71	286
B.24	Average Annual Rates of Growth of Investments in CMEA Countries, 1951–67	286
B.25	Annual Rates of Growth of Investment in CMEA Countries, 1956–73	287
B.26	Sectoral Distribution of Investments in Some CMEA and Western Countries, 1951–73	288
B.27	Sectoral Distribution of Investments in Bulgaria, 1949–73	290
B.28	Branch Distribution of Investment in Industry in CMEA Countries, 1961–67	291
B.29	Branch Distribution of Fixed Assets in Industry in Bulgaria, 1952–70	292
B.30	Indexes of Production and Inventory Dynamics in Some CMEA Countries, 1961–63	293
B.31	Share of Increase of Inventories in Distribution of National Income in Some East European and Western Countries, 1960–64	293

Table		Page
B.32	Average Annual Growth Rates of Gross Industrial Output in CMEA Countries, 1951-74	294
B.33	Fluctuating Annual Growth Rates of Gross Industrial Output in CMEA Countries, 1951-74	294
B.34	Growth and Fluctuations of per Capita Gross Industrial Output in CMEA Countries, 1951-68	295
B.35	Shares of Producer Goods (A) and Consumer Goods (B) in Industrial Output in Some CMEA Countries, 1950-72	296
B.36	Fluctuating Annual Growth Rates of Producer (A) and Consumer (B) Goods Output in Some CMEA Countries, 1951-68	297
B.37	Structure of Socialized Industry in Some CMEA Countries, 1950-67	297
B.38	Fluctuating Growth Rates of Output of Various Branches of Socialized Industry in Bulgaria, 1951-70	298
B.39	Alton-Project Preliminary Estimates of Industrial Growth in Eastern Europe Compared to Some Western Countries, Prewar to 1967	299
B.40	Indexes of Industrial Production in Eastern Europe in 1973	299
B.41	Comparisons of Alton-Project (Provisional) Independently Constructed Indexes of Growth of Bulgarian Industrial Output with Official Indexes, 1939 and 1948-65	300
B.42	Comparison of Official and Alton-Project Estimates of Growth of Production by Branches of Industry in Bulgaria, 1967-72	301
B.43	Alton-Project Estimated Index of Industrial Production in Bulgaria, 1963-74	302
B.44	Shares of Various Countries in the Output of Some Industrial and Agricultural Products in CMEA, 1950-67	303

Table		Page
B.45	Index of Agricultural Output of CMEA Countries, 1950-68	304
B.46	Index of Agricultural Output of CMEA Countries, 1960-73	304
B.47	Fluctuating Growth Rates of Agricultural Output in CMEA Countries, 1951-73	305
B.48	Alton-Project Estimates of Levels of Agricultural Output per Agricultural Worker and per Hectare of Agricultural Land in CMEA and United States, 1959, 1966, and 1970	306
B.49	Alton-Project Estimates of Average Annual Growth Rates of Outputs, Inputs, and Selected Productivity Measures in Bulgarian Agriculture, 1948-70	307
B.50	Birthrate in CMEA Countries, 1950-74	308
B.51	Rates of Birth and Natural Increase in Total, Urban, and Rural Population in Bulgaria, 1939-70	308
B.52	Sectoral Distribution of Employment in Bulgaria, 1948-70	309
B.53	Average Annual Growth Rates of Industrial Employment (1) and Labor Productivity (2) in CMEA Countries, 1951-67	309
B.54	Dynamics of Employment in CMEA Countries, 1951-67	310
B.55	Sectoral Distribution of Employment in Some CMEA and Western Countries for Selected Years, Prewar-1972	311
B.56	Index of Growth of Labor Productivity in Industry in CMEA Countries, 1965-72	312
B.57	Growth of Employment and Labor Productivity as Factors in Growth of Industrial Output of CMEA Countries, 1961-72	313

Table		Page
B.58	Changes in Labor Productivity (1), Capital Productivity (2), and Capital Intensity (3) in CMEA Countries, 1961-69	314
B.59	Incremental Gross Capital-Output Ratios in Some CMEA Countries, 1955-75	315
B.60	Alton-Project Estimated Annual Growth Rates of Industrial Output per Unit of Labor, Capital, and Combined Inputs in Some East European Countries, 1950-67	316
B.61	Alton-Project Estimated Average Annual Rates of Growth of Fixed Capital Inputs and Capital Productivity in East European Countries, 1960-72	317
B.62	Index of Nominal Nonfarm Wages in Some CMEA Countries, 1955-74	318
B.63	Sectoral Indexes of Monthly Wages in Some CMEA Countries, 1960-72	318
B.64	Index of Growth of Real Wages in CMEA Countries, 1951-67	319
B.65	Index of Growth of Real Wages in Some CMEA and Western Countries, 1965-74	320
B.66	Index of Growth of Average Annual Wages and Salaries by Sectors of Bulgarian Economy, 1948-70	321
B.67	Index of Average Annual Wages and Salaries by Sectors of Bulgarian Economy, 1948-70	322
B.68	Per Capita Consumption of Some Foodstuffs in CMEA and Some Western Countries, 1960-73	323
B.69	Retail Sales of Some Durables in CMEA Countries, 1960-72	324
B.70	Finished Housing per 1,000 Population in CMEA and Some Western Countries, 1955-73	325

Table		Page
B.71	Share of Individual Countries in CMEA Trade, 1960-72	326
B.72	Growth of Export and Import in CMEA and Some Western Countries, 1965-72	327
B.73	Structure of Export and Import According to Product Groups in CMEA Countries, 1960-72	328
B.74	Structure of Export and Import of CMEA Countries by Geographical Areas, 1950-67	330
B.75	Geographical Distribution of Bulgarian Exports and Imports by CMEA and Some Western Countries, 1939-70	331
B.76	Average Annual Growth Rates of National Income (1), Industrial Output (2), Agricultural Output (3), Investment Expenditures (4), and Labor Productivity in Industry (5) in CMEA Countries, 1956-75	332
B.77	Annual Plan and Fulfillment of Growth Rates of National Income in CMEA Countries, 1966-74	333
B.78	Annual Plan and Fulfillment of Growth Rates of Industrial Output in CMEA Countries, 1966-74	334
B.79	Annual Plan and Fulfillment of Growth Rates of Agricultural Output in CMEA Countries, 1966-74	335
B.80	Annual Plan and Fulfillment of Growth Rates of Investment Expenditures in CMEA Countries, 1966-74	336
B.81	Annual Plan and Fulfillment of Growth Rates of Labor Productivity in CMEA Countries, 1966-74	337

LIST OF FIGURES

Figure		Page
2.1	Fluctuating Rates of Growth of National Income, Industrial Output, and Investment in Bulgaria, 1951-74	39
2.2	Fluctuating Rates of Growth of National Income, Industrial Output, and Investment in Czechoslovakia, 1951-74	40
2.3	Fluctuating Rates of Growth of National Income, Industrial Output, and Investment in the GDR, 1951-74	40
2.4	Fluctuating Rates of Growth of National Income, Industrial Output, and Investment in Hungary, 1951-74	41
2.5	Fluctuating Rates of Growth of National Income, Industrial Output, and Investment in Poland, 1951-74	41
2.6	Fluctuating Rates of Growth of National Income, Industrial Output, and Investment in Rumania, 1951-74	42
2.7	Fluctuating Rates of Growth of National Income, Industrial Output, and Investment in the USSR, 1951-74	42

LIST OF ABBREVIATIONS AND ACRONYMS USED IN TEXT

AIC	Agro-Industrial Complex
AMIF	Additional Material Incentive Fund
AMS	Automated Management System
APAF	Automation of Production and Administration Fund
CC BCP	Central Committee, Bulgarian Communist Party
CC TU	Central Council, Trade Unions
CEC	Committee for Economic Coordination
CM	Council of Ministers
CMEA	Council for Mutual Economic Assistance (Comecon)*
CP	Committee on Prices
c.p.	central planner
CRIF	Capital Repairs and Improvements Fund
DF	Development Fund
EAF	Economic Assistance Fund
EEC	European Economic Community (Common Market)
FCF	Foreign Currency Fund
f.t.e.	foreign-trade enterprise
FYP	Five-Year Plan
GDP	gross domestic product
GDR	German Democratic Republic (East Germany)
GFR	German Federal Republic (West Germany)
ICOR	incremental capital-output ratio
IFTEF	Increased Foreign Trade Efficiency Fund
IRF	Inventions and Rationalization Fund
MAF	Maintenance of Association Fund
MF	Ministry of Finance
MFT	Ministry of Foreign Trade
MSSR	Ministry of Supply and State Reserves
NPF	New Products Fund
OECD	Organization for Economic Cooperation and Development
R&D	research and development
RWF	Reserve Wage Fund
SCSTP	State Committee for Science and Technical Progress
SPC	State Planning Commission
s.p.f.	special purpose funds
STE	Soviet-type economy
WF	Welfare Fund

---

*Includes Bulgaria, Czechoslovakia, the GDR, Hungary, Poland, Rumania, and the USSR. For our purposes Mongolia and Cuba have been excluded.

LIST OF BIBLIOGRAPHICAL ABBREVIATIONS

AER	American Economic Review
BP	Bulgarski Profsuyuzi
BTA	Bulgarian News Agency
DV	Durzhaven Vestnik
EEE	East European Economics
EMM	Ekonomika i Matematicheskie Metody (Moscow)
FAZ	Frankfurter Allgemeine Zeitung (GFR)
FK	Finansi i Kredit
GP	Gospodarka Planowa (Warsaw)
IM	Ikonomicheska Misul
IZ	Ikonomicheski Zhivot
JEL	Journal of Economic Literature
JPE	Journal of Political Economy
KES	Keio Economic Studies (Japan)
KhP	Khranitelna Promishelnost
Koz.Sz.	Kozgazdasagi Szemle (Budapest)
KS	Kooperativno Selo
KZ	Kooperativno Zemedelie
MO	Mezhdunarodni Otnosheniya
NA	Naruchnik na Agitatora
NArm.	Narodna Armiya
NM	Narodna Mladezh
NS	Narodni Suveti
NV	Novo Vreme
OF	Otechestven Front
OKSS	Otchestnost i Kontrol v Selskovo Stopanstvo
PE	Politicka Ekonomie (Prague)
PKh.	Planovoe Khoziaistvo (Moscow)
PP	Politecheska Prosveta
PS	Planovo Stopanstvo
PSS	Planovo Stopanstvo i Statistika
PT	Problemi na Truda
PZ	Partien Zhivot
RD	Rabotnichesko Delo
RES	Review of Economic Studies
SH	Svet Hospodarstvi (Prague)
SI	Statisticheski Izvestiya
SK	Schetovodstvo i Kontrol
SSt.	Soviet Studies
ST	Sunshan Turgoviay
Stat.	Statistika
Tanyug	Yugoslav News Agency
Tass	Soviet News Agency
TD	Tekhnichesko Delo

TG	Transporten Glas
TT	Trud i Tseni
VE	Voprosy Ekonomiki
VN	Vecherni Novini
VT	Vunshna Turgoviya
VTM	Venshnaya Torgovlya (Moscow)
ZD	Zamedelsko Delo
ZG	Zycie Gospodarcze (Warsaw)
ZhT	Zhelezoputen Transport
ZO	Zahranicni Obchod (Prague)
ZZ	Zamedelsko Zname

Growth and Reforms in Centrally Planned Economies

CHAPTER 1

HIGH-PRESSURE ECONOMY

Some of the crucial questions confronting the centrally planned (socialist) economy are the following: What are the pace, content, and path of economic growth? How efficient is the economic mechanism? How are resources distributed and utilized? On what basis are techniques chosen? What is the composition of direct and indirect production? What are the sources, benefits, and costs of economic change? And, most importantly, for whom does the economy work?[1] The choice of an appropriate growth rate and composition of national income is the problem par excellence of the central planner (c.p.). Obviously, the c.p. tends to accelerate the tempo of economic growth by setting as a target an immediate growth rate at the "highest possible" level. But such a decision encounters a number of constraints: the consumers' sensitivity to an adverse time distribution of consumption, and various technical and organizational barriers, ceilings, and rigidities.

CENTER-PERIPHERY INTERACTION

The centrally planned (socialist) economy provides an opportunity for resolving simultaneously the formidable problems of capital accumulation, effective demand, and income distribution. In principle, by setting an appropriate price-wage ratio in such a way as to achieve nearly full utilization of resources, the c.p. determines the partition of national income into accumulation and consumption and solves the financing of accumulation (investment). The fundamental decision about the size of accumulation is largely political. The c.p. usually has to resolve the major conflict between accumulation and consumption, that is, the time distribution of consumption. The short-run effects of the increase of the relative share of accumulation in national income are obvious. But it is not always the case that a higher share of accumulation leads to a higher standard of living in the long run.

The slogan, that the present should be sacrificed for the future, is a popular one. One of the key problems of development strategy is the choice of the time horizon for maximizing consumption. While the intertemporal choice is customarily presented as a resolution of the conflict between present and future consumption, the problem is complicated by the fact that the future can be interpreted as a time interval of various lengths. Since, by raising the growth rate, the level of consumption may be less favorable over a certain period and more favorable thereafter, the relative advantages of the c.p.'s maneuver become greater, the longer the time span he is considering. The choice is not only one between the present and future generation(s) but one of intertemporal allocation in the life span of the present generation. The resolution of the conflict between immediate burdens and more or less distant benefits for the present generation is the crucial choice—where the overall size of consumption during that period is not the only concern, but where the time distribution is the key issue.

Maximization of consumption during the lifespan of one generation, or even a good part thereof, seems to be an inequitable and unrealistic prescription: such a policy might lead to significant inroads into consumption during a longer span of life of that generation, or it might produce consumption paths such that the fruits are to be enjoyed only in the relatively distant future. The importance of the time horizon in consumption-maximization exercises is crucial, for, in an extreme case, it might enable the c.p. to show that an excessively high rate of investment would produce such a high growth rate that consumption in the entire plan period would be higher than it would have been if a lower rate of investment had been chosen. But even if the c.p. were higher than it would have been if a lower rate of investment had been chosen. But even if the c.p. were not concerned with the hardships on the population during the early years of the plan, he would be forced to realize that, with the assumed standard of living, the population might die of starvation before it fulfilled the plan.[2]

The c.p. should compare the benefits from raising the growth rate with the inroads into consumption entailed by raising the accumulation rate resulting in possible consumers' resistance. Broadly, the higher the accumulation rate, the stronger the consumers' opposition (the frustration barrier). There is a lower limit, or perhaps a maneuverable range, for the amount by which real wages can fall without provoking pressures to restore the previous level. A conflict arises between the desire of the c.p. to step up capital accumulation and the refusal of the workers to accept the level of real wages compatible with the postulated investment. Reduction of real wages below a certain level encounters the frustration barrier (defined as the lower limit below which real wages cannot be reduced without provoking political and social disturbances and intolerably affecting labor productivity, morale, and social consciousness). Such a barrier generally arises at the level of real wages to which the workers have become accustomed or at the "normal" increase that they have come to expect, so that the c.p.'s attempt to increase the rate of capital formation is frustrated.

An authoritarian regime has special advantages for extracting a relatively high surplus product. Yet, in practice, the imposition of planners' intertemporal preferences is not as forceful as could be expected: The powers of resistance are much stronger, and the share and size of consumption fund are much less manipulable variables. This is one of the key lessons that the c.p. is learning.[3] It raises the important question of interaction between the c.p. and the population: of alternative institutional and political arrangements for the flow of information, mutual influence, and resolution of conflicts.

DETERMINING ECONOMIC GROWTH RATE

The average annual growth rate of national income is the crucial parameter of the long- and medium-term plan.[4] In constructing the plan, the first step is to start with a growth rate that is high in the light of past experience. Thus a high variant is deliberately chosen. Already at the first testing stage it might have to be scaled down. Then a capital coefficient (the incremental capital-output ratio, or ICOR) is chosen (to relate the increment of national income and productive investment) on the basis of past experience and amended to fit changing conditions. This would still be only a very rough approximation, for the ICOR is largely influenced by the structure of the increment of output, which might not be the same as in the past.

Having chosen the ICOR, the first approximation of the annual productive investment in the beginning, middle, and end of the plan is obtained, together with the coefficient of increase in inventories. The total of consumption and nonproductive investment at the various periods of the plan is obtained by deducting productive investment and increase in inventories from national income. Consumption is separated from nonproductive investment by comparing the resultant level of consumption with the capacity of fixed assets that render consumer services. The decision thus obtained will still be largely arbitrary. By this time the plan might already prove to be untenable because the relative share of productive investment plus increase in inventories in national income impinges intolerably on consumption in the short run. Yet the variant may be entertained somewhat longer since it is as yet based on the hypothetical ICOR.

Before drawing a general outline of the industrial structure, one must make some reasonable assumption of the future structure of consumption. This can be based on the structure of consumption of more developed countries, adapted to local conditions, or on domestic income elasticities of demand derived from family budgets. Then the industrial structure may be roughly determined. It is required both for testing the balance of trade and for reaching a second approximation of the ICOR.

The foregoing roughly determines the distribution of national income into productive investment, nonproductive investment, increase of inventories, and consumption. It also determines largely the branch structure of national income and of productive investment. The branch structure further depends on the choice of both technology and the direction of foreign trade (for example, import versus domestic production of a given good). In essence, the branch structure of national income is determined by its rate of growth, the structure of consumption and its relationship to nonproductive investment, and efficiency analysis of alternative production techniques and foreign-trade variants.

Sectors and branches of industry can be roughly divided into two groups: supply-determined and demand-determined. The first includes industries whose output, owing to certain technical and organizational reasons, cannot grow in the long run beyond a certain ceiling, despite considerable increase in capital outlays. The second includes industries where such ceilings do not occur, at least within the range considered. Therefore, their output can grow in the long run in accordance with demand.

A wide range of technological and organizational factors condition the growth rates of supply-determined industries. The simple illustration is limited natural resources. The lapse of time needed to promote and to adapt new techniques is another. Difficulties in attracting labor to certain industries or in training a sufficient number of skilled personnel may constitute a third. In particular, shortages of sufficiently skilled and experienced construction staff constitute the underlying factor of the extended construction (gestation) periods hampering expansion of particular industries. A perennial problem is the overextended periods of gestation and fruition of investment. Given the growth rate in a particular sector, the volume of gestating investments is proportional to the length of the construction periods. If a period is protracted and the rate of expansion of construction activity high, the number of different projects under construction is so large that the available qualified personnel are incapable of coping with all efficiently. Consequently, the period of gestation is further lengthened, and the excessive number of construction projects results in freezing of capital rather than in more rapidly expanding the construction activity as intended.

Apart from the stressing of the "internal" ceilings to growth, of particular interest is the analysis of difficulties in foreign trade that may eventually make it impossible to exceed a certain level of the growth rate, and the relationship of growth dynamics to the problems and benefits of foreign trade.

In an economy that neither grants nor receives foreign credits (moderate credits may be assumed, but not those that are so large that they allow for the foreign-trade gap that must always be covered by credit), increase in imports has to be reciprocated by increase in exports. As soon as domestic demand for the goods of various branches has been estimated, in supply-determined industries an estimate can be made of what is left for export or what has to be imported. Imports will also include products that cannot be domestically produced. However, the import estimate will have to take into account domestic production of some substitutes. Thus, the total import requirements can be estimated. By deducting from the latter the value of exportables provided by supply-determined industries, one obtains the necessary exportables to be covered by demand-determined industries. The output of demand-determined industries is thus established so as to cover domestic demand for the given goods and their contribution to exports.

During the course of economic development, import requirements, especially for raw materials and semifabricates, are accelerated; and export earnings must rise to pay for the growing imports. The higher the rate of growth, the more rapidly must exports be accelerated, and the greater will be (in all likelihood) the problems of securing foreign markets for effective exchange. Higher rates of growth require greater exports or a limitation on imports. Large physical volume will probably be sold only at reduced prices. Export drives are likely to be accompanied by price markdowns for particular products in some markets; they will force export activity to other markets offering less advantageous terms and, in the process, will make exports less effective as foreign-exchange earners (while still preempting goods from domestic advancement or consumption). Thus, owing to powerful pressures of supply, the prices received on foreign markets may decline to such an extent that it would become impractical to obtain the amounts of foreign currency necessary to cover import requirements. Even if the foreign currency earned were sufficient to cover such needs, it would necessitate relatively high investment outlays in relation to the effect in terms of foreign currency. Therefore, a given level of investment will then yield a smaller increment of national income, and a given investment rate will yield a lower rate of growth. The dangers of unrealistic and ineffective selection of foreign-trade plan variants cannot be underrated.

Obviously, if the export plan is not realistic, the growth rate will have to be scaled down. The balance of trade is more likely to be equilibrated if the rate of growth of national income is pared down while the rate of growth of supply-determined industries is preserved. Thus, demand for imports should fall considerably, together with a likely increase in the surplus of supply-determined industries released for export. But even if the foreign-trade plan proves to be realistic, a foreign-trade-efficiency calculus is required to ascertain the influence of expansion of exports or import substitutions on investment outlays. The outlays necessary for securing a unit of imports or of domestic production of their substitutes would tend to rise because (1) they have to be imported at the sacrifice of a larger physical volume of exports, (2) a changed product mix of exports requires more inputs, or (3) the inputs required to produce import substitutes would be larger than those for producing the goods formerly exported to secure the foreign exchange to pay for the imports. Such large investment outlays may be required to obtain the required exports, that the escalated rate of investment might cause rejection of a plan that would make excessive inroads into consumption in the short run.

In fact, there may be an absolute limit to the growth rate when it appears that foreign trade cannot possibly be balanced. Autarky avoids foreign-trade difficulties but does not eliminate other growth barriers. In the absence of foreign trade there would be no possibility of widening bottlenecks by imports in exchange for exports of demand-determined industries. Import substitutes could widen bottlenecks, but often at higher costs than increase of exports.

The state of labor supply and the prevailing types of technical progress are of major importance in plan construction. For the capacity of the system to generate long-term growth and the limits to potential expansion are ultimately governed by the dynamic forces of technical progress, including organizational changes, and the rate of increase of the labor supplied. Technical progress is usually, although not exclusively, injected into the economic process via investment that, in turn, creates the potential capacity to produce. Taking advantage of the capacity depends on the availability of labor. In case of labor shortage, the potential cannot be translated into actual. Under such conditions it would be undesirable to raise the share of investment as, other things being equal, it would result merely in an underutilization of productive capacity caused by a shortage of labor to man the equipment. If, at the postulated growth rate, labor barriers are likely to occur, it may be necessary to increase the share of investment to favor mechanization as a substitution for labor.

If labor supply is unlimited then capacity is necessary to produce more and to employ the employable labor. In this case the potential growth rate of output—that is, the growth rate of capacity—can always be equal to the actual one, and the ICOR plays no role in equilibrating the two rates. When the supply of labor is limited, the growth rate allowed for by supply of labor consists of three elements: the rate of growth of labor, the rate of growth of labor productivity at a given ICOR, and relative increase in labor productivity due to the substitution effect.

Thus, the state of the labor reservoir limits only the growth rate of the national product under the assumption of constant ICOR. Within a certain range, labor and capital are substitutable inputs. Indeed, application of more mechanized production processes saves labor by substituting capital. But even though it overcomes a labor shortage, such a substitution necessarily entails additional investment per unit of national product. Or, to put the matter differently, change is achieved at the cost of raising the ICOR. A labor shortage does not inhibit acceleration of the growth rate; it makes an increase of the share of investment indispensable if the adopted higher growth-rate target is to be achieved. Briefly, under conditions of full employment the growth rate may be accelerated by one of the following: (1) raising ICOR (if technical progress is of a neutral type), (2) shortening the life span of equipment, which leads to increase in the parameter of amortization, (3) improving the system's efficiency independent of investment activity, (4) raising ICOR if technical progress is of the type encouraging capital intensity, or (5) raising the rate of increase of productivity at a given ICOR.

Before investment can be estimated, a balance of and demand for labor is drafted. The supply of labor is estimated on the basis of demographic data. The demand for labor is estimated on the basis of the industrial structure of production with allowance for increased productivity owing to technical progress. The task of an efficiency calculus is to achieve a given increment of national income with the smallest possible investment resources, consistent with ensuring equilibrium in the labor market and the balance of payments. A saving of manpower can be achieved by a variety of means, and, other things being equal, the variant with the lowest investment outlays is the one to select. No purpose is served by saving labor through costly automation if the equivalent can be achieved at smaller outlays through the primitive mechanization of "yardwork." The often-encountered opinion that the most sophisticated equipment is always preferable is fallacious. The best choice can be determined only by a calculation of investment efficiency and depends on economic circumstances. It is not necessarily true that what is profitable for a concern in the United States will also be profitable in a developing country. Effective production depends not only on a saving of labor but also on a saving of investment resources. In this context an important role is played by modernization, which often assures both savings of labor and increased production capacity, with relatively low investment outlays. But, with the assumed growth rate, an emerging labor shortage might require escalation of investment outlays to provide more mechanization.

At this point total productive investment can be estimated. The result can diverge materially from that arrived at by using the hypothetical capital coefficient. Despite the fact that the balance of foreign trade might be ensured at the assumed growth rate, the relative share of productive investment in national income so calculated might prove untenable. Then the rate of growth would have to be pared down further and the new variant reexamined as outlined above. In the final analysis, the highest rate of growth cannot be identified as a touchstone of economic advancement:

> The variant finally adopted should be characterized by the highest possible rate of growth at which there is a realistic possibility of balancing foreign trade and at which the relative share of productive investment plus the increase in inventories in national income is considered tolerable by the authorities from the point of the impact upon consumption and nonproductive investment in the near future. [5]

SOVIET GROWTH STRATEGY

The following have been among the distinguishing features of the traditional growth strategy in Soviet-type economies (STEs): stress on rapid industrialization; priority of investment and other resources for heavy industry and discriminatory allocation, with relative neglect of light industry and agriculture; growth of investment faster than that of national income (and ipso facto consumption); autarkic tendencies; and the nationalization of industry and forced collectivization of agriculture. Progress was identified not only with maximization of the "short-term" growth rate of industrial output but also with the rate of growth of specific key industries, or "leading links." Such growth strategy gave rise to striking imbalances and incongruities in the growth process and deleterious, ubiquitous disproportions between the development of agriculture and that of industry; among branches of industry (broadly, heavy versus light); between new processing capacity and the supply of raw materials; between sheer quantitative growth of output and quality, production techniques, and costs; between investments in new factories and the obsolescence of the underprivileged branches; and between productive and nonproductive activities, with the appalling neglect of the service sector. Consumption was considered largely a residue already at the planning stage, and suffered further during implementation when it was treated as the shock absorber for planning blunders, unforeseen development in unplannable activities, such as foreign trade and agriculture, and interim shifts in priorities.

The growth rate of output cannot be treated as a maximand mainly because it fails to reflect both costs and output composition.[6] Obviously, it matters to what uses the extracted surplus is put and whether the increment of output is composed primarily of growth- or welfare-promoting assortments and at what costs it was achieved. The traditional growth strategy relied primarily on quantitative or "extensive" growth, propelled largely by huge investment and employment, as contrasted with predominately "intensive" development characterized by improved quality and composition, cost reduction, and increased productivity, spurred, among other things, by diffusion of technical and organizational progress and incentives to produce. The fundamental problem of the socialist economy is not really one of quantity, but rather one of quality. Increase of output is not always beneficial, because output of useless or obsolete goods is wasteful. An interrelated process of expansion of quantity and improvement of quality occurs in a dynamic system.

Traditionally, investment was used as the principal growth propellent and accorded priority on all fronts. The institutionalized growth mania, permeating the whole system, compelled the economy to expand at a rate that heavily overtaxed its capacity, and it was not counteracted by any endogenous mechanism to reduce the size and alter the composition of investment to the economy's actual possibilities. The most sensitive spot of the plan was the stupendous investment program. The taut investment plan induced unrealistic assumptions in other activities. As a result, bottlenecks emerged, promoted by forced and abrupt overexpansion of certain activities that were bound to prolong the gestation periods of investments and fruition of output, to raise the cost of an incremental unit of output, and to lower the investment-efficiency index.[7] To sustain economic growth, with declining increase in productivity, a proportionately larger share of national income had to be channeled to investment. The other three alternatives (or combination there) available to the c.p. were to take the necessary steps to increase productivity, to reconcile himself to a lower growth rate, or to secure external financing.

Accelerated industrialization drives are usually associated with growing disparity between a rapidly expanding wage fund and sluggish supply of consumer goods. While investment leads to enhanced purchasing power, it also means cuts in production of consumer goods. The latter can be procured by imports, but increased investment strains the balance of payments by generating greater requirements for imports of producer goods. In the end, not only can industrial consumer goods not be imported, but imports of raw materials for consumer-goods production are also constrained, while some consumer goods are being exported to pay for the imports.

Usually an increase in the rate of investment is accompanied by a growth of employment in the investment (investment-activity-supporting) sector. Even without a change in wage rates, this would produce a rise in the total wage fund in this sector. The average wage rate in this sector is usually above that in the economy as a whole, so that even without a rise of overall wage rates, the reallocation of labor from lower to higher-than-average wage sectors boosts the wage fund. Intensified industrialization is usually supported by material inducements to speed up the process of transformation and by using the allocative function of wages (premiums) to lure labor to priority activities.

The tendency to increase employment is not sufficiently counteracted by commands and monetary restrictions on enterprise liquidity. During the course of plan implementation, the workers benefit from "excessive" purchasing power, as compared to the diminished production of consumer goods. Pressures to exceed the wage fund are created in both the producer- and the consumer-goods sectors, and they are intensified by the investment drive. The latter increasingly absorbs resources originally destined for current consumption; as a result there is underfulfillment of the plan for consumer goods and a deficiency of materials and equipment (domestic or imported) for the consumer-goods branches. The real test comes not only at the blueprint stage but also when the leaders are confronted with the dilemma of sacrificing part of the resources designated for growth-promoting activities (threats of inroads into the capital-formation plans).

In an attempt to mitigate the inflationary pressures that arise as a result of overheating in the economy, the c.p. endeavors to contain the rise in wages and employment. However, in this he is severely constrained by the mechanism of plan construction and implementation he has created, which not only does not oppose households' pressures to raise their living standards by means of increasing the number of gainfully employed but also provides its own pressures for expanding employment, giving rise to the phenomenon of disguised unemployment in industry.

LABOR PRODUCTIVITY

The most common form of measure is that of labour productivity. . . . Although this measure has a perfectly respectable ancestry—it is no more or less than the average physical production of conventional theory—the critics object that it does not measure anything peculiar to labour and that increased capital or materials may raise labour productivity while labour itself remains passive.[8]

The traditional STE method of plan construction views the labor force and labor productivity as the prime strategic determinants of output dynamics. Viewed from the supply side, output (Y) is conceived as the production of the number of people employed in productive activity (L) and their productivity (P). With constant P, Y expands <u>pari passu</u> with a rise in L. With constant L, the enlargement of Y is a function of the rise of P. Naturally, this raises the intricate question of theoretical or statistical attribution to or decomposition of P. As Joan Robinson correctly emphasizes, "[The] rate of increase in productivity of labor is not something given by Nature."[9] Labor productivity is a function of, among other things, quantity and quality of cooperating factors and labor's quality and motivation to produce. These in turn depend on (1) the rate and types of technical advancement and organizational progress, (2) the system's ability to generate this advancement internally or to borrow it from abroad, and to incorporate, diffuse, and disseminate it, (3) the rate of capital formation, its structure and choice of techniques, (4) improvements in working arrangements for resource allocation, and (5) working conditions and rewards.

But not all increases in labor productivity are equally beneficial. In analyzing the consequences of an increase in labor productivity, two specific cases should be distinguished: (1) growth of labor productivity leading to an increased output, with the same employment and (2) growth of labor productivity leading to reduced employment, with the same output. If, for example, in some factory the run of machinery is accelerated with a given employment, production will increase, owing to a more intensive utilization of the existing productive capacity. But if growth of labor productivity involves a smaller number of workers servicing the same machinery (working at the same speed), the same output corresponds to a reduced labor force. In the first case, national income is directly increased; in the second case, savings are obtained in manpower, which could be more or less propitious for the national economy, depending on the labor situation.

After distinguishing those two cases, one should note that the more advantageous case of increased production through better utilization of capacity and labor may also cause all sorts of disturbances in practice. An indispensable condition for such a solution is the additional supply of raw materials. Such an interaction of the production effort (and of the export effort in order to make up the shortages in foreign trade) will not always be possible, and at any rate it will not take place automatically as a result of the sloganeering that calls for increased labor productivity. If the increase of production has to be forgone as a result of a deficiency of additional raw materials, the first case is transformed into the second; that is, instead of an increase of production, a reduction of employment will take place.

Expansion of output might not always be desirable even when there are no technical limitations on expansion of supply. Difficulties in increasing production cannot be considered without regard to demand conditions. One could always counter that such a danger does not exist, for an increase of marketable output is followed by an appropriate increase of real wages through a rise in wages or a decrease in prices. But, again, this is not always a simple matter. With an increase of real wages, demand increases not only for the consumer goods that are in surplus as a result of additional production but also for those, such as food and clothing, whose supply does not change (the Allen-Hicks-Slutsky income effect of price change).

Theoretically, this difficulty could be counteracted by a price maneuver, that is, a sufficient price reduction of surplus goods in comparison to the prices of other consumer goods. However, in order to ensure partial equilibria in the consumers' market, it might be necessary to increase somewhat the prices of other goods at the same time. In this case, the increase of real wages would take place as the difference between the effect of price reductions on surplus goods and the effect of price increases on other consumer goods. However, if the price hike affected necessities, such a maneuver would meet with difficulties in practice, since the population would adversely evaluate and react to an increase in real wages that was derived in this manner. Moreover, the level of the lower real wages could actually decrease as a result of such a price maneuver. If, owing to the difficulties mentioned above, real wages were not increased in relation to the increased production of consumer goods, then stockpiling would result.

Obviously, when an economy suffers from general labor shortages, increasing labor productivity is essential for a full use of the productive apparatus. Otherwise, the additional savings of the labor force can only allow for a shortened working time, and, if working time is not shortened, the outflow from agriculture will be curbed, employment of married women and young people will decrease, and some form of unemployment will result. In such a case, the increase of productivity is hardly beneficial, as it is compensated by a lower level of employment.

However, even though the entire labor force could be completely sufficient for manning the productive apparatus, there might be labor shortages in certain areas. In such cases, pressures on savings of the labor force in the activities where there are shortages can be of significance, but should not entail any exaggerations about the necessity of such pressures in the entire economy.

The likely results are different if the assumption of a given productive apparatus is relaxed and we allow for enlargement of capacity and face the choice of techniques problem. The increased productivity obtained from an existing plant and equipment entails savings of investments necessary for carrying out a given production program. In turn, such savings in investment will result either in a reduction of the volume of investment and a corresponding increase of consumption or in the achievement of a larger increment of national income with a given volume of investment. In both cases, the procurement and sales difficulties that were discussed above will be alleviated or even eliminated, for required changes in the structure of production can be largely achieved by appropriate alterations in the investment structure. In this context, modernization of old factories is of particular significance, because it promotes a better use of the labor force and the productive apparatus, with relatively small investments. More or less capital-intensive techniques are, per se, neither superior nor inferior. The choice of a correct capital intensity of production depends on the state of the labor supply; on the prevailing trends of technical advancement, after allowance for technical limitations of substitutions; and on imposition of a condition of a certain rate of increase of real wages over time.[10]

REALISTIC PLANNING

Planners tend to fix the highest possible growth rate, underrate barriers and ceilings, and overrate their ability to overcome them. This is how an optimistic plan arises, leading to misallocation of resources and in more extreme cases to breakdowns with resultant costly shifts.[11] In general, the periodic fluctuations in activity lead to underutilization of resources and depress long-term performance. The postulation of overambitious growth rates results in a de facto lower long-term rate than what could have been achieved, had more modest goals been planned at the outset.

Internal consistency and feasibility of the plan are indispensable to practical implementation. The famous method of balances is used as a technique.[12] The plan should be so constructed that both shortages and useless surpluses can be avoided. As we have seen, the iterative process of plan construction <u>should</u> start out by assuming a certain growth rate of national income. Then, assumptions are made about the future structure of consumption and its relationship to nonproductive investments. These assumptions roughly determine the distribution of national income into productive and nonproductive investment, increase in stocks, and consumption. Essentially, the branch structure is determined by the growth rate of national income, the structure of consumption in relation to nonproductive investments, and efficiency analysis of alternative production techniques and foreign trade variants.

By implication, the traditional method that starts out by assuming target rates of expansion of various activities is incompatible with integral plan coordination. Such a method leads to shortages in some and surpluses in other activities. Often, the planner attempts to eliminate shortages merely through a spurious accounting balance, by manipulating input-output coefficients, by relying on output of capacities that are not likely to be commissioned during that period, by imprudent projections of foreign trade and agricultural targets, by expecting plan overfulfillment, without providing the necessary inputs, and by anticipating gains in efficiency from postulated reforms, without creating conditions for their implementation. This is often accompanied by an equally dangerous tendency to maintain the high indexes of "surplus" branches, especially those identified with progress. Such an approach tends to swell the inputs appropriated for such products and/or to classify the surplus as a reserve. Such reserves are wasteful at both the planning and the execution stages. Such an ambitious and unrealistic plan contains the seeds of frequent revisions. If such a plan misfires, it entails losses and costs of change. Hence, the result is worse than what could have been achieved, had a more cautious plan been adopted. This raises the questions of plan stability and impact on executants. While the plan is supposed to be regarded as law by the executant, the same does not apply to the c.p. The plan is often adopted after the beginning of the period; the annual Five-Year Plan (FYP) excerpts do not correspond to the actual annual plan, and even the annual and quarterly plans are manipulated and often retailored to fit reports. This procedure debunks the proclaimed and desired plan stability, and it should be kept in mind in discussing the conditions for reform implementation.

A realistic plan has to be set so as to minimize the risks of its unfulfillment. While the list of growth barriers and ceilings is extensive, the balancing of foreign trade, the lag of construction capacity behind investment and of raw materials behind the processing industries, the emerging labor shortages (in particular shortages of specific skills), and the frustration barrier, should be singled out. With an excessive rate of growth of national income, the output of particular industries tends to lag behind due to technical and organizational factors, and demand for imports increases, with consequent export increases and deteriorating terms of trade.

NOTES

1. On the question of objective function, see G. R. Feiwel, <u>Industrialization and Planning under Polish Socialism</u>, vol. 1, "Po-

land's Industrialization Policy" (New York, 1971), Chapter 1, where extensive references to the literature can be found.

2. M. Kalecki, Introduction to the Theory of Growth in a Socialist Economy (Oxford, 1969), pp. 28-34. Cf. G. R. Feiwel, The Intellectual Capital of Michal Kalecki (Knoxville, 1975), pp. 325-34.

3. G. R. Feiwel, SSt. (July 1974): 344-62; J. Robinson, The Accumulation of Capital (New York, 1956), passim.

4. The specifications of the terms of choice for the c.p., what he should and should not do, and the limiting factors he must take into account have been explored in the pioneering work of Michal Kalecki, which has been the inspiration of this section. See M. Kalecki, Introduction; M. Kalecki, Z zagdnien gospodarczo-spolecznych Polski Ludowej (Warsaw, 1964). Cf. Feiwel, Intellectual Capital, Part 3.

5. Kalecki, Z zagdnien gospodarczo-spolecynych Polski Ludowej, pp. 58-59.

6. Compare J. Kornai, Rush Versus Harmonic Growth (Amsterdam, 1972); National Bureau of Economic Research, Economic Growth (New York, 1972); J. Tinbergen et al., Optimum Social Welfare and Productivity (New York, 1972).

7. Compare N. Ivanov, IM 19 (1961): 80-81; T. Yordanov, NV 7 (1969): 33.

8. W. E. G. Salter, Productivity and Technical Change (Cambridge, England, 1969), p. 2.

9. J. Robinson, Collected Economic Papers, vol. 1 (Oxford, 1951), p. 101.

10. Kalecki, Z zagdnien gospodarczo-spolecznych Polski Ludowej, pp. 41-48.

11. G. R. Feiwel, Industrialization and Planning under Polish Socialism, vol. 2, "Problems in Polish Economic Planning (New York, 1971), pp. 293ff.

12. Compare E. Mateev, Balans na narodnoto stopanstvo (Sofia, 1972).

CHAPTER

2

GROWTH AND INSTABILITY

COMPARATIVE STAGES OF DEVELOPMENT

Comparisons of the relative levels of development is a very tricky task, even if the comparison is restricted to a specific social system.[1] Generally, the most commonly used indicator of a nation's level of development is the per capita national product, whatever the vexing problems of definition and interpretation.[2] All the indicators must be viewed with great circumspection (see Tables B.2-6). The notable differences in growth rates among various countries that adopted the Soviet mode of development are obviously traceable to manifold factors, among which are the past industrialization, the varied degrees of war devastation and dislocations, the speed and specificity of postwar reconstruction, specific variations of policies and arrangements, and different paths of history and national character.

To trace the history of the industrialization of Council for Mutual Economic Assistance (CMEA) countries is a task beyond the scope of this study. Suffice to say that at the threshold of their embarkation on Soviet-type development the East European countries were at disparate stages of industrialization and could be classified in three groups: most developed—Czechoslovakia and the German Democratic Republic (GDR); intermediate—Hungary and Poland; and least developed—Bulgaria and Rumania.

At the beginning of the twentieth century Bulgaria was undoubtedly one of the most backward European countries. During the first quarter of the century the government made an attempt to encourage industrialization by means of a variety of benefits conferred on private enterprises. From 1909 to 1937 the average annual rates of growth of industrial output in state-encouraged enterprises gravitated between 5 and 8 percent, with higher rates recorded in 1909-29 and rates below 5 percent in 1929-37. Presumably, growth in the less-favored enterprises was considerably lower, and productivity almost stagnated. The "growth" industries were ceramics, textiles, and energy. Structural change, that indicator of industrial development, was conspicuously absent. Agricultural production was small-scale, with considerable disguised unemployment and a declining income per unit of land even before the Great Depression. The peasantry effectively opposed attempts to encourage industry. Above all, there was considerable dearth of capital. State aid was not very strong, and in many cases it was misdirected from the standpoint of structural change. Foreign capital—what little there was—was mainly directed to food processing. All in all, the prewar attempt at industrialization floundered in its half-measures.[3]

Generally, in all East European countries at the threshold of the 1950s the productive potential apparently regained or exceeded prewar levels, with industrial output exceeding prewar levels and agricultural output below these levels.[4] According to Alton-Project estimates, by 1950 Bulgaria surpassed by 3 percent its own 1939 level of per capita GNP; Poland surpassed its own 1939 level by about 40 percent; Czechoslovakia its own 1937 level by more than 20 percent; Hungary was 7 percent below its 1938 level; and in Rumania and GDR (for which data are particularly treacherous) the level was not quite reached in the first and was almost 25 percent below in the second.[5]

As shown in Table B.2, in 1955 the Balkan countries and Spain were low on the scale of relative development levels; Czechoslovakia and Norway were relatively high; and Poland, Hungary, and Ireland were intermediate. Bulgaria was roughly at the same level as Spain, fairly ahead of Greece and Yugoslavia, but somewhat behind Rumania. Thus, a case could be made for comparing Bulgaria's growth dynamics with those of countries at roughly similar levels, rather than concentrating on intra-CMEA comparisons. Similarly, the growth performance of the more advanced CMEA countries should be compared with that of West European countries at a similar level of development: France and the German Federal Republic (GFR). But whatever the advantages of such comparisons, they are not at the forefront of attention of policy makers in Eastern Europe. Policies are shaped with a view to the relative performance of a given country within the CMEA bloc. This is not to say that Bulgarian policy (and Soviet aid) is not conditioned by its relative performance and position compared to its neighbors, and naturally East German patterns are somewhat molded by what happens in West Germany.

As indicated in Table B.3, from 1950 to 1967 the relative positions of Bulgaria and Rumania improved considerably; those of the GDR and USSR, only slightly; Poland and Hungary lost substantial ground; Czechoslovakia came out as the remarkable loser. In comparison to the USSR (100) from 1950 to 1973, the per capita level of national income in Bulgaria rose from 66 to 101 percent; in Rumania, from 56 to 80 percent; and in the GDR, from 112 to 137 percent. Czechoslovakia slipped from 157 to 107 percent; Poland, from 103 to 89 percent; and Hungary, from 114 to 83 percent. In terms of industrial output Bulgaria advanced from 43 to 88 percent; Rumania, from 33 to 67 percent, Poland, from 68 to 79 percent, and GDR from 131 to 157 percent. Czechoslovakia lost from 151 to 111 percent; and Hungary, from 77 to 71 percent. In terms of agricultural output Bulgaria gained from 97 to 112 percent; Rumania, from 77 to 94 percent; the GDR, from 99 to 111 percent; and Hungary from 138 to 143 percent. Czechoslovakia fell from 110 to 94 percent; and Poland, from 185 to 130 percent.[6] Clearly, throughout the postwar period Bulgaria and Rumania were the most consistent gainers, whereas Czechoslovakia was the most consistent loser—respectively, the most bakcward countries and the most advanced country at the time of embarkation on Soviet-type development.

A ranking of CMEA countries in 1967 in descending order, according to the relative weights of industry and agriculture in national income (index, agriculture = 1.0) shows: Czechoslovakia (5.3), GDR (4.3), Poland (3.1), Hungary (2.8), USSR (2.3), Rumania (1.8), and Bulgaria (1.5). In comparing these figures with those of the West, one finds that Czechoslovakia was substantially below the United States and the United Kingdom (both 10.9), GFR (9.4), and Sweden (7.2), and at about the level of Belgium (5.8) and France (5.5). The GDR was at about the level of Austria (4.1) and Norway (4.2). Poland, Hungary, and the USSR were close to Italy (2.6), Japan and Denmark (2.5), and Yugoslavia (2.4). Rumania and Bulgaria were somewhat below Portugal (2.0) and above Greece (1.0).[7]

According to computations of the Research Institute for Planning and Management, in Czechoslovakia per capita GNP in 1965 prices (U.S. dollars) would grow from 1970 to 1990 as follows: Czechoslovakia, from 1,932 to 4,251 (120 percent); GDR, from 1,825 to 4,309 (136 percent); USSR from 1,431 to 3,294 (130 percent); Hungary, from 1,358 to 2,985 (120 percent); Poland, from 1,252 to 3,117 (149 percent); Bulgaria, from 1,201 to 3,195 (166 percent); and Rumania, from 943 to 2,547 (170 percent).[8] By 1990 the GDR would displace Czechoslovakia as the country with the highest per capita GNP; Bulgaria would come up from sixth to fourth place; Poland and Rumania would retain fifth and seventh places, respectively; Hungary would slide down from fourth to sixth place. In terms of Czechoslovak per capita GNP = 100, the respective figures for 1970 and 1990 would be as follows: GDR, 94.46 and 101.36; USSR, 74.06 and 77.48; Hungary, 70.28 and 70.22; Poland, 64.80 and 73.32; Bulgaria, 62.16 and 75.15; Rumania,

48.8 and 59.72. Again, the fastest rates of growth were predicted for Rumania and Bulgaria; Poland, the GDR, and the USSR would make a relatively good showing, and Czechoslovakia and Hungary would trail considerably behind.

In terms of Soviet growth strategy, per capita industrial output indicates the progress achieved in the industrialization march (see Table B.5). But the pertinent question is not only what was achieved, but at what costs, and who benefited from it. A relatively high per capita aggregate product is not a sufficient criterion of the "wealth of nations." Even in consumer-oriented economies, growth of national income is not a measure of improvement in welfare.

GROWTH AND COMPOSITION OF OUTPUT

While the refractory nature of the statistics and the tentativeness and limited comparability of growth measures must be stressed, CMEA countries have recorded relatively high growth rates in the postwar period. During the period 1951-67 the countries could be ranked as follows, in descending order of the average growth rates: Rumania, Bulgaria, USSR, GDR, Poland, Czechoslovakia, and Hungary. Obviously, the ranking differs depending on the period considered, but in almost all cases Bulgaria and Rumania have recorded the highest growth rates (see Table B.7). The Alton-Project data indicate a different order of ranking, but still Rumania and Bulgaria hold the first and second place (see Table B.10). In a ranking based on per capita national income, Bulgaria moves to first place, and Rumania to second, in view of a higher population growth rate in the latter. In addition, the GDR moves ahead of the USSR, due to a decline of population (see Table B.9). However, the gap between the output growth leaders and the other countries is somewhat narrowed when per capita growth rates are compared.

The range of growth rates is fairly large. While there have been other exceptions, Hungary appears to have been the major deviator, from the observed rough pattern that the growth rate is more rapid with a less advanced level of development. These growth rates are quite impressive if compared with those of the same period in Western countries, but they do not fare quite as well in comparison with those of Western countries at comparable levels of development.[9]

The dynamics of the Bulgarian index of national production, as depicted by GNP, are very impressive, but the available measures are tentative. The Alton-Project's latest estimates indicate that GNP virtually doubled in the turbulent 1950s, and that it expanded by nearly 80 percent in the 1960s, registering a fairly respectable average annual growth rate of 6 percent during 1950-70 (see Table B.12). This trims down the official claims but does not invalidate them.

If economic progress is to be measured by growth rates, we must know the extent to which growth rates have been sustained and were proceeding along an ascending trend. What matters is a long-term increase in the capacity to produce and effective utilization of resources over time. Among the remarkable features of postwar growth in CMEA countries, all countries registered in the 1960s and 1970s a retardation of the growth momentum of the 1950s, and throughout the period all countries were subject to fluctuations of the growth rates and underutilization of resources (see Tables B.7-9).

Structural Change

Sweeping structural breaks occurred in the USSR before the East European countries adopted the Soviet pattern. Some of these countries were then at a level of development similar to the Soviet level of the early 1930s, but the imitation of the Soviet model was mechanically implemented without understanding that the Soviet economic structure could not be transplanted without grave repercussions.[10] A distinguishing feature of this type of industrialization was the speed and character of the structural transformation, that is, the striking rapidity of industrial growth and the disproportionate rates of sectoral advance, with priority of heavy industry as a hallmark, as reflected in the shifting composition of the structure of production by sectors of origin over time. The drastically shifting sectoral contribution to national output reflects the economy's structure at the time of the take-off: the strength of the industrialization push and the stormy drives to sustain and accelerate it, followed by periods of consolidation and retrenchment; the discriminatory and shifting allocation policies; the economy's limited absorption capacity; and a certain system-made proclivity to perpetuate the existing structure.

Again, the most backward countries registered the highest increase in the share of industry in national income (in Bulgaria, from 37 percent in 1950 to 51 percent in 1972). Among the most advanced countries, the share of industry rose in the GDR and declined or was barely maintained in Czechoslovakia. There were changes in prices, and the data for the 1970s are not quite comparable, particularly in the case of Hungary (see Table B.13). Industry exhibited the highest or second highest (after construction) growth rates for most of the postwar period in almost every country. With some notable exceptions, construction recorded high rates, and welfare-oriented activities tended to trail behind. The growth of agricultural output was considerably below that of nonfarm sectors (see Tables B.14-15). Since industry and agriculture are the principal sectoral

GROWTH AND INSTABILITY

determinants of the size of national output, their relative rates of growth tend to have a commanding influence over the velocity of the aggregate rate. Other things being equal, the higher the level of industrialization in the base period, the heavier the weight of industry's performance in the overall growth rate, and the smaller is the depressing effect of lagging agricultural performance. Ipso facto, a country like Bulgaria, with a weighty but slowly growing agriculture, must outpace other countries in industrial growth in order to produce the same aggregate rate.

The findings of the Alton-Project generally corroborate the official picture of the growth pattern (see Tables B.16-17). The differences pertain mainly to the varied assessments of the speed of the sectoral rates and their impact on the shifting structure of production. Alton's latest estimates for 1965-74 are only partly comparable with his estimates for earlier periods.[11] On the whole, the contribution of agriculture is reduced in favor of industry, but the process takes place at varied rates intertemporally and among countries. By their very approach, the Alton-Project estimates imply the upward bias of the official measure of industry's contribution.[12] Toward the end of the 1960s the Alton-Project's ranking according to contribution of industry to GNP (percentages) was as follows: GDR, 41.2; Czechoslovakia, 39.7; Poland, 35.3; Hungary, 33.7; Bulgaria, 33.3; Rumania, 30.6. The ranking according to agriculture was as follows: Rumania, 37.3; Bulgaria, 29.2; Hungary, 25.6; Poland 23.9; Czechoslovakia, 19.3; GDR, 15.8.[13]

In comparing GNP growth data for Bulgaria, one should be particularly circumspect of the earlier Alton-Project estimates, partly due to the estimate of the agricultural component. The latter was calculated in 1939 prices, which had a depressing effect on the index. Subsequently, a new index for agriculture, using a 1968 price base, was estimated, thus revising the earlier measures. According to the recalculated overall index, the relative share of industry jumped from 13.9 percent in 1950 to 23.3 percent in 1960 and to 33.7 percent in 1970, while agriculture and forestry dropped from 47.9 percent to 40.7 and 28.5 percent, respectively. Despite the relatively high growth rates, the share of construction increased from 4.3 percent in 1950 to only 5.7 percent in 1960 and 6.9 percent in 1970. Throughout the two decades, both transportation and trade almost doubled (reflecting their small initial share), while government services declined sharply (from 26.4 percent in 1950 to 16.4 percent in 1970). The structural transformation was also accomplished at the sacrifice of housing and other welfare-oriented activities. There was a noticeable retardation in the dynamics of sectoral rates in the 1960s (compared with the preceding decade), with the exception of housing and transport (see Table B.17).

One of the major characteristics of modern economic growth in market economies is the rapidity of structural change, including the shift from agricultural to nonfarm activity, and recently the shift from industry to services.[14] In the postwar period the East European countries displayed on the whole a greater change in the structure of production than did the West. In addition to the forced industrialization rush and investment priority for heavy industry (collectivization of agriculture and emphasis on autarky), the specific characteristic of the Soviet mode is the neglect of the service sector (see Table B.55).

Shifts in Bulgarian Industrialization

During 1945-48 the Bulgarian communist regime consolidated its power. In the economic sphere a series of institutional changes took place. In 1946 the already small landholdings were further subdivided by a land reform. Initial nationalization of industry took place also in 1946, but it was fully accomplished only at the end of 1947, together with nationalization of the financial apparatus. The year 1948 saw the nationalization of wholesale trade and the establishment of foreign-trade monopoly. The initial plan of reconstruction (1947-48) aimed at achieving the 1939 production levels and at preparing for the intensive industrialization rush.* The accent was on the electric power and railroad networks. Cement factories and plants producing basic chemicals were built. Some heavy industry factories were also started.

The first industrialization rush spanned 1949-53, supported by collectivization of agriculture. Its spearhead was the development of heavy industry, especially metallurgy and machine building. Further development of building materials, basic chemical production, electric power, and transport took place. All this involved very high investment outlays. The share of accumulation in national income was considerably above the quarter mark. The annual rates of growth of industrial output were the highest of the postwar period, and per capita industrial output increased by 57 percent from 1950 to 1953 (see Tables B.33-34). In 1954-55 raw materials shortages appeared. Supply of raw materials did not keep pace with the commissioning of new processing facilities. Investment outlays for widening the raw-materials base were considerably increased, but without interrupting the development of heavy industry: hence, the high capital intensity of investment in 1953-56. Concurrently, the need to supply

---

*Industrial production (1938 = 100) reached 175 in 1948 and 227 in 1949, but agricultural output (1934-38 = 100) hovered at about 95 in 1948-49.[15]

the operating factories with raw materials caused shifts in imports at the cost of machinery. This resulted in further dispersal of investments, reduction in their effectiveness, and, consequently, a drop in growth rates. Though the share of investment in national income was still high, the relative increases of investment outlays fell considerably.

In 1957-60 the newly built raw-materials facilities began operating. The intensified collectivization and mechanization of agriculture released considerable manpower for industry. The second phase of Bulgarian industrialization rush at the threshold of the 1960s upheld the further development of heavy industry, but with increasing stress on development of chemical, light, and food-processing industries, as well as agriculture. The gap between growth rates of producer goods (A) and consumer goods (B) was somewhat narrowed, but the principle of faster growth of A was maintained.

Industrial Growth Performance

From the standpoint of the Soviet growth strategy, the growth rates of industrial output are the pulse of economic performance. The ranking of CMEA countries in descending order of average annual growth rates of industrial output in 1951-67 was as follows: Bulgaria, Rumania, Poland, the USSR, Hungary, the GDR, and Czechoslovakia (see Table B.32). This ranking differs somewhat from that according to national product growth rates, with the variances obviously influenced by the performance in agriculture and other sectors. But once again, the most backward countries exhibited the highest growth rates, with Czechoslovakia trailing considerably behind (there is a spread of five percentage points between Bulgaria and Czechoslovakia). This general pattern holds whether measured by official or recalculated data (see Tables B.39-40). In comparing growth of per capita industrial output from 1950 to 1968 we find that Bulgaria far outdistanced the other CMEA countries, so that the spread between it and Czechoslovakia (1950 = 100) was in 1968 from 853 to 371 (see Table B.34).

Whatever the shortcomings of the data, in comparing the average annual per capita growth rates of industrial output in CMEA countries we find considerable deceleration even from the 1951-68 average to the 1961-68 average, with the notable exception of Rumania, which in the latter period outdid Bulgaria. The respective figures were as follows: Bulgaria, 12.6 and 10.9 percent; Rumania, 11.9 and 12.3 percent; the GDR, 9.4 and 6.1 percent; Poland, 9.3 and 7.3 percent; the USSR, 8.7 and 7.3 percent; Hungary, 8.4 and 6.9 percent; Czechoslovakia, 7.6 and 5.0 percent. In the first period Japan did better than any CMEA country (13.4 percent), but in the second period (12.1 percent) it was slightly outdistanced by Rumania.[16]

According to the Alton-Project estimates, in the 1960s growth rates of industrial output in Czechoslovakia and the GDR compared unfavorably with those in the EEC, whereas those of Poland and Hungary were slightly above those of the EEC, and Bulgaria and Rumania maintained their leading positions.[17]

Growth rates of A tended to exceed those of B, with the notable exceptions in Bulgaria in 1951, 1953, 1955, 1957, and 1961; Czechoslovakia in 1954, 1955, 1957, 1961, and 1965; Poland in 1955, 1957, 1958, 1971, and 1972; Rumania in 1954, 1955, 1960, and 1972; and the USSR in 1953 and 1971. In Hungary there was a sharp reversal of this policy during the New Course. Also during the 1960s and 1970s, B grew at a slightly higher rate than A. Moreover, in all countries since the mid-1960s the discrepancy between the growth rates of A and B has been less pronounced (see Table B.36).

A rough indicator of the progress of industrialization of CMEA countries is the relative share of A in total industrial output, but the ambiguous nature of this measure of priority accorded to heavy industry is well recognized.[18] In 1950 Bulgaria and Czechoslovakia were the only countries whose share of A was below 50 percent (38.2 and 48.6 percent, respectively). By 1970 this distinction had disappeared, but whereas the share of A in Bulgaria was somewhat more than 50 percent, it was far more than 60 to 70 percent in other countries, indicating the greater importance of consumer-goods production in Bulgaria than in the other CMEA countries (see Table B.35). But this picture is not complete without the relative dynamics of A and B output. The priority accorded to production of A was more pronounced in Bulgaria and Rumania than in other countries; taking 1950 as 100, A grew 18- and 15-fold, respectively, and B was 806 and 719. In comparison, in Poland A and B grew to 957 and 548, respectively; in Czechoslovakia, to 608 and 373; in the USSR, to 781 and 538.[19] According to the Alton-Project estimates, during the period 1950-65 the ratios of growth of heavy industry (excluding mining) to all industry were as follows: Bulgaria, 1.9; Rumania, 1.5; Poland, 1.4; about 1.2 for Hungary, Czechoslovakia, and the GDR. The ratio for the two indexes in the EEC was about 1.3, so that the first three countries were above, and the last three were below. However, the West European ratio cannot be taken as a yardstick.[20]

Another commonly used indicator is the share of machine building in total industrial output.* By that standard, in 1950 the ranking

---

*The criteria adopted in a Polish comparative statistical study of the modernization of industrial structure were the respective shares

in descending order was Hungary, Czechoslovakia, Rumania, Bulgaria, and Poland. By 1967 Poland had displaced Rumania, and Rumania had displaced Bulgaria. On the whole, although Bulgaria seems to have made considerable progress (from 9.1 to 18.8 percent), the strongest effort in that direction seems to have been made by Poland and Czechoslovakia, in both cases primarily at the expense of the food industry (see Table B.37).

However, looking at the dynamics of machine-building industry in the various countries, Bulgaria again has the lead. Taking 1950 as 100, by 1970 this industry grew 36-fold in Bulgaria, 30-fold in Rumania, 26-fold in Poland, 13-fold in the USSR, in Czechoslovakia to 973, in Hungary to 799, and in the GDR to 400.[22] The rates of increase are strikingly high, even if recalculated indexes are used or if only developments in the 1960s are considered so as to account in part for the relatively small machine-building industries in Rumania and Bulgaria in the early 1950s. For 1950-67 Alton estimated average annual growth of machinery at about 18 percent in Bulgaria, 16 percent in Rumania, 13 percent in Poland, 10 percent in the GDR, 8 percent in Hungary, and 7 percent in Czechoslovakia. In 1961-67 the recomputed and official figures were respectively as follows: Bulgaria, 17.2 and 18.9 percent; Rumania, 12.2 and 16.6 percent; Poland, 8.4 and 13.3 percent; Hungary, 4 and 9.6 percent; the GDR, 3.9 and 4.6 percent; Czechoslovakia, 0.5 and 7.7 percent.[23] The decline and the variance between recalculated and official indexes were particularly sharp in Czechoslovakia.[24]

Although in Bulgaria the shares of textiles and clothing and food industries dropped significantly from 1950 to 1967, the fall was not so large as in some other countries (see Table B.37). Whereas textiles, most light industries, and food processing in most CMEA countries registered slowest rates of increase, the relative sluggishness was less pronounced in Bulgaria, where these were the traditional industries. Bulgaria maintained its lead in the growth of food processing. With 1950 as 100, by 1970 this industry had grown to 598 in Bulgaria; in Rumania, to 513; in the USSR, to 444; in Hungary, to 379; in Poland, to 337; in Czechoslovakia, to 253; in the GDR, to 198.[25]

---

of processing industry, and particularly electro-machine-building, chemical, and energy branches. By this yardstick, Poland's industrial structure was considered close to that of Czechoslovakia and Hungary, but it was superior to those of Bulgaria and Rumania, where in 1967 the share of mining, electro-machine building, and chemicals was considerably lower than in Poland, but the share of food processing was higher than in Poland.[21]

Looking at the growth rates of various industries in Bulgaria throughout the two decades, we find it obvious that the most consistently high rates have been maintained (although declining) for electric power, machine building, and chemical industries, whereas the rates of fuels, textiles and food industries have fluctuated more widely (see Table B.38).* During the 1961-67 period of intensive industrialization in Bulgaria, important structural changes took place that have been to

---

*It would take too much of a digression to trace and compare the rates of expansion of industrial branches in individual countries through time, and even that is not illuminating without further disaggregation into components.[26] Briefly, Alton's estimates indicate the following growth rates in the 1950-67 period:[27]

	Bulgaria	Czechoslovakia	GDR	Hungary	Poland	Rumania
Electric power	18.2	9.0	6.7	9.3	10.5	15.6
Mining and fuels	11.2	4.7	2.8	4.1	3.2	6.3
Ferrous metals	11.4	n.a.	9.2	n.a.	8.7	n.a.
Nonferrous metals	16.9	n.a.	n.a.	n.a.	7.3	n.a.
Machine building	18.5	6.5	9.6	8.2	12.7	15.6
Chemicals	18.1	10.5	9.0	18.6	10.6	18.7
Building materials	11.0	11.1	6.4	8.7	7.8	11.2
Lumber and woodworking	3.7	10.1	4.9	10.7	8.7	3.2
Textiles	7.8	3.0	5.5	5.4	4.9	6.3
Clothing	6.5	8.0	7.1	12.3	5.1	10.7
Food processing	9.1	4.6	4.8	9.7	6.2	8.9

n.a. = not available.

The above data indicate that both Bulgaria and Rumania were leading in the growth rates of electric power, mining and fuels, machine building (with Poland a close third), chemicals (with Hungary recording a similar rate), building materials (with Czechoslovakia at a similar rate, and textiles. In food processing they were outdistanced by Hungary, but they kept a close second and third position. In clothing Bulgaria recorded the second lowest rate, with Hungary, Rumania, and Czechoslovakia in the lead. In the 1960s the growth retardation proceeded at varied rates, but it was pronounced in almost all branches (least appreciable in chemicals) in most CMEA countries.[28]

GROWTH AND INSTABILITY   27

some extent perpetuated in the 1970s. Ranking the industries by the
highest average annual growth rate (percentages), we obtain the following: ferrous metallurgy, 25.8; glass and china, 19; machine building,
18.9; chemicals and rubber, 18; fuels, 16.4; building materials, 16.2;
cellulose and paper, 15; electric power, 14.8; clothing, 12; nonferrous
metallurgy, 10.2; food processing, 9.7; leather and footwear, 9.5;
textiles, 6.7.[29]

Bulgarian industrial expansion was estimated by two current
Alton-Project weighted sample-commodity indexes. The first (see
Table B.41) covers the years 1939 and 1948-65, and the second
(see Table B.42) is the new provisional index for 1963-74. The
first avoids some of the major shortcomings of the notorious gross
value of output, but it seems to suffer from limited coverage and a
particularly spurious value frame. The second relies on a fairly larger, but admittedly still small, sample index. It is constructed with
a completely revised system of weights. Its weaknesses notwithstanding, it is probably a better measure than the official indexes.[30]
Whether measured by the official or recalculated indexes, the industrialization process was very rapid. Despite the differences in approach between official and recalculated indexes, the variations in
quantitative measures obtained for postwar dynamics of industrial
production are not overly striking. However, the differences in assessing the growth of production from prewar to postwar periods are
sharp, thus determining the base for measuring postwar performance.
The average growth rate for 1848-65 computed by the first index
(13.9 percent) falls only slightly below the impressive gross value
(14.5 percent). A comparison of industrial growth in 1963-74 computed by the second index indicates an expansion by roughly 2.25 to
2.5 times, as against an official claim of 3.5 times (see Table B.43).
Bulgarian performance could be contrasted with that of other CMEA
countries (except for Czechoslovakia), where substantial differences
between official and recalculated rates were registered in the initial
period, and where the rates converged somewhat in later periods,
due partly to improvement in East European statistics and price
revisions.

In analyzing the structure of industry in Bulgaria, the Alton-Project estimates indicate a somewhat different picture than the official one. Table B.42 indicates a higher rate for some branches
than the official (electric power and woodworking), practically the
same rate for others (food), and a much lower rate than the official
in still others (strikingly in fuels, but also in machine building, chemicals, and others). For some branches the intertemporal pattern differs sharply; for example, in the chemical industry the recalculated
index outpaces the official one in 1963-66 and trails behind thereafter.
The typical pattern of highly unbalanced growth is corroborated by

both the official and recalculated indexes. Again, we observe, with some shift in priorities, the industrial progress focused on electric power, ferrous metals, machine building, and chemicals, with strikingly disparate rates at which the various branches of heavy industry have developed.[31]

Agricultural Performance

The dynamics of agricultural production in the CMEA have been far below those of national income and industrial output and have generally depressed the growth rates of national income.* By 1968 the ranking in descending order of growth of agricultural output was as follows: Bulgaria, Rumania, the USSR, the GDR, Poland, Hungary, and Czechoslovakia. Whereas from 1950 to 1968 Bulgarian agriculture was supposed to have increased by 128 percent, that of Czechoslovakia grew only by 42 percent.[32] Agricultural performance seems to have been much better in the first decade in Bulgaria than in the other countries (81 percent growth by 1960 in Bulgaria and 20 percent in Czechoslovakia). However, in the second decade Bulgarian agricultural growth was slower than in the first and about the same as in the other countries. From 1970 to 1973 Bulgarian agricultural output apparently grew at the slowest rate in the CMEA (see Tables B.45-46). The prewar backwardness and inefficiency of Bulgarian agriculture contributed to the relatively good showing of the immediate postwar period. Moreover, investment allocation favored this sector more in Bulgaria than in the other countries.

The Alton-Project estimates bear out the better performance of Bulgarian agricultural production in the 1950s, compared to that in the 1960s. The average annual rates of growth for 1965-70 were particularly low, yet inputs continued to grow at high rates. Whereas in 1948-54 and 1960-65 crop production grew faster than livestock, in 1954-60 they grew at about the same rates; in 1965-70 crop production declined on the average by 0.8 percent, and livestock grew by 3.3 percent. On the whole, from 1948 to 1970 livestock grew slightly faster than crop production. With the exception of corn, almost all per hectare yields showed a decline over time (see Table B.49). Whereas in all Eastern Europe agricultural output per hectare is considerably above the USSR and the United States, Bulgaria ranks

---

*One should point out that only in Poland is the bulk of agriculture in private hands; in the other countries it is primarily socialized, with 8 to 16 percent of agricultural land in private hands.

GROWTH AND INSTABILITY 29

second to last, followed by Rumania. In the GDR, Hungary, Czechoslovakia, and Poland agricultural output per person employed in agriculture is higher than in the USSR, but way below that in the United States, whereas in Bulgaria it is somewhat below the USSR, and in Rumania it is less than half that of the USSR (see Table B.48). The Alton-Project estimates indicate that in 1963-68 Bulgaria ranked third in Eastern Europe in the use of commercial fertilizers (78.8 kilograms per hectare), behind the GDR (200.9 kilograms) and Czechoslovakia (117.2 kilograms), whereas before the war the respective figures were 0.5, 93.4, and 12.2 kilograms. With 1950-54 as 100, the dynamics of fixed agricultural investment in Eastern Europe were as follows by 1965-67: Rumania, 859; Poland, 529; Bulgaria, 407; Czechoslovakia, 351; Hungary, 266; the GDR, 130. But whereas the share of agriculture in investment expenditures increased in almost every country from 1950-54 to 1965-67 (in some considerably: in Poland, from 9.1 to 15.7 percent, and in Rumania, from 10.1 to 18.5 percent), it dropped in Bulgaria from 10.0 to 9.3 percent (having recorded a high of 14.5 percent in 1955-59).[33] In Bulgaria total agricultural production almost doubled between 1939 and 1970. Inputs from nonagricultural sectors increased fivefold. Thus, value added increased at a slower rate (62 percent) than agricultural production (71 percent).[34]

CAPITAL FORMATION

Comparisons of the shares of national product allocated to capital formation between the CMEA and the West are difficult, if not treacherous. Nevertheless, the available evidence seems to point to significantly higher shares in the CMEA. Throughout the postwar period Bulgaria exhibited extraordinarily high rates of capital formation, considerably above the very rapid rates of output expansion (see Tables B.7, B.14, B.19, B.24, and B.32). Moreover, for most years the variances between these rates were higher in Bulgaria and Rumania than in most other CMEA countries.

Consumption and Accumulation

In the statistics on distribution of national income in CMEA countries (see Table B.18) the most conspicuous is the upward trend in the share of accumulation and ipso facto the downward trend in that of consumption. For example, in Bulgaria the index of national income (1952-67) increased 3.4-fold and distributed 3.6-fold (accumulation 5.3-fold, and consumption only 3.1-fold).[35] The average annual rates of growth of accumulation in 1952-60 were 11.4 percent (consumption, 8.7 percent); in 1960-65, 8.25 percent (consumption, 6.6 percent); in 1965-70, 10.85 percent (consumption, 7.1 percent); in 1970-72, 3.5 percent (consumption, 6.85 percent). In general, there appears to have been a tendency toward bridging the gap between those rates (with the uncommon excess of the growth rate of consumption over that of accumulation in 1970-72), with the 1971-75 FYP postulating slightly higher growth rate of consumption (6.5 percent) than accumulation (6.2 percent).[36] The remarkable features were the spectacular speeds and aftermaths of recurring investment drives, partly reflected in the notable fluctuations in the shares of accumulation in Bulgaria.

Even if the comparisons of the partition of national income into consumption and accumulation are confined within the CMEA or intertemporally to one country, the customary warning about the distortion of shares is in order. The shares calculated in current prices tend to differ markedly from those in constant prices, particularly if substantial changes in relative prices of investment and consumer goods took place. The bulk of turnover tax is levied on consumer goods, whereas investment goods tend to be exempted and benefit from various subsidies. The prices of the latter were greatly understated, particularly in the earlier period. Thus, the share of consumption was inflated, and accumulation was understated.

Turning to the record, in 1951-67 all East European countries, for which data were available, registered a marked increase in the shares of accumulation; this, at first sight, showed that these rates of growth outpaced those of national income. Indeed, the average rate of growth of consumption was substantially lower than that of accumulation, though the spread varied greatly among countries and periods (see Table B.19). Obviously, the averages offer no more than a very broad indication of the dynamics during the period considered, and they obscure significant intertemporal fluctuations. *

---

*According to a Soviet source, accumulation increased in Bulgaria in 1953 by 54 percent, and declined by 35 percent in 1954, registering sharp variations in annual rates of change thereafter. In the GDR the rate of growth was 30 percent in 1951, 36 percent in 1952, 19 percent in 1953, -19 percent in 1954, 24 percent in 1955, 33 percent in 1956, 18 percent in 1957, 34 percent in 1958, 8 percent in 1959, and -5 percent in 1960 and 1961. In Czechoslovakia it was 122 percent in 1951, 41 percent in 1952, 16 percent in 1953, -42 percent in 1954, and 69 percent in 1955. In Hungary the amplitudes were very wide: 71 percent in 1951, -23 percent in 1952, 28 percent in 1953, -30 percent in 1954, and 58 percent in 1955.[37] The changes reflected in Table B.28 for 1956-58 seem to be grossly understated.

The "leakage" from national income in the form of redundant stocks and protracted and dispersed investment effort can be related to the strength and rapidity of the investment push. The unreliability and aggregation of data make comparisons of distribution of accumulation difficult and the dimensions of maldistribution and stock redundancy hard to ascertain. Apparently, Bulgaria surpassed most CMEA countries in inventory additions, and its rates seem to have been extraordinarily high, not only in comparison with the West but also within the CMEA (see Tables B.22 and B.31). In Bulgaria investment in inventories and unfinished construction fluctuated markedly.[38] For example, this share in accumulation grew from 60.3 percent in 1952 to 76.7 percent in 1953, reflecting considerable pressures on the investment front, wide dispersal of resources, overextended gestation periods, and failures in commissioning new capacities. This share receded to 35 percent in 1954, 20.5 percent in 1955, and 8.1 percent in 1956, corresponding to a decline in the rate of increase of investment and the consequent relaxation of pressures. However, by 1957 (though investment declined by 3 percent) the share rose to 39.1 percent; this was partly due to the replenishing of inventories depleted by the increase in consumption in 1954-56. As could be expected, the unfinished construction and additions to stocks jumped following the resumption of the investment rush (and deterioration and expectation of further worsening in supply) to 43.5 percent in 1958 and 53.1 percent in 1959.[39]

Investment

The data on relative movements of investment outlays and fixed capital put into operation are inaccurate and sometimes conflicting. Roughly, the greater the acceleration of investment outlays, the slower the additions to the stock of fixed capital, and the greater the accumulation of unfinished construction. While many factors are at work (including the arbitrary cut-offs of FYPs and planners' anxieties to start as many projects as possible at the beginning of the period and to finish as many as possible before the start of the new period, time lags, and so on, the strategic factor in the rate of improvements of additions to capital stocks is the slowing down in the investment push (the number of projects started).

CMEA countries generated remarkably high rates of capital investment, generally surpassing those of OECD countries by a wide margin and only occasionally equaled in the West. Tentative Alton-Project estimates suggest that the shares of GNP (at factor cost) allocated to gross fixed investments increased almost continuously in most European countries in 1950-63. But the increase tended to be considerably higher in the CMEA. The shares in Western Europe rose from 20 percent (or below) in the early 1950s to about 25 percent in the early 1960s, whereas in Eastern Europe (except the GDR) they rose from a low 20 percent to around 30 percent. In 1950-54 Hungary was the leader (with 25.9 percent), but by 1955-59 Bulgaria shifted to the forefront and retained this rank (far outdistancing the others) in 1960-63 (with 41.5 percent).[40]

Japan's star performance in postwar growth experience cannot be excluded from these comparisons. Whatever other factors there are in this context, it is noteworthy that Japan had the highest investment (and small defense) shares.[41]* Similarly, West Germany (another front-ranking postwar performer) channeled a relatively high share of GNP to investment (and little to defense), but less than Japan. Apparently, Japan's share of gross investment in 1955-66 (32.3 percent) surpassed that of the USSR (26.6 percent). Among CMEA countries, the Soviet share was above the Czechoslovak (25.2 percent), Polish (25.5 percent), Hungarian (23.9 percent), and GDR (20.7 percent) shares. Bulgaria's share (34 percent) seems to have exceeded even the Japanese.[43] But given the reliability of the data, it would be venturesome to suggest by how much.** According to the average annual growth rate of investment in 1951-67, CMEA countries could be ranked in the following descending order: Bulgaria, Rumania, the GDR, the USSR, Poland, Czechoslovakia, and Hungary (see Table B.24).[45] However, the burden of investment on consumption was partly alleviated in Bulgaria by Soviet and other CMEA credits and preferential treatment, as will be seen in Chapter 3.

Distribution of Investment

The distinctive features of the Soviet growth strategy can be gauged not only by the growth of the share of national product allocated

---

*Norway registered nearly as high investment shares, but Simon Kuznets contends that this might have been a "statistical illusion."[42]

**According to UN recalculations, to mitigate incomparability of data, the share of gross fixed investment in GDP amounted to about 33 percent in most East European countries in 1965, with the highest (35 percent) in the USSR, followed by Bulgaria, Czechoslovakia, Poland, and Rumania (32 percent), Yugoslavia (30 percent), and Hungary (28 percent). The "comparable" shares in the West were as follows: Spain, 28 percent; the GFR, Italy, and Norway, 27 percent; Japan, Sweden, and Austria, 26 percent; France and Canada, 25 percent; Greece, 24 percent; the United Kingdom, 22 percent; the United States, 21 percent; New Zealand, 17 percent.[44]

to accumulation but also by its destination, and especially by the structure of investments. The growth rate, structure, and path depend, among other things, on (1) distribution of national income into consumption and accumulation, (2) apportionment of accumulation into investments in fixed assets and increases of inventories, (3) distribution of investments between so-called spheres of material and nonmaterial production (including administration, education, science, health services, and housing), (4) allocation of productive investments between sectors and branches (subbranches) of material production, and (5) choice of technology.

Generally, investment was allocated by according a strikingly discriminatory preference to growth-forcing, rather than to welfare-oriented, activities. This policy was periodically tempered. The rise in physical capacity to produce depends directly on productive investments alone—that is, on expenditures on capital goods involved in production of goods. Nonproductive investments refer to outlays on such new fixed capital goods as do not directly contribute to production of goods (such as housing, education, health-care centers, cultural, sport, and recreational facilities, and so on). However, the share of nonproductive investment is often inflated by including all kinds of defense and production-supporting activities. The preferential treatment of productive investment is obvious at the allocation stage, but even more revealing during execution, when resources are diverted from nonproductive investments (especially the welfare-oriented ones) to bolster up execution of productive investments.

The data on the shares of productive and nonproductive investments in CMEA countries are of limited comparability. In ranking the countries in ascending order according to the share of nonproductive investments, it is significant that the least developed countries, which place the greatest stress on investment as a growth-forcing factor, have the lowest share of nonproductive investments: Rumania, Bulgaria, Hungary, Poland, Czechoslovakia, and the USSR. In Bulgaria the share of nonproductive investments was about 37 percent in 1950. By 1951-55 it declined to 19.7 percent. Despite the period of consolidation, this share continued to drop in 1956-60, to 15.6 percent, and even further, to 13.8 percent, in 1961-65. However, in the latter part of the 1960s it reached again about 20 percent, climbing to about 25 percent in the early 1970s. In Rumania the share of nonproductive investment was only 13 percent in 1951-55 climbed to 16.6 percent in 1956-60, but reentrenched to 15 percent in 1961-65, and gravitated around that figure in the latter 1960s and early 1970s. In Hungary the share was 21.3 percent in 1951-55 and rose slightly, to 23.3 percent, in 1956-60, only to fall back (to 21.7 percent) in 1961-65. But in the latter 1960s it climbed to about 30 percent, exceeding that mark in the early 1970s. In Poland the share was 25.6

percent in 1951-56, rose to 29.7 percent in 1956-60, returned to the previous level in 1961-65, and dropped slightly (to about 23 percent) in the latter 1960s and early 1970s. In Czechoslovakia the share dropped from 30.9 percent in 1951-55 to 27.2 percent in 1956-60 and 23.9 percent in 1961-65; it returned to about 30 percent in the latter 1960s and early 1970s. In the USSR the share was 32.4 percent in 1951-55, climbed to 37.8 percent in 1956-60, decelerated to 35.3 percent in 1961-65, and gravitated around 30 percent in the latter 1960s and early 1970s.[46] Whereas it was the general tendency in the CMEA for the share of nonproductive investments to decelerate in 1961-65, compared to the previous quinquennium, in the post-1965 period this share accelerated markedly in Bulgaria, Hungary, and Czechoslovakia, remained at its low level in Rumania, and fell somewhat in Poland and the USSR.

In almost all countries, in most years, the rates of growth of productive investment outpaced those of total and nonproductive investments. For example, in Bulgaria productive fixed investment increased by 436 percent and nonproductive by only 176 percent from 1952 to 1966. From 1950 to 1966 the respective increases were as follows: Hungary, 517 and 198 percent; Rumania, 292 and 193 percent; Czechoslovakia, 251 and 168 percent; the USSR, 439 and 345 percent.[47]

In the CMEA industry and construction absorbed a considerably higher share of investment than in Western Europe. With large variations, the shares in the former were between 40 and 50 percent, and in the latter about one-third (see Tables B.26-27). The absorption by agriculture also surpassed Western countries (even those with a significant agricultural sector). Although one cannot generalize on this basis, it appears that the problems of agriculture in the CMEA cannot be largely attributed to discriminatory allocation of investment. But the record of allocation would look considerably worse if meaningful adjustments could be made for the often lower quality of resources for agriculture. The share of housing in the CMEA (10 to 20 percent) was considerably below that in Western Europe (20 to 30 percent), but here again quality is a major factor.

In Bulgaria the combined share of industry and construction in total investments was less than 40 percent in the 1950s and about 45 percent in the 1960s and early 1970s. With the exception of the USSR and Hungary in post-1956, other CMEA countries allocated a larger share of investment to industry and construction than Bulgaria (Rumania exceeded the 50 percent mark in the latter 1960s and early 1970s). Since the mid-1950s Bulgaria has allocated to agriculture a considerably larger share of investment than have other CMEA countries. From the mid-1950s to the mid-1960s the share of investment for agriculture was largest in Bulgaria as a support for the collectivization drive in the second half of the 1950s.[48] This share also increased in the same period in Czechoslovakia and Rumania, whereas it began rising only in the early 1960s in Hungary and in the latter 1960s in Poland. In Bulgaria in the 1960s and early 1970s the share of investments allocated to industry grew at the cost of those for agriculture and housing. A somewhat larger share of investment was channeled to housing than in other CMEA countries in the early 1950s, but this share began to fall off in the late 1950s and, together with Rumania, Bulgaria had one of the lowest shares in the late 1960s and early 1970s.

Major structural shifts in allocation of investment were not restricted to countries like Bulgaria where the investment rates increased most sharply in the 1960s and redistribution could rely considerably on the increment. Specific factors were at work, but in Czechoslovakia the sectoral shifts were very pronounced, while the investment rate ascended only relatively modestly.

Investment in Industry

In the period 1951-67 in all CMEA countries the average rate of growth of investment in industry was higher than the average rate of growth of total investment. Classifying the countries in descending order of the growth rate of investment in industry we get the following: Bulgaria, Rumania, the GDR, Poland, the USSR, Hungary, and Czechoslovakia (see Table B.24). Alton-Project estimates of industrial fixed capital expansion suggest that in 1952-67 Bulgaria's average rate of growth was nearly 12 percent, almost double that of Czechoslovakia in 1948-67 (6.3 percent), and 1.5 times as large as Hungary in 1949-67 (7.7 percent). This limited intra-CMEA comparison was extended for 1960-67, indicating a top rank for Bulgaria (14.3 percent), followed by Rumania (10.7 percent), Hungary (7.5 percent), Poland (6.9 percent), and Czechoslovakia (6.7 percent).[49] Despite the retardation of the growth rate of industrial output in the 1960s, there was no corresponding decline in the rate of increase of capital inputs in industry. On the contrary, except for Hungary, the average rate of growth of capital rose, indicating declining returns.

The average growth rates tend to blur the considerable fluctuations of investment outlays in industry, which not only correspond to the fluctuations of investment per se—to be discussed in the next section—but also reflect periodic relative priorities, and the large swings of the pendulum of priorities. Thus, in Bulgaria, with a 2-percent growth of investment in 1956, investment in industry declined by 16.7 percent, and with a total decline by 3 percent in 1957, industrial investment grew by 17.9 percent. The upsurge of investment in 1958 (22 percent), 1959 (63 percent), and 1960 (18 percent) seems to have accorded preference to nonindustrial investments, for the respective growth rates of investment in industry were -4 percent, 40.4 percent, and -0.6 percent. However, the next two resurgences of investment momentum gave clear priority to industrial investment. Thus, with a total growth of 15 percent in 1963 and 10 percent in 1964, industrial investments grew by 96.6 and 27.2 percent, respectively, and with total growth of 22 percent in 1966 and 25 percent in 1967, industrial investments grew by 20.6 and 25.6 percent, respectively (see Table B.25).[50]

The tenet of investment priority for heavy industry was influenced by the highly discriminatory allocation of industrial investments though the actual degree of priority accorded between and within groups shifted over time. In 1951-55 most CMEA countries allocated 85 to 90 percent of industrial investments to group A. In 1956-60 this share receded to less than 85 percent (with the exception of Rumania and the USSR) and increased to more than 85 percent in 1961-65 (with Rumania and the USSR at about 88 percent). In Bulgaria the share of investment in group A rose from 78.6 percent in 1949 to 87 percent in 1950, declined to 84.3 percent in 1952, rose again to more than 90 percent in 1956, receded to 83 percent in 1957 and 79.5 percent in 1960, and returned to 85 percent in 1965 and 84 percent in 1966-67.[51] In Bulgaria, even during the period of consolidation and partial placating of the consumer (the mid-1950s), no striking shifts occurred in the allocation of industrial investment to group B, although those are hardly accurate guides, partly because group A includes producer goods for the consumer sector, and heavy industry produces consumer durables.

While the composition of industrial investments did change considerably over time, it is remarkable that the shifts were not more pronounced. By and large there were rather limited modifications of the leading-links approach. The factors favoring perpetuation of the existing structure must have been strong. In the earlier period a considerable share of industrial investment in Bulgaria was allocated to "electrification"; there was a decline of this share thereafter (an experience similar to those of the USSR and Rumania). In Bulgaria this share was about 37 percent in 1950, 23 percent in 1952, 28 percent in 1956, and receded to about 15 percent in the 1960s. Electric power and fuels (primarily coal) absorbed some 45 percent of industrial investment in 1950, 31 percent in 1952, 49 percent in 1956, 30 percent in 1960, 29 percent in 1965, and 21 percent in 1970. In the second half of the 1950s the share of investments in fuels was reduced. Metallurgy was accorded high priority in the USSR, Rumania, and Bulgaria; in the last it was the second-largest beneficiary. The share of investment allocated to metallurgy rose from 4 percent in 1950 to 28 percent in 1952, dropped to 15 percent in 1956, rose again to 18 percent in 1960 and 25 percent in 1965, and fell to 11 percent in 1970.

Ferrous metallurgy was forced vigorously, with insufficient consideration for factor endowments. Although Bulgaria is much better endowed with nonferrous ores, investments in nonferrous metallurgy were on the rise only in the latter part of the 1960s. The chemical industry was the third-largest recipient of investments, but the reordering of priorities was considerable, with drastic shifts in the 1960s, influenced by various attitudes of different Soviet leaders. Because of its costly commitments to capital-intensive basic-material projects, such as coal and ore mining (extracted under adverse geological conditions with heavy investment in equipment) and metallurgy in the early 1960s, Bulgaria was falling behind the USSR and Rumania in allocating investments to chemicals. From 5 percent in 1960 the share of investments allocated to chemicals almost doubled by 1965, and again nearly doubled (19 percent) in 1967; it fell off to 15 percent in 1968 and 13 percent in 1969 and rose to nearly 16 percent in 1970. Naturally, machine building is among the favored branches, with a constantly rising share (from 5 percent in 1952 to 10 percent in 1960, 12 percent in 1965, and 20 percent in 1970), but this share is apparently insufficient, given the extensive investments in metallurgy. In 1967 the share of investments allocated to machine building in Bulgaria approached the average in the CMEA but exceeded the share in Rumania by a wide margin (see Table B.28).

Among the incongruities of investment allocation in the CMEA throughout most of the postwar period was a relatively very high share (about 75 percent) of industrial investments absorbed by basic-material-supplying industries and relative underinvestment in some processing branches. With some variations, in the 1960s there was a moderate decline in this share in the USSR, Hungary, and possibly Poland, whereas Bulgaria returned to its earlier extraordinarily high share, and Rumania maintained its high share practically unchanged.[53]

In the CMEA the disparity between construction capacity (including building materials) and investment activity plagued the planners throughout most of the postwar period. In Bulgaria various measures were adopted, including increased allocation of investment to building materials. The share of investments in this branch rose from 3 percent in 1952 to 5 percent in 1956 and 6 percent in 1960, receded to 4 percent in 1965, increased somewhat thereafter, but fell back to 4 percent in 1970. In Bulgaria the share of investment absorbed by food processing declined sharply in the first half of the 1960s, compared with the preceding quinquennium: This share was 6 percent in 1950, 7 percent in 1952, 5 percent in 1956, 13 percent in 1960, 8 percent in 1965, and 10 percent in 1970. The corresponding shares in textiles were 3 percent, 4 percent, 3 percent, 6 percent, 4 percent, and 6 percent.[54] These shares tended to be somewhat higher than in Rumania and in some other CMEA countries. Nevertheless, despite their export potential, the c.p. often neglected food processing and some branches of light industry, though he treated them somewhat more favorably in the 1960s. The shares of investments allocated to food processing and light industry exhibited a sustained increase only in Hungary. In these branches output grew mainly through increase in employment. Whatever other problems beset food processing and light industry the key one seems to be deficiency of fixed capital and its antiquated and sometimes dilapidated state. More palpable incentives to produce the desired assortment and quality are needed, but they are insufficient without substantial enlargement and rejuvenation of the capital stock.

FLUCTUATIONS AND DECELERATION OF GROWTH RATES

Figures 2.1 to 2.7 indicate two basic features of development in Eastern Europe: fluctuating and declining rates of activity. The relatively lower rates of activity in the 1960s, compared with the 1950s, were generally accompanied by relatively less-pronounced fluctuations. At first sight it would appear that the CMEA countries could be ranked as follows, in descending order according to their experience of intensity of fluctuations in economic activity: Bulgaria, Rumania, Hungary, Czechoslovakia, the GDR, Poland, and the USSR. The fluctuations were particularly pronounced in the first four countries listed, and especially in terms of the most volatile macrovariable considered (investment). In almost all cases the fluctuations in the growth rates of investment were more striking than those of national income, which in turn were more marked than those of industrial output, partly because of the agricultural component. In the boom periods growth rates of investment far exceeded both industrial output and national income, and in the downturns investments grew at much slower rates than either industrial output or national income. Czechoslovakia and the GDR followed the general pattern, but fluctuations in national income and industrial output evinced roughly similar amplitudes, partly due to the relatively lesser weight of agriculture. While both Bulgaria and Rumania exhibited appreciable fluctuations of industrial output, their amplitudes seem to have been less sharp than those in Czechoslovakia, especially in view of the deceleration from 1960 to 1963 and the absolute decline in 1963, which (with the exception of the tragic events of 1956 in Hungary) was unprecedented in the postwar CMEA.[55] Fluctuations of industrial output appear to have been weakest in the USSR and Poland.

In countries that rely predominantly on the domestic supply of capital goods, such as Czechoslovakia and the GDR, the wide ampli-

FIGURE 2.1

Fluctuating Rates of Growth of National Income, Industrial Output, and Investment in Bulgaria, 1951-74

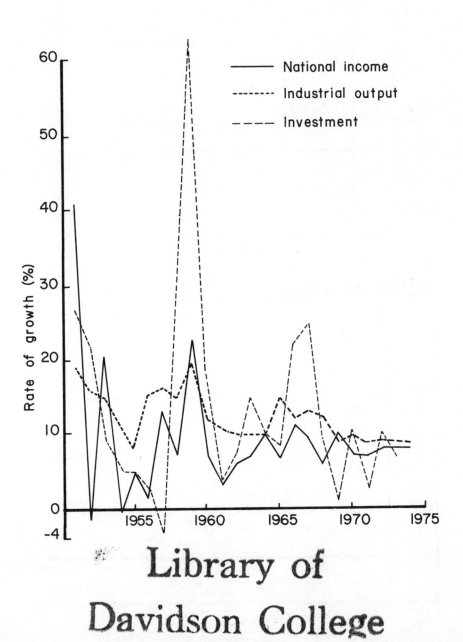

FIGURE 2.2

Fluctuating Rates of Growth of National Income, Industrial Output,
and Investment in Czechoslovakia, 1951-74

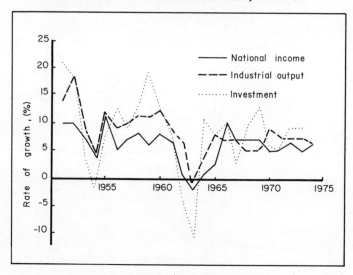

FIGURE 2.3

Fluctuating Rates of Growth of National Income, Industrial Output,
and Investment in the GDR, 1951-74

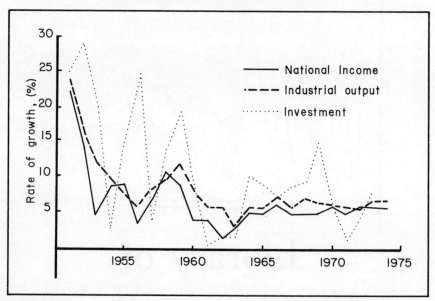

FIGURE 2.4

Fluctuating Rates of Growth of National Income, Industrial Output, and Investment in Hungary, 1951-74

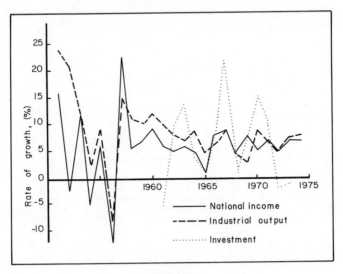

FIGURE 2.5

Fluctuating Rates of Growth of National Income, Industrial Output, and Investment in Poland, 1951-74

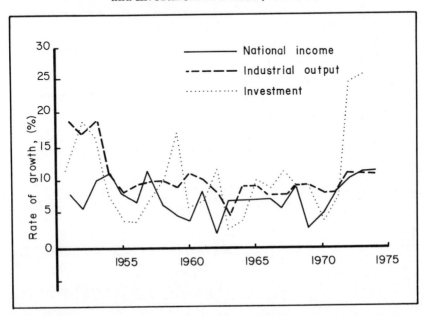

FIGURE 2.6

Fluctuating Rates of Growth of National Income, Industrial Output, and Investment in Rumania, 1951-74

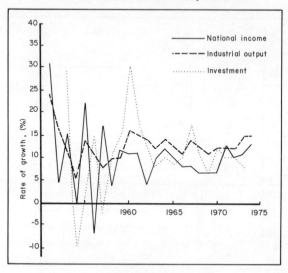

FIGURE 2.7

Fluctuating Rates of Growth of National Income, Industrial Output, and Investment in the USSR, 1951-74

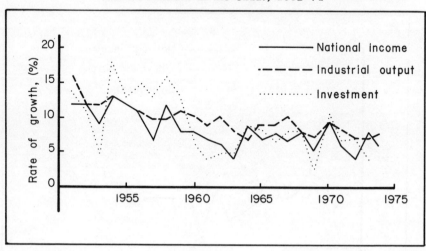

tudes in fluctuations of investment affect significantly the fluctuations of industrial output. In countries that rely primarily on import of capital-goods, balance-of-payments tensions and the offsetting effect of aid are crucial factors affecting fluctuations. Other factors are the erratic supply of imported and domestic machinery, the supply of raw materials, the absorptive capacity of construction, and the availability of building materials.

Poor harvests—whether attributable to nature or to mismanagement—contributed to amplifying fluctuations in national income. Their influence was similar in case of industrial output, especially in countries, such as Bulgaria, where a weighty share of industrial output is derived from agricultural raw materials. Both Rumania and Bulgaria suffered from ample fluctuations in agricultural output; sharper fluctuations were evident in the former. In both countries the fluctuations were less pronounced in the 1960s than in the 1950s (see Table B.47). Hence, in the latter period the impact of agriculture on fluctuations of overall production was weaker, also because of the relatively lower share of agriculture in total output.

It is difficult to judge to what extent the booms and downturns in activity coincided among the countries. With the exception of the USSR, and allowing for certain variations in time span and considerable differences in amplitudes, it would appear that from 1950 until 1965 in most East European countries the periods of booms and downturns were more or less synchronized, with the booms occurring in ca. 1951-53 and 1958-60 and the downturns in ca. 1954-56 and 1961-63. Thereafter, the fluctuations became less pronounced and less synchronized throughout the bloc.

Postwar Reconstruction and Industrialization

In broad terms the first boom was attributable to rapid recovery from war devastation to the reaping of quick returns on investments of relatively low capital intensity. The policy of industrialization pursued was not devoid of a measure of success: Heavy industry formed the spearhead of rapid development. However, it gave rise to serious disproportions and dislocations that hindered further expansion of certain branches, claimed resources that had to be withdrawn from underprivileged areas of activity, produced disparate and deleteriously unbalanced rates of expansion of various components of the economic process, and led to sheer waste of resources in high-priority areas (including negative marginal overproductivity in expanded branches).

Despite the overall fulfillment of industrial output plans in 1951-52, in many countries output of some basic industries, such as steel, machine building, fuels, and power, was below plan targets.[56] In all countries even the modest goals set for consumer goods and housing were unfulfilled. The industrialization rush was supported by collectivization of agriculture. In Bulgaria 15 percent of farmland was collectivized in 1949, and 50 to 60 percent in 1950.[57] But this was often involuntary and achieved by means of threats and pressure that, together with increased compulsory deliveries, at prices way below market levels, had disastrous effects on agricultural output. Moreover, the establishment of large farms was not accompanied by adequate mechanization.

The overheating of the economy, together with the growth of nonagricultural employment, resulted in increased demand for food by the urban population. The pressure of demand on food supply was also aggravated by bad weather and mismanagement of agriculture.* Moreover, the outflow of population from agriculture was not accompanied by a shift in food supplies but by a higher per capita consumption of the remaining peasants. The flow of food supply to urban centers was aggravated by speculation.

From 1951 to 1953 real wages were barely maintained, or they even fell. Per capita real incomes of the urban population fared somewhat better because of the larger number of employed household members. Deficiencies in consumer goods and the threatening decline of per capita food consumption of the urban population tended to decelerate increases in labor productivity and to disrupt industrial production.[59]

THE NEW COURSE

This first boom was followed by a downturn in 1954-56 that has gone down in East European history as the New Course (in some countries it started as early as 1953 and spread through 1957). The new policy adopted featured mainly a relaxation of investment push and consequent slackening of pressures, scaled-down growth rates of national income and industrial output (with certain changes in composition, while growth rates of A were slower than those of B, in marked contrast to the preceding period). In general, the New Course featured policies more palatable to the consumer-worker, who had been given a rough time during the reconstruction and early industrialization period.

The aim of the New Course was to eliminate some of the disproportions that had previously developed. To that effect it was postulated

---

*In Bulgaria, while agricultural production was planned to increase by 57 percent from 1949 to 1953, it only grew by about one-third.[58]

that within the next two to three years the output of consumer goods, agriculture, and certain lagging raw materials would be increased consequentially. To support this an attempt was made to rechannel resources from investment to consumption, while agriculture and light industry were allocated more investment. Wages were raised and higher prices were paid to farmers, while prices of some consumer goods were reduced. Substantial progress notwithstanding, the shift from high investment rates and priority of heavy industry was both difficult and costly in terms of considerable dislocations. The flow of materials and semifabricates through the productive process was often disrupted, resulting in accumulation of stocks and bottlenecks. The increased purchasing power was not matched by a supply of consumer goods, and shortages of many consumer goods continued.[60]

In 1953-54 retardation of growth rates was probably the most crucial outward demonstration of the need for remedial action. The death of Stalin was only one and probably not the key factor in the adoption of the New Course. This is not to say that the extent and timing of the shift in economic policy were not influenced by the political situation and social pressures. But I would venture the opinion that the pivotal factor in altering the course was the marked deceleration of growth rates that could be discerned virtually throughout the area. The timing and intensity of the retardation varied significantly among the countries of the bloc, and the attribution of the proximate causes of the phenomenon differed somewhat. The remedial measures adopted were, in turn, conditioned by the relative severity of the disproportions and decline in growth rates, the relative stage of development, dependence on foreign trade, and a host of other factors. The New Course was more strongly emphasized in some countries than in others.

The Second Industrialization Rush and Downturn

In almost all countries the New Course was terminated in 1957. The following year saw an upsurge of investment activity that could be termed a second wave of investment. To be exact, in Czechoslovakia, the GDR, and Rumania investment activity had picked up momentum already in 1955-56, but it decelerated again in 1957. In Poland a slight acceleration started in 1957 as a prelude to a new wave of investment, which resembled that of the early 1950s but was less extreme.[61] Similarly, Hungary embarked on a new industrialization drive—a considerably toned-down version of the extremely lopsided investment rush of the early 1950s.[62] In 1958 investment rose by 28 percent (from a reduced base); in 1959, by 41 percent; in 1960, by

17 percent.[63] In Bulgaria investment activity reached its nadir in 1957 (-3 percent). None of the other countries could match the pace of the second wave of investment in Bulgaria that began in 1958, reached its peak in 1959, and bottomed out in 1961. This was indeed the Big Leap Forward from which the Bulgarian economy has not yet fully recovered. In view of its sheer magnitude and abruptness of takeoff, the Leap considerably magnified the barriers, ceilings, and inefficiencies that usually follow a spurt of investment and extended the period of retrenchment necessary to overcome them.

The Leap Forward was motivated by a number of economic and political factors. Among them were (1) the pressing need to create jobs for those displaced by amalgamation and mechanization in agriculture and the clamping down on handicrafts and the artisanat, (2) the desire to catch up with the more-developed CMEA members who were also starting on a new investment wave, (3) the pressures for raising agricultural output to supply raw materials and foodstuffs and notably as a source of export earnings essential for import of equipment, and (4) intraparty strife.[64] By far the most startling feature of the Leap was the rise in investment activity, accompanied by a 60-percent rise in imports of machinery and a 50-percent increase in output of metal-using industries. Acute difficulties in construction developed already in the second half of 1959, manifested in the well-known frittering away of resources over numerous projects, shortages of skilled construction workers, tie-ups in design offices, and shortages of building materials.[65] Despite the relative slowing down of the investment momentum, these problems continued to plague Bulgarian construction in 1960 and 1961, with a growing disparity between investment outlays and completion of projects. For example, whereas unfinished investment was 2.4 percent of national income distributed in 1958, it grew to 4.3 percent in 1959 and to 7.6 percent in 1960.[66] The strain of the Leap on the balance of payments was indicated by the new credit received from the USSR in 1960 and the postponement of a $40-million debt service due to the USSR in 1961.[67] Fortunately, the impact of the Leap on consumption was reduced by good harvests, sizable import surplus, and inventory disinvestment.

The second waves of investment also reached "relatively respectable" proportions in Rumania (where the speed-up continued until 1960), the GDR, Czechoslovakia, and Poland; with the exception of Poland, were not to be equaled by future investment waves in the 1960s and early 1970s. In all countries this period featured considerably lower growth rates of B.

The second downturn in economic activity was most pronounced in Czechoslovakia and the GDR. In both countries, after the peak of investment activity in 1959, investment decelerated until it reached a nadir in 1963 (in the GDR it actually hit bottom in 1961 and leveled

GROWTH AND INSTABILITY 47

off in 1962-63). In Czechoslovakia growth rates of national income and industrial output began decelerating in 1961 and reached their lowest point in 1963; in the GDR the deceleration process set in in 1959-60, and the low point was also reached in 1962-63, but it was not quite so disastrous as in Czechoslovakia.

In Bulgaria the second downturn was of shorter duration. After the 1959 adventure the rate of growth of investment began declining until it reached its lowest point in 1961. Similarly, growth of national income and industrial output peaked in 1959 and did not bottom out until 1961-62. In Rumania the second downturn in growth of national income and industrial output began in 1960-61 and bottomed out in 1962-63. The rate of growth of investment reached a peak in 1960, after which it declined until 1963. In Poland, after a considerable decline in 1960, the rate of growth of investment picked up momentum again, peaked in 1962, and declined in 1963. After a peak in 1960-61, the growth rates of national income and industrial output began to fall, reaching a trough in 1962-63. In Hungary, too, a low point in the growth rate of investment was recorded in 1961 (-5 percent), but another peak was achieved in 1963. Growth rates of national income and industrial output recorded a peak in 1960 and bottomed out in 1962-63. The modest rate of increase of investment was supported by substantial imports, contributing to marked foreign-trade deficits in 1962-64. This was one of the reasons for scaling down the investment targets.

As in the previous downturn, growth rates of B came to the forefront again. In Bulgaria group B grew faster than A in 1961. In Czechoslovakia the rates converged in 1961 and B grew faster than A in 1963. In Rumania the rates converged in 1960. However, in Poland rates of growth of A declined from 1961 to 1963, and those of B from 1962 to 1963, almost at the same velocity. In Hungary there was some convergence of growth rates of A and B in 1959-60, with a slight excess of A over B that was eliminated in 1961, and B exceeded A by one percentage point in 1962 (see Table B.36).

This second downturn in economic activity was primarily traceable to magnified pressures and imbalances, rising growth barriers, and encountered ceilings that accumulated during the second wave of investment. It proved to have been inordinately costly, not only in its short-term adverse effects but for the years to come. Among its notable side effects was the rise in almost all CMEA countries of economic reformism. Reform of the system of functioning was sought as a panacea for the decelerating growth rates. As we shall see later, the system was blamed not only for its own sins but also for many of the shortcomings inherent in the growth strategy pursued.

Decelerating Growth Rates

In Bulgaria investment activity again picked up momentum in 1962-63, but the peaks of activity in national income and industrial output were not reached until 1965-66. Investment grew sharply from a low point in 1965 to a peak in 1966-67 and decelerated at a rapid rate in 1968-69, in order to allow for completion of the widely spread investment front. This retrenchment was also due to accumulated difficulties in foreign trade. From the low point in 1969 (1 percent) the acceleration of investment in 1970 was not spectacular, exceeding only slightly the trough in 1965; it fell off again in 1971 and grew in 1972 to slightly below the 1970 rate. This was an indication of considerable difficulties at the end of the 1960s and early 1970s. Although national-income and industrial-output growth were subject in the post-1965 period to slighter fluctuations than in the past, they were also subject to an obvious deceleration. However, the retardation of growth of industrial output in the 1960s was less pronounced in Bulgaria than in other CMEA countries, except for Rumania. Similarly, in Rumania peaks of investment activity occurred in 1967 and 1970, but the troughs were not so low as in Bulgaria and, on the whole, Rumania maintained a more sustained investment drive in post-1965. Also, though the rates of national-income and industrial-output growth fluctuated slightly more than in Bulgaria, they were maintained at considerably higher levels in the post-1965 period than in Bulgaria.

Both in Czechoslovakia and the GDR the post-1965 growth rates of national income and industrial output were fairly steady but considerably below those reported in the previous period. Czechoslovakia exhibited spurts of investment in 1964-66 (with a trough in 1967) and in 1968-69; it had a deceleration in 1970-71 and a pickup in 1972-73. In the GDR a fairly steady climb of investment that started in 1964 (with a slight deceleration in 1965) reached a peak in 1969 and declined thereafter, with a trough in 1971 and some acceleration in 1972-73.

The decline of growth rates in the post-1965 period was not so pronounced in Poland and Hungary as in Czechoslovakia and the GDR. In Hungary both national-income and industrial-output growth rates continued to fluctuate considerably, whereas in Poland the growth of industrial output was fairly steady, but national income fluctuated widely (due to the erratic performance of agriculture). In Hungary, after a low point in 1965, the investment momentum picked up again in 1966 and peaked in 1967, only to decelerate abruptly in 1968. This was followed by another surge of investment that lasted from 1969 to 1971. Investment recorded negative growth rates in 1972-73. During the post-1965 period, Poland seems to have made the most consistent investment effort. Fairly steady and relatively high investment rates were maintained from 1965 to 1969, resulting in a rising consumer-

GROWTH AND INSTABILITY

frustration barrier and the tragic explosion of December 1970.[68] A deceleration occurred in 1970, but Poland embarked on a new wave of investment in 1971-73.

In Bulgaria from 1962 to 1965 growth rates of B were considerably below those of A; there was a slight convergence in 1966-67 and a continuing disparity in 1971-72. The situation was similar in Rumania, with the exception of 1971, when these growth rates were quite close, and 1972, when B actually grew by one percentage point faster. In Czechoslovakia, after a convergence in 1965-66, these rates diverged in 1967, but for 1966-70, on the average, and for 1971-72 the convergence held, with only one-percentage-point excess of A over B. In Poland there was a more restricted divergence of growth rates of A and B from 1964 to 1968, but rates converged in 1971, and B exceeded A by three percentage points in 1972. Hungary, in which growth rates of B exceeded those of A in 1963 and 1967, on the average throughout the 1966-70 period, and in 1971-72, appears to have been the only CMEA country where the primacy of A over B has been more consistently rejected in the most recent years (see Table B.36).

It is difficult to compare the fluctuations in the USSR with those of other CMEA countries. This difficulty is partly due to the Soviet Union's relative self-sufficiency and greater ability to contain the rise in the consumer's frustration barrier, making the country less subject to erratic movements. However, as shown in Figure 2.7, the USSR presents an obvious example of declining growth rates over about a quarter century. Its fluctuations were less abrupt and erratic than those in other CMEA countries, nevertheless a certain regularity of fluctuations can be discerned. A trough in growth rates was reached in 1953; they then picked up in 1954, began decelerating considerably in 1959, and reached a trough in 1962-63. The pick-up in 1964, which lasted till 1968, was at a considerably lower level. A deceleration occurred again in 1969, with a much shorter upswing in 1970-71 and a deceleration again in 1972-73.

Generally, two distinct periods in the development of CMEA countries can be detected: (1) postwar to 1960, marked by relatively high and fluctuating growth rates, and (2) post-1960, distinguished by considerable retardation of growth rates, especially in the more developed countries, and some tendency for the fluctuations to subside.

HUMAN RESOURCES AND EMPLOYMENT

Broadly speaking, industrialization in Eastern Europe has proceeded by drawing the unemployed or employable of productive age into the labor force and by shifting disguised unemployment in agriculture into the productive process. Once this is accomplished, further increases in employment have to rely on the natural increase of the population and the removal of sources of inefficiency in the use of labor. To a varying degree, during the postwar period, the three main sources of labor supply were the exit from agriculture, youngsters coming of working age, and increasing labor participation rate of women.

GROWTH OF THE LABOR FORCE

At the threshold of the 1950s the STEs could be divided into two distinct groups: (1) those with considerable disguised unemployment in agriculture, a relatively small industrial labor force, and a fairly sizable artisanat (Bulgaria, Rumania, and Poland) and (2) those with a relatively fully employed rural population and a well-trained industrial labor force (Czechoslovakia, the GDR, and, partly, Hungary, where there was still some unemployment among the rural population). The Soviet pattern was more easily adaptable in the initial period in the first group than in the second, where it placed a severe strain on employment (the strain was partly mitigated by increasing the labor participation rate of women and use of pensioners).

The CMEA countries could be ranked in descending order as follows, according to their dynamics of employment during the period 1950-68: Bulgaria, Rumania, the USSR, Hungary, Poland, Czechoslovakia, and the GDR (see Table B.54). The difference between Poland and Hungary is inconsequential, so it is the countries of the first group that have exhibited the highest growth dynamics in employment. According to official statistics, in all CMEA countries increase of employment outpaced that of population. Countries with the highest population growth (the USSR and Rumania) tended to have the highest growth rates of employment. Bulgaria is a notable exception: The population increased only by 15 percent, and employment by 221 percent. In contrast, Poland, with a population growth at about the Soviet rate (30 percent), increased employment at about the same rate as Hungary, where the population grew by 10 percent. In the GDR, with a decline of population by 7 percent, the employment rose by nearly 30 percent. In Czechoslovakia, where the population grew at about the same rate as in Bulgaria, employment expanded at about one-third of the Bulgarian rate. But these statistics are highly suspect. The United Nations attempted to reduce the incomparability of data (partly due to change in ownership), and we can deduce the following ranking in descending order of employment growth from 1950 to 1967: the USSR (39.6 percent), Poland (36 percent), Hungary (18.2 percent), Rumania (18 percent), Czechoslovakia (9.8 percent), Bulgaria (8.5 percent), and the GDR (7.6 percent).[69] According to these data, employment grew faster than population in the USSR, Poland, Hungary, and the GDR, but slower in Bulgaria, Rumania, and Czechoslovakia.

A more interesting picture emerges if the increase in employment is analyzed by periods, but such an analysis points out the deceptiveness of averages (see Table B.53). During the initial period (1951-55), Bulgaria registered a 5.2 percent average annual rate of growth of employment; this was exceeded by Hungary, Poland, and Rumania. In Hungary and Poland this was the period of highest employment rates, as the exodus from agriculture was being encouraged. The highest employment rates (11.5 percent), registered by Bulgaria in 1955-60, far exceeded those in other countries, as well as those in Bulgaria in any other period. This was the result of the Leap Forward, partly motivated by the need to provide jobs for the rising unemployment, which was rumored to be at least 350,000 (an astounding two-thirds of industrial employment in 1957, when about 3 million people were still employed in agriculture)—a great embarrassment to the regime.[70] This was partly due to the highly capital-intensive techniques adopted. But it was also affected by the unprecedented rapidity with which manpower was shifting away from agriculture, far in excess of the needs of the other sectors, creating a "migration loss." Forced collectivization and initial imposition of extremely harsh methods of collecting the agricultural produce (such as discriminatory low prices for high compulsory deliveries) pushed peasants to seek a livelihood elsewhere; the exodus was in addition to the "normal" release resulting from reorganization and mechanization of farms. Improper choices of unduly capital-intensive techniques reduced demand for labor, while pressures on the supply side were intensified by low real wages and the strife of families to earn a "living wage" for the household. Aside from the incentives for the rural population to enter industrial employment, the industrial managers were induced to hire the rural workforce in preference to the urban workers because the former had fewer demands for working and living conditions and even wages. The urbanization of the rural population created the well-known problems of housing shortages, overloading of utilities, shortages of schools and hospitals, and so forth. The successive plan revisions raised the target for new jobs to 400,000 by 1962. Other measures to reduce unemployment included encouragement of early retirement, prohibition of holding several jobs simultaneously, and "export" of workers mainly to Czechoslovakia and the USSR (where about 10,000 youths were sent mostly for agricultural work).[71]

In 1961-65 growth of employment receded to 4.4 percent, coinciding with lower increases in investment. However, in 1966-67 growth of employment picked up considerably. In Bulgaria the periods of high growth rates of investment and relatively lower growth rates of consumption have coincided with those of high growth rates of employment and relatively lower growth rates of labor productivity: hence, the increasing strains (excess of spending power over goods supplied) that have required immediate relaxation and have given rise to considerable fluctuations.

The intensive rural-urban migration reached exceptional proportions in Bulgaria between 1953 and 1963. Between 1948 and 1964 the number of farm workers declined by nearly 50 percent, with the slowdown thereafter partly achieved by "administrative means." The exist from agriculture primarily benefited industry and construction. On the whole, in Eastern Europe the agricultural work force was estimated to have decreased from 23.4 to 18.6 million (by 20.5 percent) during 1950-68. The countries with the sharpest decrease were the GDR (39.2 percent), Bulgaria (38.8 percent), and Czechoslovakia (38.5 percent); in the other countries the decline was relatively smaller.[72] For example, the average annual rates of decline of agricultural labor force were as follows: 3 percent in Bulgaria in 1952-57, 2.8 percent in Czechoslovakia in 1950-67, 2 percent in the GDR in 1952-57, 2 percent in Hungary in 1950-67, almost nil in Poland in 1951-67, and 0.9 percent in Rumania in 1951-57.[73] In these countries the non-farm workforce increased from about 22 to 34 million during 1950-68 (by almost 56 percent). The growth was most rapid in Bulgaria (123 percent), Rumania (84 percent), and Poland (80 percent), followed by Hungary (60 percent) and Czechoslovakia (51 percent), with the GDR trailing (at 12 percent), not only due to its relatively higher industrialization level at the initial stage but also due to the depletion of its population.[74]

Sectoral Distribution of Labor Force

If the sectoral shift of employment is to be a rough yardstick for measuring industrialization, Tables B.52 and B.55 indicate the progress accomplished by CMEA countries.* But such a yardstick is

---

*The sharp decline in the private sector and the absorption by the socialized—sometimes by rather ficticious formation of cooperatives—inflates growth of employment in socialized industry, particularly during periods of significant ownership shifts. According to the United Nations, in Bulgaria employment in socialized industry and handicrafts increased from 262,000 in 1948 to 589,000 in 1958, but during these years employment in private handicrafts and self-employ-

not particularly reliable, since it gives relatively greater weight to
the undermechanized (and hence often the less-privileged) sectors
and branches. Here again the lead was taken by Bulgaria and Rumania,
with Bulgaria (where in 1948 industry and construction accounted for
9.9 percent of employment and agriculture and forestry for 82.1 percent, and in 1972 for 40.1 and 32.8 percent, respectively) outpacing
Rumania (where in 1950 the respective figures were 14.2 and 74.3
percent, and in 1972, 34.5 and 44.2 percent). Czechoslovakia and
the GDR were the countries with the least structural change in employment; the change was somewhat more pronounced in Hungary than in
Poland. Thus, Bulgaria had the highest rate of expansion of industrial
employment, and the GDR had the lowest.

In almost all East European countries heavy industry claimed
the predominant share of the workforce. In Czechoslovakia the share
was already about 60 percent in 1950, and it rose to nearly 70 percent
by 1967 (with a high proportion employed in steel production). By
1967 this share was about two-thirds in the GDR, somewhat below
that in Hungary, and it had risen to 60 percent in Poland and to 55
percent in Rumania.[76] Bulgaria registered the highest rate of expansion of employment in heavy industry. Within Bulgarian industry the
shifts in employment structure tended to favor ferrous metallurgy
(whose share in total employment in state-owned industry rose from
0.2 percent in 1948 to 2.9 percent in 1970), nonferrous metallurgy
(from 1.9 to 3.8 percent), machine building (from 12.1 to 24 percent),
and chemical industry (from 1.9 to 5.8 percent); the declining branches
were electric power (from 2.1 to 1.7 percent), fuels (from 9.8 to 5.6
percent), woodworking (from 19.4 to 7.2 percent), textiles (from 18.7
to 10.9 percent), and food processing (from 19.9 to 16.5 percent).[77]

During the early 1950s the increase of employment in Bulgaria
was predominantly in the traditional industries (woodworking, textiles,
and food processing), where considerable unused capacity existed.
But investment allocations favored heavy industry, making it much
more realiable as a gauge of state priorities than increase of employment. By 1970 the more traditional industries (woodworking, food
processing, glassware and china, textiles, garments, and leather
and footwear) still accounted for 42.8 percent of the workforce in
state-owned industry, as against 48.5 percent in producer-goods industries (electricity, fuels, metallurgy, machine building, chemicals,
and building materials).[78]

The declining share of agriculture in total employment is a general feature of industrialization. The STEs can be distinguished from

---

ment declined from 81,000 to 41,000 (in 1956). Industrial employment
was estimated to have risen by 85 percent over 1948-58.[75]

the West by the use of labor released from farming. In the West, though industry has benefited, the most obvious increase of the share of employment has occurred in services (see Table B.55).* The ranking of CMEA countries in the 1970s in descending order, according to the share of services in the labor force (the GDR, Czechoslovakia, Hungary, Poland, Bulgaria, and Rumania), might represent a fair picture of their comparative levels of development. Whereas in the West the share of services in employment has been increasing steadily in the postwar period, in the CMEA countries the most palpable increase seems to have occurred in the 1960s and thereafter. In comparing CMEA and Western countries at similar levels of development in the 1970s (Bulgaria and Rumania with Spain and Greece, Poland with Spain, Hungary with Italy, Czechoslovakia and the GDR with Austria), in almost every case the share of employment in services is lower and in industry higher in the former than in the latter. In fact, the least developed Western countries (Greece and Spain) have almost the same share of employment in services as the most developed CMEA countries.

Emergent Labor Shortages

In Bulgaria, as in the more industrialized CMEA countries, the increase in the labor-participation rate of women has also been an important factor, especially during the 1970s. The ascent was characterized by appreciable fluctuations. For example, after 1957 a fall occurred, indicating a slight increase in unemployment among women.[80] This was partly due to a deliberate attempt to restrict employment. Subsequently, these policies were reversed. The share of women in the labor force has grown as follows from 1960 to 1970: Bulgaria, from 32.1 to 42.5 percent; Hungary, from 32.5 to 42.8 percent; Poland, from 33.1 to 39.7 percent; Czechoslovakia, from 39.8 to 45.7 percent; the GDR, from 44.4 to 48.6 percent; Rumania, from 26.9 to 30.2 per cent.[81]

Increasing labor shortages have plagued Czechoslovakia and (especially) the GDR since the mid-1950s, and echoes of labor shortages have been picked up in other East European countries, even in Bulgaria. To some extent these impending shortages are exaggerated as they reflect rather partial shortages of particular skills, age, and sex groups, and in specific locations.[82] To a large extent the prob-

---

*Though this is true of the postwar period, it has not necessarily been a consistent experience, especially during the period of ebullient industrialization.[79]

GROWTH AND INSTABILITY 55

lem is structural and aggravated by "extensive-type" performance. To the extent that significant changes in macro and micro policies can be ruled out, the situation is likely to deteriorate. But, again, the state of labor supply is not an entirely exogenous variable, even if in the short run the c.p. has little influence on the demographic situation. In Bulgaria recently some attempts were made to encourage an increase in the birthrate, centering primarily on increased family allowances, longer leaves for mothers of infants (at minimum-wage payments), and other measures. The main stress was on daytime child care and education, so as to draw mothers into the workforce and encourage increased childbearing by working women.[83]

Undoubtedly, there is a significant decline in the birthrates of East European countries, as shown in Table B.50, that will endanger the future labor supply. It would appear that in Bulgaria agriculture can no longer be used as a reservoir of additional labor. Table B.51 indicates the rapid "aging" of the agricultural labor force. As a rule, the rural birthrate has been higher than the urban, but this tendency was reversed in 1966. In addition, since 1967 the natural increase of the rural population has fallen to alarmingly low levels. Generally, in Eastern Europe the agricultural workforce now consists mainly of women, the aged, and the least skilled. Moreover there are grim prospects for improving this situation or even maintaining the status quo.

The policy of "overfull" employment, coupled with institutional arrangements conducive to misallocation of labor and overemployment in enterprises, was hardly a vehicle for promoting efficiency and improved performance. Unquestionably, such a policy created "schools or laboratories of industrialization," provided vast opportunities for learning by doing, converted peasants into industrial workers, and brought housewives into the labor force, with profound and enduring social and economic repercussions. But the massive growth of the labor force was not devoid of adverse consequences that still haunt the c.p. Among them was the difficult process of adapting to the industrial production regime, which was not accompanied by the discipline of the industrial regime. Perhaps the absence of a "reserve army of workers" offers, at least in part, an explanation of the phenomenon of generally lax discipline and poor performance. Experience has shown that overinvestment is likely to have an adverse effect on productivity and that there is a high correlation between consumption standards and performance (at least within a certain range). Employees also tend to develop a defense and adaptation mechanism that consists of holding several jobs and performing poorly on each (partly in order to conserve energy). Absenteeism is rife, and there is inordinately high labor turnover. The system is also conducive to corruption, embezzlement, speculation, and theft of state property.

It should be noted that—although a temporary decline of efficiency or retardation of its rise could be clearly expected as a cost of launching and sustaining the vast industrialization effort (rapidly expanding employment on such a scale tends to have a low or even negative marginal productivity within a certain range) and although it is possible to argue that there is likely to be overcompensation for the immediate effect in terms of productivity by the training of the labor force in the process—the manpower policy, combined with specific system-created features has had an enduring effect on attitudes and motivation to work and on social consciousness. It demoralized the managers and the managed, lacked an effort-releasing effect, and has plagued the system to this day. Measures to subject the industrial worker to some "reasonable" standards of performance, in a certain comparative sense, have encountered great and predictable resistance.

NOTES

1. E. Denison, Why Growth Rates Differ (Washington, D.C., 1967), pp. 11ff; United Nations, Economic Survey of Europe in 1959 (Geneva, 1960), p. 8.

2. S. Kuznets, Population Capital and Growth (New York, 1973), p. 263.

3. A Gerschenkron, Economic Backwardness in Historical Perspective (Cambridge, Mass., 1962), pp. 198-234. Cf. N. Spulber, The State and Economic Development in Eastern Europe (New York, 1966).

4. United Nations, Economic Survey since the War (New York, 1953), Chapters 3, 4.

5. U.S., Congress, Joint Economic Committee, Economic Developments in Countries of Eastern Europe (Washington D.C., 1970), p. 47; G. R. Feiwel, ed., New Currents in Soviet-Type Economies (Scranton, Pa., 1968), p. 82.

6. R. Constantinescu, Era Socialista (Bucharest) 24 (1974): 24.

7. Przemysl w Polsce i wybranych krajach 1950-1968 (Warsaw, 1970), p. xxvii.

8. Der Spiegel, January 20, 1975, p. 59.

9. Compare S. Kuznets, Modern Economic Growth (New Haven, Conn., 1966).

10. G. R. Feiwel, Industrialization and Planning under Polish Socialism, vol. 1, "Poland's Industrialization Policy" (New York, 1971), Chapter 3.

11. Feiwel, New Currents, pp. 84-85; U.S., Congress, Joint Economic Committee, Economic Developments, pp. 53-57; U.S.,

Congress, Joint Economic Committee, Reorientation and Commercial Relations of the Economies of Eastern Europe (Washington, D.C., 1974), pp. 255-60.

12. See G. R. Feiwel, AER December 1966, pp. 1300-02.

13. T. P. Alton, E. M. Bass, L. Czirjak, and G. Lazarcik, Statistics on East European Economic Structure and Growth, OP-48 (New York, 1975), pp. 8-57.

14. Kuznets, Population Capital, p. 168.

15. United Nations, Economic Survey of Europe in 1949 (Geneva, 1950), pp. 3, 16.

16. Przemysl w Polsce i wybranych krajach, p. xxxiii.

17. U.S. Congress, Joint Economic Committee, Economic Developments, p. 437.

18. See G. R. Feiwel, The Soviet Quest for Economic Efficiency (New York, 1972), pp. 516ff.

19. Statisticheski Yezhegodnik Stran-Chlenov Soveta Ekonomicheskoy Vzaimopomoshchi 1971 (Moscow, 1971), p. 58.

20. U.S., Congress, Joint Economic Committee, Economic Developments, p. 244.

21. Przemysl w Polsce i wybranych krajach, pp. xxxiv-xxxv.

22. Statisticheski Yezhegodnik . . . 1971, pp. 61-68.

23. U.S., Congress, Joint Economic Committee, Economic Developments, pp. 447-51; Przemysl w Polsce i wybranych krajach, p. 198.

24. For marked variances between the official and the Lee-Montias indexes of Rumanian industrial growth (1948-63), see J. M. Montias, Economic Development in Communist Rumania (Cambridge, Mass., 1967), p. 13.

25. Statisticheski Yezhegodnik . . . 1971, pp. 61-68.

26. See ibid., pp. 61ff; Przemysl w Polsce i wybranych krajach, Part 3.

27. U.S., Congress, Joint Economic Committee, Economic Developments, p. 447.

28. Ibid., p. 451. See United Nations, Economic Survey of Europe in 1971, Part 1 (New York, 1972), pp. 48, 66ff.

29. Przemysl w Polsce i wybranych krajach, p. 53.

30. G. J. Staller, Bulgaria: A New Industrial Production Index, 1963-1972, with Extension for 1973 and 1974, OP-47 (New York, 1975), p. 13.

31. For an analysis and quantitative appraisal of industrial expansion in Poland, Czechoslovakia, and the GDR in the 1950s and the early 1960s, see A. Zauberman, Industrial Progress in Poland, Czechoslovakia and East Germany (London, 1964). For Rumania, see Montias, Economic Development.

32. For a description of postwar development in Bulgarian agriculture, see B. Dobrin, Bulgarian Economic Development since World War II (New York, 1973).
33. U.S., Congress, Joint Economic Committee, Economic Developments, pp. 515-17.
34. G. Lazarcik, Bulgarian Agricultural Production, Output, Expenses, Gross and Net Product, and Productivity at 1968 Prices, 1939 and 1948-70, OP-39 (New York, 1973), pp. 12-13.
35. Statisticheski Godishnik na Narodna Republika Bulgaria 1968 (Sofia, 1968), p. 102.
36. I. Khadzhiivanov, VT 9 (1974): 4; Razshireno Sotsialistichesko Vozproizvodstvo v NRB (Sofia, 1973), p. 91. Compare E. Kudrova, Statistika natsionalnogo dokhoda europeyskikh sotsialisticheskikh stran (Moscow, 1969), p. 163.
37. Kudrova, Statistika, p. 134.
38. Statisticheski Godishnik na Narodna Republika Bulgaria 1973 (Sofia, 1973), pp. 98-102.
39. For comparative statistics on the distribution of accumulation into stocks and fixed-capital formation from 1950 to 1966 in CMEA countries (except Rumania), see Kudrova, Statistika, p. 144.
40. Feiwel, New Currents, p. 93. Compare Denison, Why Growth Rates Differ, p. 118.
41. A. Maddison, Economic Growth in Japan and the USSR (New York, 1969), pp. 39, 57. See K. Ohkawa and H. Rosovsky, Japanese Economic Growth (Stanford, 1973).
42. Kuznets in conversation.
43. A. Bergson, World Politics, July 1971, p. 595; A. Bergson, Soviet Post-War Economic Development (Stockholm, 1974), p. 70.
44. United Nations, Economic Survey of Europe in 1969, Part 1 (New York, 1970), pp. 143-45. Compare Denison, Why Growth Rates Differ, pp. 118-20; Maddison, Economic Growth, p. xix.
45. L. Zienkowski, Dochod narodowy wytwarzanie podzial (Warsaw, 1970), pp. 49-50.
46. Statisticheski Yezhegodnik . . . 1971, p. 141; Statisticheski Yezhegodnik Stran-Chlenov Soveta Ekonomicheskov Vzaimopomoshchi 1973 (Moscow, 1973), p. 157; United Nations, Economic Bulletin for Europe 18, no. 1 (November 1966): 41.
47. Kudrova, Statistika, p. 155.
48. Ibid., p. 157.
49. U.S., Congress, Joint Economic Committee, Economic Developments, p. 438.
50. Handel zagraniczny a wzrost krajow RWPG (Warsaw, 1969), p. 264; Statisticheski Godishnik . . . 1968, p. 234.
51. Statisticheski Godishnik . . . 1968, p. 235; United Nations, Economic Bulletin, pp. 42-43.

52. Statistical Yearbook of Bulgaria 1971 (Sofia, 1971), p. 89.

53. United Nations, Economic Bulletin, p. 42. Compare Statisticheski Yezhegodnik . . . 1971, pp. 155-59.

54. Statistical Yearbook of Bulgaria 1971, p. 89.

55. For an analysis of the Czechoslovak developments, see Feiwel, New Currents, pp. 112-22; J. Goldmann and K. Kouba, Economic Growth in Czechoslovakia (White Plains, N.Y., 1969); G. R. Feiwel, New Economic Patterns in Czechoslovakia (New York, 1968), Chapter 1. For fluctuations of industrial output in Rumania, see Montias, Economic Development, p. 16. For the record of fluctuations in Yugoslavia, see B. Horvat, Business Cycles in Yugoslavia (White Plains, N.Y., 1971); B. Horvat, Ekonomist (Belgrade) 1-2 (1974). For a comparative study of Yugoslavia and other East European countries, see A. Bajt, JEL, March 1971, pp. 56-59. On comparisons of fluctuations in the East and in the West, see G. J. Staller, AER, June 1964, pp. 385-95.

56. United Nations, World Economic Report 1951-52 (New York, 1953), p. 44.

57. United Nations, World Economic Report 1950-51 (New York, 1952), p. 34.

58. United Nations, World Economic Report 1952-53 (New York, 1954), p. 40.

59. Ibid., pp. 41-42.

60. United Nations, World Economic Report 1953-54 (New York, 1955), pp. 46-52.

61. See Feiwel, Industrialization and Planning under Polish Socialism, vol. 1, "Poland's Industrialization Policy" (New York, 1971), p. 115.

62. J. Kornai, Rush Versus Harmonic Growth (Amsterdam, 1972), p. 99.

63. United Nations, Economic Bulletin, p. 59.

64. The political factors behind the Leap Forward are well described by J. F. Brown, who also shows the connection with China and the emulation of the policy of that famous name. Brown also provides a glimpse of the fervor and pitch to which Bulgarians were exhorted to fulfill the Third FYP in three to four years. See J. F. Brown, Bulgaria under Communist Rule (New York, 1970), pp. 83-95. Compare J. Kalo, Survey 39 (December 1961): 86-95.

65. RD, October 27-29, 1959; United Nations, Economic Survey . . . 1959, Chapter 2, p. 21.

66. United Nations, Economic Survey of Europe in 1962, Part 1 (New York, 1963), Chapter 1, pp. 17, 21, 31. Compare Handel zagraniczny, p. 271.

67. United Nations, Economic Survey of Europe in 1961, Part 1 (Geneva, 1962), Chapter 2, p. 33.

68. Compare Feiwel, "Poland's Industrialization Policy," pp. 333-476.
69. United Nations, Economic Survey . . . 1969, Part 1, p. 10.
70. Compare The Economist, January 3, 1959.
71. Compare RD, March 12, 1957.
72. U.S., Congress, Joint Economic Committee, Economic Developments, p. 150.
73. United Nations, Economic Survey . . . 1969, Part 1, p. 17.
74. U.S., Congress, Joint Economic Committee, Economic Developments, pp. 160-61.
75. United Nations, Economic Survey in Europe in 1960 (Geneva, 1961), p. 23.
76. U.S., Congress, Joint Economic Committee, Economic Developments, p. 444. Compare Kudrova, Statistika, p. 85.
77. Statistical Yearbook, 1971, p. 93.
78. Ibid.
79. Kuznets, Modern Economic Growth, pp. 105-11. Compare Denison, Why Growth Rates Differ, Chapter 6.
80. A. Dobrev, PT 1 (1970): 78.
81. Statisticheski Yezhegodnik . . . 1971, pp. 386-87.
82. Dobrev, PT, p. 78.
83. R. Gocheva, IM 2 (1973): 9ff.; T. Zhivkov, RD, December 14, 1972; I. Vasilev, NA 24 (1972): 3.

CHAPTER

3

RESOURCE UTILIZATION

PRODUCTIVITY

In general CMEA countries have been falling behind the West in productivity gains. Bergson's research on comparative productivity has led him to conclude that productive efficiency in the USSR "may well be low by Western standards."[1] For CMEA countries productivity increases "have always been distinctly more costly" than for Organization for Economic Cooperation and Development (OECD) countries.[2] The c.p. is particularly concerned with the relative position of his country within the CMEA, and drastic measures are called for if the relative productivity deteriorates.

The relative levels of productivity, as measured by national product per employee, are indicative of the countries' relative stages of development. The varied growth rates of output per employee are considered rough indicators of relative advance made by different countries in growth of productivity. Estimates differ, but in terms of GDP per employee in 1950 (index CMEA = 100), the levels in Rumania (49) and Bulgaria (59) were the lowest. The USSR and Poland were about 100; Hungary was 109; GDR, 139; Czechoslovakia, 161. By 1967 (CMEA = 100) Rumania's and Bulgaria's relative positions (62 and 77, respectively) improved; Bulgaria did slightly better than Rumania: the average annual growth rates in 1950-67 were 6.9 percent in the former and 6.6 percent in the latter. Czechoslovakia deteriorated sharply, to 132; the GDR fell only slightly, to 134; Hungary fell to 87; Poland, to 82; the USSR improved slightly (103).[3]

Labor Productivity

Despite their wide divergencies, both official and recalculated data place Bulgaria and Rumania at the top in terms of average growth rates of aggregate output per employee in 1952-67. Both these countries seem to have done better in the 1960s than in the 1950s; Rumania (with 4.8 and 7.6 percent, respectively) outdid Bulgaria (with 5.7 and 7 percent, respectively), a notable source of disquietude in Bulgaria in the 1960s and 1970s. With the exception of Hungary (with its depressed performance in the 1950s), all other CMEA countries registered lower growth rates in the 1960s.

In comparing the growth rates of labor productivity for the same period, it appears that Bulgaria (with 6.4 percent) and Rumania (with 6.2 percent) have done better than the Western countries at a roughly comparable level of development (Greece, 5 percent; Spain, 5.3 percent; Yugoslavia, 5.4 percent). However, the record looks less impressive if recalculated data are applied (Rumania, 5 percent). The comparison of CMEA countries at higher levels of development with their Western counterparts indicates lower growth rates of labor productivity in the former, and far worse performance, if recalculated data are used.[4]

As shown in Table B.53, in terms of growth of industrial labor productivity, again, the least industrialized CMEA countries held the lead, though some data suggest that Rumania considerably outpaced Bulgaria. Although Bulgaria and Rumania have outperformed their closest Western counterparts (Greece, 6.5 percent; Spain, 5.3 percent; Yugoslavia, 6.6 percent), the differences are much smaller than in the comparison of economy-wide labor productivity. Moreover, if recalculated data are used (see Table B.60), both Bulgaria and Rumania trail their Western counterparts. Such comparisons are especially disadvantageous to the more industrialized CMEA countries.[5]

In comparing the dynamics of labor productivity for the 1960s in the economy as a whole and in industry, it is apparent that the least-developed CMEA countries recorded the quickest advance, and the most-developed the slowest. Thus, by 1967 (1963 = 100) labor productivity in the economy and industry was, respectively, in Bulgaria 141.9 and 126.6, and in Rumania 139.5 and 139. The significantly faster rise of labor productivity in industry in Rumania must have been of considerable concern to the Bulgarians. The respective figures for Hungary were 122.2 and 122.3, and for Poland 121.1 and 118.9; the GDR (with 119.2 and 117.8) and Czechoslovakia (with 118.2 and 115.6) trailed.[6]*

---

*Alton-Project estimates (Bulgaria and Rumania excluded due to data deficiency) indicate that in 1967 the GDR had the highest level of

Obviously, the labor-productivity indicator depends on the values of the numerator and denominator. It cannot be overemphasized that comparative measures of productivity must inquire into the content and valuation of output. Only if the available statistics could be adjusted for qualitative differences in output would the comparison be meaningful, whatever the other tricky problems. The issue is that in quality, variety, modernity, and durability of output CMEA countries are inferior to the West. But there are also significant intra-CMEA variations.

Statistical differences in sectoral-growth patterns of productivity are particularly suspect, especially because of the peculiarities of pricing and deficiencies of employment statistics. Estimates are tenuous, despite the crude adjustments.[8] Thus, the results on productivity in agriculture are at best very tentative. For example, farm-labor productivity clearly depends on the size of the agricultural population, so the index must be depressed in countries with excessive absorption of labor in farming (particularly Poland and Rumania). Be that as it may, in countries with ample labor supply, agricultural policy should not be swayed by the issue of the highest possible productivity per farm worker; the guidepost should be the highest possible productivity per hectare of agricultural land.[9]

Taking all this into account, it might be of some interest to look at the UN estimates of labor and capital productivity in agriculture. In the 1950-52 to 1967-69 period Bulgaria and the USSR had the highest average growth rates of labor productivity (4.7 percent), followed closely by the GDR (4.5 percent), and Rumania (3.9 percent). In the other countries the rates were considerably lower: Hungary, 3.1 percent; Czechoslovakia, 2.3 percent; Poland, 1.9 percent. All CMEA countries registered negative growth rates of output per unit of fixed capital. The worst results were obtained by Bulgaria and Czechoslovakia (about -5 percent), followed by the USSR (-3.4 percent), Hungary (-3.2 percent), and the GDR (-2.8 percent); Rumania (-0.9 percent) and Poland (-0.5 percent) registered relatively the best results.[10]

---

industrial labor productivity in Eastern Europe, but its index was only 63 (the GFR = 100); Czechoslovakia's was 54; Poland's, 50; Hungary's, 35. The Czechoslovak index reached only two-thirds of the French level, and that of Hungary only three-quarters of the Austrian level.[7]

Capital Productivity

Bulgaria seems to have performed fairly well, judged by the yardstick of labor productivity, but a different picture emerges when attention shifts to sources of productivity growth and to returns to investment in particular. For example, even given the tentativeness of the recalculations, in the 1960s in Bulgaria there was a modest retardation in the rate of expansion of industrial output (from 12.7 percent in the 1950s to 11.9 percent in 1960-68), a marked speedup in the rate of growth of fixed capital in industry (from 12.6 percent in the 1950s to 14.3 percent in 1960-67), and a substantial deceleration in the rate of expansion of industrial employment (from 8.3 percent in the 1950s to 5.1 percent in 1960-68).[11] The index of labor productivity registered an acceleration, but the speedup in the rates of capital formation clearly suggests that productivity cannot be accounted for by labor alone; allowance must be made for different quantities and qualities of investment goods, incorporating changes in production techniques and other cooperating factors.

The ICOR plays a prominent role in development planning, but the ambiguity of the coefficient and the pitfalls of measurement have long been recognized.[12] As with many other measures it is probably more important to point out the elements this one omits (for example it is not a meaningful indicator of capital productivity; it neglects labor's contribution and tends to underrate that of technical progress), the assumptions on which the available estimates are based, and the irksome problems of identifying and associating the increment in capital with that of output. For example, the usually assumed one-year time lag between commissioning of capacities and flow of output underrates the protracted period of fruition of output, especially in heavy industry and particularly in capital-intensive branches. The results are affected by the variable rates of utilization of capacities over time. Some increases in output occur primarily due to increased rate of utilization of capacity (economies of scale) rather than because of investment. Obviously, the yields are affected by the coverage and methods of computation (prices used for aggregation, gross or net of depreciation and allowance for obsolescence, gross value or value added and so forth). Naturally the yields would differ, whether investment in fixed capital series were confined to new capacities commissioned (whether fully utilized or not) or, as in the Alton-Project indexes of Bulgarian investment, both finished and unfinished construction were included as part of investment.[13] The shorter the period under consideration, the more spurious is the relation likely to be, and the more pronounced the neglect of earlier investments. The short-term variations in ICORs indicate the fluctuations in investment activity and the instability in the rate of resource allocation.

RESOURCE UTILIZATION

The size of the coefficient is affected by structural changes some of which are profound and enduring. Measures aimed at rising dynamic efficiency of the system may entail rising the overall coefficient (investment embodying capital-using technical progress, infrastructure, housing, and so forth). Such returns are likely to be reaped only in the long run. System changes constitute only one of the determinants affecting the movements of the coefficient, and caution must be exercised not to infer success or failure of reform from changes in the coefficient. Whatever its limitations, the coefficient is important because it is one of the basic criteria used by the c.p. for evaluating the efficiency and costs of development. Its movements and intra-CMEA variations matter to the c.p.[14]

Throughout the 1960s the c.p. in most CMEA countries was anxious about the deterioration of the ICORs. From the second quinquennium of the 1950s to the first of the 1960s this coefficient increased 4.5-fold in Czechoslovakia, about twofold in Bulgaria and Hungary, by 50 percent in the USSR, and by 25 percent in Poland (see Table B.59).* It is difficult to reduce plausibly to quantitative terms the weight of different factors affecting the returns on investments. Some of the forces previously mentioned were pulling in different directions. The structural shifts are emphasized, but this seems to underrate the retrogressive effect of the working of the economic mechanism itself. It should be recalled, for example, that in Bulgaria in the early 1960s a large share of industrial investments was allocated to expansion of capital-intensive branches, such as coal, metal-ore mining, and metallurgy. Often, the extraction was undertaken under adverse geological conditions, supported by very costly equipment; moreover, capacities were vastly underutilized, partly due to difficulties of recruiting skilled personnel and failures to handle production problems. The quality of Soviet equipment and training were also handicaps, whatever other factors were at work. The notably higher returns to investment in Rumania in the 1960s must have been a source of considerable concern to the Bulgarian c.p.[17] Bulgarian difficulties in containing investment costs were aggravated by unfavorable comparisons with Soviet developments. In the USSR, after rising for some 15 years, ICORs fell in the second half of the 1960s, whereas in Bulgaria they rose. Bulgaria must have been under Soviet pressure to lower investment costs.

Tentative Alton-Project estimates suggest that investment costs in Eastern Europe tended to be significantly higher than those in the

---

*Other estimates indicate a relatively sharper increase in Bulgaria.[15] Moreover, the UN estimates of the ratio of productive investment to increase in income show a similar—slightly higher—tendency to increase.[16]

West. On the average East European countries required about 25 percent more gross fixed investments to achieve one unit increase of GNP (at factor cost) in 1951-64. The sectoral differentials were substantially higher: 45 percent more in industry, considerably larger in agriculture, but similar in services. Czechoslovakia exhibited the highest ICORs, followed by the GDR, Hungary, Bulgaria, and Poland. The coefficient in the GDR was considerably higher than that in the GFR; Czechoslovakia was above France; in Bulgaria it was some 70 percent above that in Greece. The differences are even more striking if investment costs in industry are compared, for the Bulgarian ICOR was twice as high as the Greek. Not only did the East European countries show worse yields on investment but the difference between them and the closest comparable Western countries grew in the 1960s.[18]*

Attribution of yields to productive factors is a controversial theoretical question, beset with grave measurement problems.[20] As tenuous as the estimates are, in the 1950s the rates of increase of industrial output per unit of fixed capital were positive in Bulgaria, Czechoslovakia, and Hungary. But for the 1961-67 period these rates became negative for all East European countries, with wide discrepancies in the intensity of change (see Table B.60). The negative changes of capital productivity were particularly striking in heavy industry, mining, and electric power in Czechoslovakia, the GDR, and Hungary. In Bulgaria the most striking changes were in mining, ferrous metallurgy, building materials, woodworking, paper, glass, and textiles.[21] In addition, GNP per unit of capital indicated negative growth rates in all East European countries in 1960-72, as shown in Table B.61.

According to UN estimates of changes in capital productivity, there was considerable slowdown in the rate of decline in the second half of the 1960s in Czechoslovakia, the USSR, the GDR, and Bulgaria. This partly resulted from an acceleration of the growth rate of national output (with the exception of Rumania) and a deceleration of the growth rates of capital stock (except for Bulgaria and Rumania, where an acceleration was recorded). In 1966-69 in Rumania the rate of fixed capital accretion in industry markedly outpaced that for 1961-65; industrial output increased more slowly. Bulgaria trailed Czechoslo-

---

*On the basis of Alton-Project data, Bergson adjusted the ICOR reducing it by an amount proportional to the increase in employment, roughly obtaining the incremental capital-productivity ratio. The comparisons of this ratio indicated that costs of productivity growth were generally higher in CMEA than in OECD countries. The disadvantage of CMEA countries was particularly notable in comparisons of countries with lower productivity.[19]

vakia, the USSR, and the GDR in this improvement; it was only slightly ahead of Hungary, where the improvement was negligible.[22] In Bulgaria capital productivity in industry improved from -2.3 in 1961-65 to -0.9 percent in 1966-69. Fuels, the woodworking, paper, printing, glass, and china category, and food processing still had negative rates of productivity (fuels improved from -4 to -0.5; food processing deteriorated from 2.5 to -3). Considerable improvements were recorded in ferrous metallurgy, chemicals, and textiles and clothing, where the negative rates became positive. There was also a slowdown in the positive rates in electric power, building materials, and a slight slowdown in machine building.[23]

Comparisons of productivity in terms of the admittedly arbitrary measure of "factor productivity" show that in Bulgaria the slowdown in industrial output growth was accompanied only by a negligible slowdown of the composite of labor and capital taken together (from 8.5 percent in the 1950s to 8.4 percent in 1960-67).[24] Thus, the growth retardation was attributed to retardation in the pace of productivity gains (see Table B.60), granting the difficulties in accounting for quality improvements in inputs, technical progress, economies of scale, and derivation of reasonable factor-income shares as weights for labor and capital inputs, and so forth and deficiency of data.[25]

FOREIGN TRADE

Foreign trade reflects the benefits, costs, and shifts in industrialization policy. The question is not so much the size as it is the composition and efficiency of external trade. Autarky—a feature of past Soviet-type development—is in the process of erosion. There are demonstratively compelling empirical reasons for drawing on the theory of comparative advantages and exploiting the advantages of backwardness, which are obviously limited by political considerations. Since the death of Stalin, the fairly indiscriminate pushing of import substitution has been relaxed and foreign trade has grown beyond the principle of limiting exchange to the extent that was mandatory for carrying out the industrialization rush. The foreign trade of CMEA countries has increased considerably, but it would be difficult and hazardous to estimate how undersized it still is. Starting at a very low level (1953 = 100), the indexes of imports and exports of centrally planned economies were by 1971 466.7 and 454.4, respectively (for the world as a whole, 429.9 and 420.8, respectively); these were still lower than those of the developed market economies (476.6 and 465.7, respectively). Thus, from 1953 to 1971 the structure of world imports was altered in favor of the developed-market economies (65.3 and 72.4 percent, respectively) and to the detriment of the developing-market economies (25.5 and 17.6 percent, respectively), leaving the centrally planned economies practically unchanged. A similar situation emerged in exports.[26]

It would seem that Bulgaria (whose 1964 share of world industrial output was 0.3, and world trade 0.6 percent) is the most "open" of the CMEA countries, followed by Hungary (0.6 and 0.8 percent, respectively) and Czechoslovakia (1.5 and 1.6 percent, respectively), with Rumania (0.6 and 0.6 percent, respectively) and the GDR (1.7 and 1.7 percent, respectively) tying for fourth place, followed by Poland (2 and 1.2 percent, respectively), with the proverbially "self-sufficient" USSR (19.9 and 4.5 percent, respectively) trailing far behind.[27]*

---

*It is difficult to ascertain to what extent CMEA countries underutilize their foreign-trade potential. For example, Zauberman offered interesting estimates on underdevelopment of foreign trade in Poland, Czechoslovakia, and the GDR by examining the relationship of foreign trade to industrial output and GNP in aggregate and per capita. He warned that the results should be used with great circumspection.[28] Following Chenery's research,[29] Pryor found in eight West European countries a strong correlation between per capita trade and per capita industrial output (Chenery used national income per capita). Pryor used the results of these correlations for estimating "trade potential" for CMEA countries. The divergence between actual and potential was strikingly sharp: In 1955 Bulgaria and Rumania apparently utilized only 46 percent; Czechoslovakia, Poland, and the GDR, 35 percent; and Hungary, 30 percent of their respective potentials.[30] Pryor admitted that the results were only tentative. Wiles suggested that while income per capita is universally regarded to be a key determinant of trade volume, it is not clear how it works. He offered the "participation rate," defined as the proportion of national income earned by or spent on foreign trade. By this method he found that the STEs (Bulgaria included) were indeed very autarkic.[31] The rapid progress of their trade in the 1960s notwithstanding, the East European countries showed considerable "trade lag," as measured by the criterion of per capita foreign trade and its share in GNP. Brzeski attempted to estimate the "import lag" (the difference between the exhaustion of opportunities for advantageous exchange under actual market conditions and actual per capita imports); he found that in 1968 this was lowest in the GDR (18 percent), followed by Bulgaria (23 percent), Poland (25 percent), Czechoslovakia (26 percent), Hungary (38 percent), and Rumania (46 percent). The difference between potential and actual per capita import multiplied by population renders the total import lag. Brzeski concluded, however, that no

RESOURCE UTILIZATION 69

Within the CMEA the USSR has by far the largest share of trade (nearly 40 percent in 1972), followed by the GDR (with only 14.7 percent); Poland and Czechoslovakia contend for third place, and Hungary, Bulgaria, and Rumania are together in the last place (see Table B.71). The great discrepancy between the shares of the USSR and of other countries indicates that CMEA trade is largely concentrated in trade between the USSR and the others, rather than among the smaller CMEA countries.

Growth Rates

In comparing average annual rates of growth of industrial output and exports in various CMEA countries, the following differences stand out: Both in 1950-60 and 1960-64 Bulgarian exports (16.6 and 15.9 percent, respectively) grew faster than industrial output (14.8 and 11 percent, respectively). The same was true in the GDR in 1950-60, when exports grew by 18.4 percent and industrial output by 11.4 percent, but by 1960-64 industrial output and exports grew at about the same rate (6 and 6.6 percent, respectively). Both in Rumania and the USSR the rates of growth of industrial output and exports were about the same in both periods. In Rumania in 1950-60 output increased by 13.1 percent and exports by 13 percent, and the respective figures for 1960-64 were 14.2 and 13.8 percent; in the USSR the respective figures for 1950-60 were 11.6 and 12 percent, and for 1960-64, 8.5 and 7.1 percent. In Czechoslovakia, Hungary,

---

"quantitative revolution" in CMEA market was in the offing for the West.[32] The assessment of shares of foreign trade in national income is fraught with grave difficulties, including the formidable problem of divergent domestic and foreign trade prices and arbitrariness and discrimination in exchange rates.[33] By itself, the indicator of per capita foreign trade is of limited significance. In 1968 the CMEA countries could be ranked as follows in terms of per capita exports: Czechoslovakia, $206; the GDR, $202; Bulgaria, $193; and Hungary, $174; Rumania and Poland ($90); and the USSR ($45) trailed far behind. In terms of imports, Bulgaria assumed the lead ($210), followed by Czechoslovakia ($205), the GDR ($192), Hungary ($177), Rumania ($97), Poland ($88), and the USSR ($39).[34] Comparing these data to national production (see Table B.4) Bulgaria seems to have underutilized its potential least, followed by the GDR and Hungary, Czechoslovakia, Rumania, Poland, and the USSR (see Table B.5). Clearly, not too much should be read into these statistics. The notion that STEs are motivated by autarky has been questioned by Holzman.[35]

and Poland the growth rates of export exceeded those of industrial output in the second period, whereas in the first period they were about the same in the first two countries, and the reverse was noted in Poland. In Czechoslovakia in 1950-60 industrial output grew by 10.9 and exports by 9.5 percent, and the respective figures for 1960-64 were 4.7 and 8.3 percent. In Hungary the respective figures for 1950-60 were 10.4 and 10.3 percent, and for 1960-64, 8.4 and 12.0 percent. In Poland the respective figures for 1950-60 were 12.9 and 7.7 percent, and for 1960-64, 8.2 and 12.9 percent.[36] In the second half of the 1960s growth rates of export outpaced those of industrial output only slightly in Bulgaria (11.3 and 11.1 percent, respectively) and Czechoslovakia (7 and 6.5 percent, respectively). In Rumania exports (10.9 percent) grew at a slower rate than industrial output (11.8 percent). In the other countries exports grew much faster than industrial output. The spread was largest in Hungary (8.9 and 6.2 percent respectively), followed by the GDR (8.5 and 6.5 percent), Poland (9.7 and 8.3 percent), and the USSR (9.5 and 8.5 percent).[37]

It appears that during the 1960s the fastest rate of growth of trade was registered in Bulgaria, with exports growing considerably faster than imports.* In Rumania the reverse trend was noted: Imports grew much faster than exports. However, Rumania outperformed Bulgaria in trade with the West, enabling it to modernize its industry. Rumania probably outdid Bulgaria in terms of long-term gains in dynamic efficiency. By 1972 the CMEA countries could be ranked as follows, in descending order of their foreign-trade-growth dynamics: Bulgaria, Rumania, Poland, Hungary, the USSR, the GDR, and Czechoslovakia. The exceptionally high growth dynamics of Bulgaria were surpassed by Spain, Japan, and Italy; Rumania was outdone by Yugoslavia, Portugal, the GFR, France, and Belgium. The rates of growth of foreign trade were faster in Greece than in Poland and Hungary, and they were faster in Austria Norway, Sweden and the United States than in the the USSR, the GDR, and Czechoslovakia (see Table B.72).**

---

*In 1961-67 in Bulgaria the growth rate of foreign trade 13.6 percent, surpassed that of growth of industrial output per capita, 11 percent. The respective rates were as follows: Czechoslovakia 6.3 and 5 percent; Poland, 8.4 and 4.7 percent; Hungary, 8.4 and 4.9 percent; and the USSR, 6.6 and 5 percent; the GDR fell slightly behind with 5.5 and 6 percent.[38]

**According to Montias apparently the key factor determining the level of trade among CMEA partners is the net import demand for machinery and industrial consumer goods from the USSR and the less-industrialized East European countries. Sharp intensification

Structure

From the standpoint of restructuring foreign trade in the CMEA during 1960-72 (see Table B.73), one of the most important feats was accomplished in exports of machinery in the less-developed countries. Ranking these countries in descending order, according to their share of machinery in exports, we obtain the following for 1960: the GDR, Czechoslovakia, Hungary, Poland, the USSR, Rumania, and Bulgaria. By 1972 the first two places remained unchanged, but Poland and Bulgaria displaced Hungary for the third and fourth places, respectively, with Rumania and the USSR in the sixth and seventh places, respectively. The share of machinery in total imports still exceeds its share in exports in Bulgaria, Rumania, and the USSR. But while the ratios of these shares have not varied much in Rumania and USSR, in Bulgaria the ratio of the shares of import to export was 3.4 in 1960, and by 1972 it was only 1.3. In contrast, in the GDR, Czechoslovakia, and Poland the shares of machinery in total exports still exceed those in imports. But while not much variation has occurred in Poland, in the GDR the ratio of the shares of exports to imports has dropped from 3.8 in 1960 to 1.6 in 1972, and in Czechoslovakia the corresponding figures were 2.1 and 1.4. The position of Hungary has shifted from a ratio of the share of exports to imports of 1.3 in 1960 to a ratio of the share of imports to exports of 1.05 in 1972. All this indicates a trend toward equalization in trade of machinery among CMEA countries. However, there continue to be formidable difficulties in penetrating Western markets. Thus, in 1967 the traditional machinery exporters (Czechoslovakia and the GDR) succeeded in exporting to developed-market economies some 5 percent of machinery; Poland and Hungary exported 3 percent, and Bulgaria exported merely 1 percent.[40] The failure is particularly striking when viewed against the major price concessions made to the West.[41]

There seem to have been no dramatic shifts in the shares of fuels and metals in the exports and imports of the individual countries

---

of industrialization is accompanied by rapid trade expansion and deceleration by flagging trade. The most industrialized CMEA countries are heavily dependent for their boom on the strength of their partners' demand for their wares. At times of slump these partners tend to discriminate in favor of their own machine-building producers and to reduce imports. The demand for machinery by Bulgaria, Poland, and Rumania largely determines their trade in raw materials, semi-fabricates, and foodstuffs (rather than the reverse) and is influenced by the credits they receive from the USSR and Czechoslovakia. The imports of capital goods are related to the vicissitudes of the industrialization path and the capacity of the home-produced machinery to expand.[39]

in 1960-72, with the exception perhaps of the GDR and the USSR, where the shares of fuels and metals in total imports have declined considerably; Poland, where the share in exports has declined markedly; and Rumania, where the shares in both exports and imports have dropped significantly, indicating larger domestic refinement and use of oil.*

During 1960-72 there was considerable realignment in the shares of industrial consumer goods in total exports of individual countries. In 1960 the ranking of these countries in descending order was as follows: Czechoslovakia, Bulgaria, Hungary, the GDR, Poland, Rumania, and the USSR. By 1972 Hungary moved out to first place, followed closely by the GDR Czechoslovakia Rumania, and Poland, with Bulgaria relegated to sixth place, and the USSR trailing. The greatest reductions in this share occurred in Bulgaria and Czechoslovakia, whereas the largest increases were recorded in Rumania, Poland, the GDR, and Hungary. The shares of consumer goods in total imports also fell in Bulgaria and Rumania; they rose significantly only in Czechoslovakia and Poland. With the exception of the USSR, in all countries the shares of exports exceed those of imports, with the most obvious positive difference in Hungary and the GDR.**

Agricultural products retained the highest place in total exports of Bulgaria in 1960-72.† They have usurped that place in Rumania, held by fuels, in 1960. The ranking according to the descending order of these shares was as follows in 1960: Bulgaria, Rumania, Hungary, the USSR, Poland, Czechoslovakia, and the GDR. By 1972 the order remained virtually unchanged, with only Czechoslovakia ceding its place to the GDR, but the relative magnitudes of the shares fell sharply, with the exception of Czechoslovakia and Hungary, where the decline was only slight, and the GDR, where there was a slight increase. Most countries reported a fall in the share of agricultural products in total imports (the sharpest falls were in Poland, Czechoslovakia, and GDR), with the exception of Hungary and the USSR, where this share went up slightly.

---

*The share of ores and metals in overall Bulgarian exports fell from about 25 percent in the mid-1950s to 9 percent in the late 1960s, and those of imports, from 23 to 16 percent, respectively. The share of fuels in exports declined from 2.5 percent to almost nil, whereas that in imports increased from 7 to 11 percent.[42]

**The share of textiles, clothing, and leather goods in Bulgarian exports dropped from 25 percent in the mid-1950s to 10 percent in the late 1960s; imports also declined from 6 to 4 percent.[43]

† Processed foods, beverages, and tobacco's share in Bulgarian exports rose from 28 percent in the mid-1950s to 35 percent in the late 1960s; imports increased only from 1 to 4 percent.[44]

RESOURCE UTILIZATION

On the whole there seems to be a tendency for the pattern of foreign trade of the various CMEA countries to converge. Although Bulgaria and Rumania remain the largest exporters of agricultural products, their respective shares of machinery plus industrial consumer goods have increased from 30.8 and 22.5 percent, respectively, in 1960 to 47.4 and 43.3 percent, respectively, in 1972. In 1972 in the GDR and Czechoslovakia the respective shares of that category were 72.1 and 68.1 percent, and in Hungary and Poland, 56.3 and 55.7 percent. The USSR remained by 1972 the largest exporter of fuels, raw materials, and chemicals (56.5 percent), while its exports of machinery and industrial consumer goods amounted only to 26.7 percent (see Table B.73).

Intra-CMEA Trade

One of the key features of CMEA countries is that by far the largest share of trade remains within the bloc, leaving these countries' industrial output rather insensitive directly to the more sophisticated technology and higher quality of goods imported from the West or to the more stringent requirements of Western buyers. As shown in Table B.74, with the exception of Rumania, more than two-thirds of foreign trade of CMEA countries was with the socialist countries. Moreover, with the exceptions of Rumania and Czechoslovakia, the geographical trade orientation of CMEA countries did not change drastically, except for a modest increase in the share of trade with nonsocialist countries that began in the mid-1960s.*

In the past, trade within the CMEA consisted mainly of bilateral agreements (with tie-in arrangements limiting gains from exchange), with the USSR as the predominant trade partner of all countries. By 1971 trade with the USSR represented more than 30 percent of overall trade in Czechoslovakia, the GDR, Hungary, and Poland, in contrast

---

*Bulgaria's share in exports to the industrialized West was the lowest in 1967 (15 percent), followed by the GDR and Czechoslovakia (about 17 percent), Hungary 23 percent, Poland 27 percent, and Rumania 30 percent. The corresponding shares of exports to developing countries in 1967 were about 10 percent, with Bulgaria the lowest (7 percent) and Rumania the highest (14 percent). The lowest imports from the industrialized West were registered by Czechoslovakia, the GDR, and Bulgaria (about 20 percent), followed by Hungary and Poland (about 25 percent), and the remarkable upsurge in Rumania to 46 percent (from 23 percent in 1965). Imports from developing countries were 5 percent in Bulgaria and Rumania, and between 7 and 9 percent in the other countries.[45]

with about 54 percent in Bulgaria and about 25 percent in Rumania.
Trade with the remaining CMEA countries gravitated at about 30 percent in Czechoslovakia, the GDR, and Hungary, and it was about 26 percent in Poland; such trade was only about 22 percent in Bulgaria and Rumania.[46]* During the 1950s and 1960s, despite the rhetoric, there was very little integration and specialization in the CMEA.[48] Each country attempted to develop the largest possible variety of types of heavy industry and followed a similar pattern of industrialization. By 1968 only 10 percent of the trade within the CMEA was a result of specialization agreements.[49] Kynziak notes: "As a rule the countries choose the same or similar directions of specialization."[50] In this connection it would seem that the newer industries (such as computers, automation equipment, and so on, those that have come into being in CMEA countries within the past ten years) would be more susceptible to specialization, for it would not entail as much restructuring of existing capacity and arrangements. However, it seems that the CMEA countries still continue the policy of "duplicating investments," so that their exports still consist of more or less identical commodity groups. Botos remarks: "The opportunities for a division of labor within the framework of economic integration are not yet utilized adequately"; such opportunities are seriously handicapped by a lack of criteria for "judging unambiguously the effectiveness of an investment decision from the viewpoint of all partners."[51]

The threatening energy crisis has promoted the pooling of resources for development within the CMEA, which is relatively independent of outside sources of supply for energy. Intra-CMEA energy prices were to remain unchanged until 1976, but they rose sharply in 1975. Joint prospecting and building of capacities in Komi (oil) and Orenburg (gas) in the USSR and coal in Poland were promoted. Some of the important joint projects under construction in other fields are: steel (Kursk, in the USSR), asbestos (Kiembaev, in the USSR), copper (in Poland and Bulgaria), sulphur (in Poland), potassium salt (in the GDR), and cellulose (Ust-Ilimsk, in the USSR). Specialization is being promoted in machine building to reduce costs of research and production of costly electronics and machinery for atomic power plants. Three CMEA associations were set up (Intertextilmach, Interatomenergo, and Interelectro) to oversee research, planning, production, sales, servicing, and technical assistance.[52]

In 1950 Czechoslovakia and Hungary were the countries with the lowest share of trade with socialist countries; Bulgaria had the

---

*Rumanian imports from the USSR were 26 percent in 1967 (a drop from 38 percent in 1965), from Eastern Europe, 19 percent (no change), from advanced market economies; 46 percent (double the 1965 share), and from underdeveloped economies 5 percent (8 percent in 1965).[47]

RESOURCE UTILIZATION

highest, followed closely by Rumania. By 1955 Czechoslovakia had reoriented its trade toward the socialist countries, whereas Hungary's shares remained virtually unchanged. By 1960 Bulgaria and the USSR recorded a slightly lower share of trade with the socialist countries and Rumania a significantly lower share, whereas the shares of Czechoslovakia and the GDR went up somewhat, and Hungary's share of trade with the socialist countries shot up. From 1960 to 1967 the shares of trade with socialist countries decreased somewhat in Bulgaria and considerably in Rumania, but in both countries they went up in 1972. Throughout the period (1960-72) there was evidence of a slight, steady decrease of this share in Hungary and a more rapid decrease in the GDR and the USSR, and there were relatively unchanged shares in Czechoslovakia and Poland. In the latter the shares went up somewhat in 1965 and 1967 and receded in 1972 (see Table B.74).

Trade with the West

It is not easy to identify the combination of political and economic forces that affect the extent to which the various CMEA countries trade with the West.[53] Obviously, political loyalty to the USSR and the difficulties imposed by some Western countries have affected East-West trade in the past and will continue to do so, albeit to a lesser extent.[54] However, in the absence of long-term Western credits, one of the primary economic reasons is the country's potential ability to earn hard currency through exports and tourism.[55] As trade with the West increased, so did CMEA indebtedness in hard currency. Decisions to import industrial products from the West must be made with an eye to the balance of payments. As could be expected, the highest indebtedness was recorded for the countries most active in trade with the West. Thus, the estimated negative balances of payments with the industrial West (in U.S. dollars) were as follows in 1971: Romania, -1.8091 billion; the GDR, -1.2373 billion; Poland, -1.1577 billion; Czechoslovakia, -1.0212 billion; Hungary, -927.9 million; and Bulgaria, -787.3 million.[56] During 1971-75 the positive balance of payments of industrial market economies with the CMEA amounted to (current) U.S. $13.313 billion.[57] In the past CMEA exportables for hard currency have been primarily raw materials, semifabricates, and foodstuffs. As a country becomes more industrialized its domestic requirements for these products (whose supply is relatively inelastic) impinges on export potential. This could be counteracted by increasing export of high-quality industrial products. Such exports already play some role in Czechoslovakia and the GDR, but they are as yet nonexistent in Bulgaria and Rumania.

In the 1970s there was relatively rapid progress in East-West trade. The rates of increase of CMEA countries' exports to and imports from the industrial market economies were respectively as follows: in 1970, 12 and 20 percent; in 1971, 9 and 6 percent; in 1972, 16.6 and 32.8 percent; in 1973, 38.9 and 45 percent; in 1974, 42 and 40 percent; in the first nine months of 1975, 6.5 and 36.5 percent. But these rates are distorted by the upward price movement so that the rates of growth by volume in 1973 were 14.1 and 21.3 percent; in 1974, -11 and 10 percent; and in 1975, nil and 10 percent.[58] The industrial market economies have an increasing positive trade balance with the CMEA (in 1974 and 1975, respectively, U.S. $3.28 and 6.164 billion.[59] The established pattern of East-West trade continued into the 1970s. The negative commodity balances of the CMEA with Western Europe concentrated on metals and metal products, chemicals and rubber, machinery and other manufactured goods, whereas the positive balances were in energy materials, agricultural products, and timber and pulp.[60]

Some of the novel forms of East-West cooperation that emerged were the 33 long-term agreements for scientific and technical cooperation signed since 1972 between the USSR Committee for Science and Technology and U.S. companies. Other examples include the agreement between the respresentatives of the Soviet Ministry of Coal Industry and Kaiser Resources, Ltd. (Canada) and Mitsui Mining Co. (Japan) for cooperation in hydraulic mining technology; a five-year contract for technical cooperation between the USSR and Mayer & Co. (GFR) for design and production of textile equipment; and the agreement between the USSR and Siemens (GFR) for medical equipment. Hungary pioneered in the creation of mixed companies. In addition, joint production facilities were to be established by the Polish tractor industry and Massey-Ferguson and Perkins of the United Kingdom.[61] Many more examples could be cited.

Bulgarian Trade Relations

On the whole, the share of Bulgaria's trade with socialist countries declined from 84.7 percent in 1956 to 77.8 percent in 1970 (with the CMEA, from 81.9 to 74.4 percent), whereas that with developed-market economies increased from 12.2 to 16.6 percent, and with underdeveloped economies from 3.1 to 5.6 percent.[62] Germany was the main trading partner of prewar Bulgaria. In the postwar period this role was assumed by the USSR, whose share in Bulgaria's foreign trade has been steadily increasing, while those of the other CMEA countries have been decreasing. In fact, the shares of Poland, Hungary, and Rumania in Bulgaria's foreign trade are not much

RESOURCE UTILIZATION

higher than those of France, the GFR, Italy, or the United Kingdom (see Table B.75).

The Bulgarians are fond of speaking of economic integration with the USSR. Bulgarian dependence on the Soviet source of supply for raw materials is in many cases overwhelming. For example, in 1970 the share of the USSR in total Bulgarian imports of coal was 100 percent; that of coke, 30 percent; of crude oil and diesel oil, 84 percent; of lubricating oil, 82 percent; of iron ore 91 percent; of pig iron, 72 percent; of iron sheets, 66 percent; of iron wire, 80 percent; of construction steel, 79 percent; of tin plate, 79 percent; of aluminum, 85 percent; of cellulose, 76 percent; of cotton, 70 percent; and of wool, 77 percent. The USSR has also found in Bulgaria a good nondiscriminating customer for its machinery and industrial consumer goods. Thus, in 1970 the share of the USSR in total Bulgarian imports of lathes was 40 percent; of milling machines, 36 percent; of light-industry equipment, 32 percent; of chemical-industry equipment, 21 percent; of road-construction machinery, 63 percent; of equipment and materials for complete factories, 55 percent; of tractors, almost 100 percent; of passenger cars, 75 percent; of motorcycles, almost 100 percent; of cotton fabrics, 46 percent, of home sewing machines, 94 percent; of bicycles, 79 percent; of cameras, 70 percent; of radios, 74 percent; of vacuum cleaners, 83 percent; and of television sets, nearly 100 percent.[63] Since the 1950s the USSR has helped to build about 240 Bulgarian industrial projects, 130 of which are already in operation, accounting for 65 percent of Bulgarian chemical output, for more than 90 percent of ferrous metals, for 85 percent of nonferrous metals, and for more than 70 percent of power production.[64]

In return, Bulgaria exports to the USSR a good share of its machinery and heavy-industry products. For example, in 1970 the share of the USSR in total Bulgarian exports of electric engines was 52 percent; of electric trucks, 68 percent; of electric pulley blocks, 56 percent; of ensilage harvesters, 100 percent; of fodder mills, 100 percent; of ships, 57 percent; of large section steel, 74 percent; and of electric truck tubes and tires, 60 percent.[65] By 1975 there were nine specialization, subcontracting, and cooperation agreements between the Bulgarian and Soviet ministries of machine building; the Ministry of Electronics and Electrical Engineering had entered into six such agreements. These agreements concentrate on electric trucks (Bulgarian production apparently covers almost all Soviet needs), automobile parts and assemblies (the Bulgarians produce these for the Zhiguli passenger cars and in return receive the finished product), assemblies for tractors produced in the USSR, and electronic computer equipment (in which Bulgaria ranks second, after the GDR, as supplier to the USSR).[66]

The USSR is also a bulk customer for Bulgarian agricultural products. Thus, in 1970 the share of the USSR in total Bulgarian exports of tobacco was 63 percent; of poultry, 53 percent; of canned meat and vegetables, 64 percent; of cheese, 44 percent; of fresh vegetables, 44 percent; of canned vegetables, 70 percent; of fresh fruit, 49 percent; of canned fruit, 60 percent, and of jams, 80 percent. Reciprocally, the USSR is not a particularly discriminating customer for Bulgarian industrial consumer goods. In 1970 it bought 84 percent of Bulgarian exports of carpets, 90 percent of ready-made clothes, 90 percent of leather footwear, and 92 percent of furniture.[67]

Bulgarian trade with the other CMEA countries concentrates to a large extent on reciprocal exchange of machinery and the exporting of Bulgarian food products. But trade with the developed-market economies is of a very specific nature, highly representative of trade between a backward country with industrialized ones: Bulgaria imports mostly machinery from these countries and sells them raw materials, semifabricates, and agricultural products. For example, in 1970 Austria, the United Kingdom, the GFR, Italy, France, and Switzerland supplied almost 10 percent of Bulgarian imports of light-industry equipment and 24 percent of chemical industry equipment; Austria, Belgium, Denmark, Sweden, the United Kingdom, France, the GFR, and Italy (with the last four representing the bulk of the transactions) supplied 31 percent of Bulgarian purchases of equipment and materials for complete factories. In return, in 1970 Bulgaria sold 48 percent of its exports of plywood to the United Kingdom; 32 percent of cellulose to Italy and France; 34 percent of steel plates (3 mm. and thicker) to the GFR, Italy, and France; 60 percent of lead to Austria, Italy, and Switzerland; 58 percent of zinc to the United Kingdom, Italy, and France; 71 percent of peppermint oil to the United Kingdom, the GFR, and France; 71 percent of attar of rose to France and the United States 95 percent of calves to Italy; 65 percent of weaned lambs to Greece and Italy; and 68 percent of pork to Italy.[68]

Generally, East European countries import a considerably larger share of fuels, mineral raw materials, and metals from the CMEA (70 to 80 percent) than they export to the CMEA (about 60 percent). While in 1967 Bulgaria imported more than 75 percent of its raw materials from the CMEA, it sold only 42 percent to the CMEA. Bulgaria's flexibility is much less than implied by these figures, for that group constitutes only about 7 percent of Bulgaria's exports, whereas in other East European countries it ranges from about 15 to 25 percent.* The largest exporters of foodstuffs are Rumania, the

---

*In 1972 Bulgaria further discriminated against imports of raw materials from the West. Imports of considerable quantities of goods

RESOURCE UTILIZATION   79

USSR, and Bulgaria (constituting about 12 percent of total exports in 1967). The USSR and Rumania exported only about 44 percent to the CMEA; Bulgaria sold nearly 70 percent of foodstuffs exports to the CMEA. The other CMEA countries (with a share of about 5 percent of foodstuffs in total exports) sold only 15 to 28 percent of foodstuffs to the CMEA.[70]

Bulgaria has not been so prompt as its CMEA partners to jump on the bandwagon of detente and expand its trade with the West. Whereas in 1972 CMEA imports from the industrial market economies grew by 32.8 percent (with the largest increases in Poland, with 55.2 percent; the USSR, with 44.9 percent; and Rumania, with 31.5 percent), in Bulgaria they increased by only 9.6 percent. But by 1973 the respective figures were 45 and 39.3 percent. By 1974 the growth rates of Bulgarian imports from the West were the highest among CMEA countries and well above the average (71 and 40 percent, respectively). This tendency continued into 1975, when the respective figures were 39.2 and 36.5 percent.[71] Bulgaria's closest trade ties in the West are with the GFR. It was reported that in 1971-75 trade between the two countries tripled. Increasingly, the trade agreements are on a cooperative basis. For example, in 1975 Bulgaria signed contracts to cooperate in production with Siemens for communications equipment, medical instruments, and automators; with Lurgi for processing Bulgarian oil shale; with Daimler-Benz for battery-driven trucks and automobiles; and with Krupp for metallurgy, heavy-machine building, and chemical equipment.[72] In view of its particularly close ties with the CMEA, Bulgaria has come out increasingly on the side of CMEA-European Economic Community (EEC) relations, rather than bilateral ones.[73] This might be interpreted as an attempt on Bulgaria's part not to be left out of the East-West trade picture by its more adventurous CMEA partners. Moreover, its recent Western shopping splurge for industrial goods has created considerable balance-of-payments problems, for Bulgaria continues to sell the West mainly industrial goods of agricultural origin and foodstuffs and these, partly due to EEC protectionist tariffs, are not overly profitable exports.[74]

For most of the postwar years the Bulgarian balance of trade (in million leva) was negative: 1950, -18; 1955, -16; 1956, +59.6; 1957, +44.4; 1958, +7.7; 1959, -131.4; 1960, -71.5; 1961, -4; 1962, -14.5; 1963, -116.1; 1964, -96.8; 1965, -2.2; 1966, -202.7; 1967, -133; 1968, -195.6; 1969, +52.8; 1970, +202.2; 1971, +73.4; 1972, +64.8.[75] When these data are compared with Figure 2.1, it is obvious that the

---

were rechanneled from capitalist to socialist countries, probably in an attempt to alleviate the growing foreign trade deficit with the West.[69]

largest negative balances occurred in periods of particularly steep investment effort. In fact, until the 1970s positive balances were recorded only in periods of considerable investment slack. This is a good indication that Bulgarian imports are geared primarily to sustain the growth effort and that, without substantial long-term credit or with the burden of repaying previous credits, the negative trade balance is one of the causes for the periodic shifts to lower gear.

As an exporter of primary products, Bulgaria's terms of trade deteriorated after the war (1948 = 100; 1952, 65; 1957, 62; 1958, 64), but because of the recent improvement in terms of trade for foodstuffs, the Bulgarian position should improve in the 1970s. The Bulgarian industrialization effort has been greatly enhanced by Soviet aid. For obvious reasons, data on this facet are extremely fragmentary. One source reports that during the first 16 postwar years, the USSR has granted Bulgaria credits amounting to $500 million, principally for capital goods.[76] During 1947-51, Bulgaria was supposed to have received credits of $237 million, mainly from the USSR. No substantial new credits were reported in 1952-55. From mid-1955 to 1958 new credits of $213 million were negotiated, and in 1961 a credit of $163 million. It was reported that the USSR financed 25 percent of Bulgarian state investments in 1947-59.[77] According to Bulgarian sources, credits received from the USSR until 1956 amounted to about 8 billion leva, and in 1958 alone Bulgaria received more than 335 million leva in Soviet credits.[78] Another source claimed that in the 15 years after the end of World War II Soviet credits amounted to more than 2 billion rubles.[79] According to the deputy finance minister, from 1949 to 1958 Soviet loans to Bulgaria exceeded 9.6 billion leva.[80]

To recall, during 1950-55 Bulgaria undertook massive investments in heavy industry that were particularly capital intensive and of long gestation periods. During this period, imports of capital goods grew more rapidly than those of raw materials. In the post-1955 period the enterprises built in the first period began production, necessitating rapid increase of raw materials. At the same time, the underinvested raw-materials base (including agriculture) received increased attention—that is, investment. In addition, the need to placate the neglected worker-consumer intensified strains by cutting down on the export of foodstuffs and industrial consumer goods and by necessitating investments in food and light industry. All this constantly swelled the import needs, while exports merely played the passive role as a means of payment for imports. But due to the deteriorating terms of trade, increasing exports were required to pay for the same imports. The withdrawal of an ever-larger quantity of goods from domestic consumption was also a brake on the pace of development.

THE TRADITIONAL PLANNING SYSTEM

The role of the planning system as a growth propeller or barrier is a controversial one. An attempt to quantify the actual effect of the traditional planning system in Czechoslovakia indicated positive effects on the growth rate in the early 1950s and negative effects thereafter. While the estimates should be approached with great circumspection, the shortcomings of the traditional system are obvious.[81] The problems created thereby intensify with the advance of industrialization, but many of the traditional system's drawbacks are a source of waste even at the early stage of industrialization. It is an untenable argument that the traditional system suited the early stage. In order to discuss the reforms of the functioning mechanism, it is necessary to analyze what is to be reformed. Literature is replete with comprehensive analysis of the traditional STE system. But in this context one would be remiss not to present briefly the Bulgarian view of its shortcomings. Understandably, many of the statements made are either incomplete or tendentious, but the author has purposely let the Bulgarians speak, rather than provide his own interpretation, which will be found in the subsequent chapters. Some of the shortcomings listed are causes and some are effects. Each of them is important enough to merit closer scrutiny, not only because they occur in the traditional system, and during the protracted process of transition, but especially because they are not eliminated by the "final" version of the reform.

1. The plan and its chief assignments have been accused of having been "turned into a fetish." The only performance criterion of the enterprise was whether and by how much it has fulfilled its plan (especially gross value of output). The primary conflict between the interests of the enterprise and those of the c.p. then became the strong drive of the former to obtain the lowest targets and the highest inputs possible, which would permit easy plan overfulfillment, and of the latter to foist on the former as taut a plan as it could.[82] Management's strategy was to conceal as much productive capacity as it could as insurance for fulfilling a continuously raised plan whose targets exceeded the revealed capacity.[83] Aware of the tendency to conceal reserves, the c.p. constantly presumed that higher targets and lower limits could be imposed thatn those warranted by the information received. In the end, many enterprises received assignments impossible to fulfill, and others enjoyed a slack plan. Moreover, management was careful not to overfulfill the plan too mucy, since overfulfillment could provoke an excessive raising of the following period's assignments.[84] Thus, plan fulfillment as a performance criterion suffers from the strife between plan givers and plan recipients that in the final analysis is influenced by the former's knowledge

of local conditions and his ability to wield authority. Rewards are much more dependent on cunning and knowing "how to play the game" and cheating than on good work. The result is underutilization and mismanagement of resources, and the user is victimized, for the producer produces for the plan and not for the user.

2. Although fulfillment of the gross value of output was the primary premium condition (and yardstick for allocating the wage fund), in time its obvious shortcomings became more pronounced, and such side conditions as costs, labor productivity, and so forth were imposed. Instead of constraining malpractices, these conditions only stimulated management to seek wider margins of tolerance.[85] Moreover, due to weak coordination of production and financial planning, management was able to secure premiums for almost all indexes without satisfactorily fulfilling any.[86]

3. Although the traditional STE system can foist technical progress from above (by direct orders and full financing), it fails to stimulate it from below, and it has built-in obstacles for implementation of orders. C.p. tends to perpetuate the existing industrial structure and techniques.[87] New plants are often equipped with obsolete equipment.[88] Technical-progress implementation disrupts plan fulfillment by upsetting the regular flow of activity, by absorbing labor and materials away from the production schedule, and so on. Then, as the enterprise is modernized, the c.p. foists on it increasing production demands, without much consideration to the kinks that have to be ironed out in the introduction of new technology. In view of the problems involved, the additional capacity, and the consequent additional assignments, the enterprise is not only indifferent but downright hostile to new ways of doing things.[89]

4. The entire system stresses quantity at the detriment of quality and drives the producer to turn out obsolete and poor-quality goods.[90] The bill of goods produced is subordinated to gross-output plan fulfillment. The economy is overheated and the system is devoid of the exigencies of a buyer's market that would subject the producer to an effective check by the buyer. In the final analysis, the user-consumer is the loser. The system is further handicapped by the ineffectiveness of the quality-control apparatus. The quality-control inspector, who relies on the enterprise's manager for his job, promotions, and premiums, is reluctant to make him hostile by vigilant surveillance of quality standards. In the early 1960s complaints of poor quality and lack of variety of consumer goods (especially goods such as shoes, children's and men's clothing, and canned goods) poured in.[91] By the end of 1962 the stock of unsold inventories was twice as large as that planned.[92] Producers aimed at producing "plan-satisfying" goods, and the various violations of assortment plans were undertaken to report the largest possible production-plan fulfill-

ment and not to satisfy changes in consumer demand. Moreover, the price structure effectively insulated producers from information about consumers' demand.[93]

5. The traditional supply system was inefficient, cumbersome, rigid, and wasteful. The supply sources were centrally assigned to users, preventing them from seeking out the more economical suppliers. In view of the spreading tentacles of the bureaucratic system, not only key materials (or subsistence necessities) in short supply were rationed but also readily available goods. In practice, users were required to stipulate their needs about a year in advance. Many production plans were changed during the course of the annual plan, by which time it was too late to rescind or alter orders for inputs. Unneeded goods often came in and piled up, while the needed inputs were unavailable, giving rise not only to frantic scrounging for all sorts of inputs, but also to the accumulation of unnecessary goods (hoarding of inputs for the future).[94] In view of plan changes, manipulations, and so forth, goods were often not delivered, or were supplied only with considerable delays and not in the composition ordered. Buyers could exact damages and fines, but these were insignificant and not worth applying, and they antagonized the supplier.[95] Furthermore, since the financial result was only of marginal importance, these fines (irrespective of their size) neither made up for the losses to the buyer nor detracted from the benefits of the supplier.

6. At every level there was a tendency toward self-sufficiency and unnecessary duplication. Because of the chronically failing supply and subcontracting system, every enterprise tended to produce component parts, to have its own repair and maintenance shops, and so on—all this at enormous costs and loss of efficiency. Regional management was fraught with local patriotism and was disruptive to the flow of output and technical advance.[96]

7. The system induced waste of fixed and working capital. Fixed capital was subsidized by the budget, and the manner of using capital did not affect the principal performance criterion of management.[97] Moreover, the expansionary impulse, generated by the c.p., pushed the manager "to keep as many machines as possible because this places his factory in a higher category, while he himself receives higher pay, more bonuses, etc."[98]

8. The wage system neither stimulated parsimony in the use of labor nor motivated the wage earner to perform better. Wages were generally paid according to a state-established wage schedule and were unrelated to performance. The base wage accounted for 60 to 70 percent of the wages of piece-rate workers, for 75 to 80 percent of hourly workers, and for 85 to 90 percent of management and white-collar workers.[99] This left little room for maneuvering the variable share (premium) of take-home pay as to induce extraordinary

performance. Some enterprises were rewarded for wage-fund savings, computed as the difference between the planned and reported wage funds. These savings reflected mainly the extent to which the planned figures were inflated.[100] Aside from the main premium for production-plan fulfillment, some poorly coordinated and insignificant rewards were given out for cost reduction and other partial tasks.[101]

9. In practice the "budget constraint" was lax, and the state budget made up most shortfalls, irrespective of flagrant inefficiencies. Subsidies in various forms permeated the system. The number of subsidized enterprises was constantly growing, and among them were the newest and technically most sophisticated ones.[102]

10. The c.p. was preoccupied with siphoning off "profits," rather than with their incentive and allocative functions; thus, the lion's share of profits went to the budget, and only about 5 percent was used for incentive purposes, mainly for welfare measures. Managers were not particularly interested in the enterprise's financial results.[103] About 74 percent of an enterprise's profits went to the budget; the remaining 26 percent was channeled into 27 various appropriations— none of very much significance.[104] More specific data indicated that between 1955 and 1960, 60 to 80 percent of profit went to the budget; 40 to 17 percent was used to finance activity; 2 to 4 percent was appropriated for personnel, of which less than 0.1 percent went for premiums.[105]

The profit plan was actually a financial recomputation of the physical plan, with allowance for projected cost reduction. Due to the various shortcomings of the planning process, followed by continuous changes, the final profit plan was often set only in the second half of the planned year, so as to accord with the expected results. The part of profit retained by the enterprise was independent of the rate of profit achieved, of the tautness of the plan fulfilled or of the enterprise's contribution to satisfying domestic market and earning foreign currency. With deficient prices and costing, profit was hardly an index of efficiency, and statistical records seem to indicate spurious results, particularly if related to the capital employed.[106]

11. In order to obviate the intensifying conflicts between the various echelons, increasing centralization was practiced, enmeshing the system in an ever-growing and constricting network of indexes. For example, the Council of Ministers (CM), the National Assembly, the State Planning Commission (SPC), the branch committees, and the various ministries confirmed more than 1,800 indexes for industry alone. The SPC alone confirmed about 400; the CM adopted eight labor and cost indexes, and the ministries and branch committees 12 to 24 of the same. Such an overgrown system obviously led to contradictory, inconsistent, and unrealistic indexes that were frequently changed during the course of the plan (adjusted to actual implementation), reducing their effectiveness and power to "mobilize."[107]

RESOURCE UTILIZATION   85

12. Overcentralization led to an overgrowth of the administrative apparatus, an increasing number of decisions being made on the basis of incomplete or distorted information, with accompanying time lags and procrastination. Overstaffing at central offices was accompanied by a similar tendency at enterprises.[108] All these shortcomings were reinforced by the inherent conservatism of the bureaucracy and other employees who opposed or obstructed changes in daily routine. This was a "serious obstacle to the further perfecting of state and economic management."[109]

In the early 1960s the economy was beset by many problems. The deterioration of performance following on the heels of the Leap Forward was probably one of the main reasons for turning the leadership's attention to reform of the functioning system as a considerable source of growth, a way of improving the position on foreign markets, and a means of exploiting the advantages of backwardness. Moreover, the winds of reform were sweeping throughout Eastern Europe, and the imitative spirit was strong. Although there was evident need for reform, the compulsion was weaker in Bulgaria than in some other CMEA countries; this is a clue to the shifting content of Bulgarian reform. Bulgarian industry was still relatively underdeveloped and unsophisticated; performance by "extensive" growth indicators was relatively satisfactory in comparison with other CMEA countries (an important performance criterion for the c.p.), Bulgaria continued to benefit from "external" accumulation; finally, the pressures from the population to raise living standards were not so strong then as in other CMEA countries or as they later became in Bulgaria. The various inefficiencies of the traditional functioning system, as viewed by the Bulgarians, were the obvious reasons given for the imperative of reform.

NOTES

1. J. Tinbergen et al., Optimum Social Welfare and Productivity (New York, 1972), p. 92.
2. A. Bergson, World Politics, July 1971, p. 599. Compare U.S., Congress, Joint Economic Committee, Economic Developments in Countries of Eastern Europe (Washington, D.C., 1970), p. 63.
3. United Nations, Economic Survey of Europe in 1969, Part 1 (New York, 1970), p. 10.
4. United Nations, Economic Survey of Europe in 1971, Part 1 (New York, 1972), p. 6; U.S., Congress, Joint Economic Committee, Economic Developments, p. 62.
5. United Nations, Economic Survey . . . 1971, Part 1, p. 13.
6. United Nations, Economic Survey . . . 1969, Part 1, pp. 50-51.

7. U.S., Congress, Joint Economic Committee, Economic Developments, p. 267.
8. Compare United Nations, Economic Survey . . . 1969, Part 1, pp. 50-51; U.S., Congress, Joint Economic Committee, Economic Developments, pp. 521-22.
9. Compare G. R. Feiwel, Industrialization and Planning under Polish Socialism, vol. 1, "Poland's Industrialization Policy" (New York, 1971), p. 355.
10. United Nations, Economic Survey . . . 1971, Part 1, p. 18.
11. For comparable data on other East European Countries, see U.S., Congress, Joint Economic Committee, Economic Developments, pp. 437-38.
12. Feiwel, "Poland's Industrialization Policy," p. 48; United Nations, Economic Survey of Europe in 1968 (New York, 1969), p. 146.
13. G. Lazarcik and A. Wynnyczuk, Bulgaria: Indexes of Construction, Investment, Housing, and Transportation and Communications, 1939 and 1948-1965, OP-30 (New York, 1968), p. 4.
14. J. Kleer, Ekonomista (Warsaw) 1 (1973).
15. U.S., Congress, Joint Economic Committee, Economic Developments, p. 64.
16. United Nations, Economic Bulletin for Europe 18, no. 1 (November 1966): 39. Compare Problemy ekonomiczne (Cracow) 3 (1971): 10.
17. Compare United Nations, Economic Bulletin, p. 54.
18. G. R. Feiwel, ed., New Currents in Soviet-Type Economies (Scranton, Pa., 1968), pp. 95-96. Compare U.S., Congress, Joint Economic Committee, Economic Developments, pp. 268-70.
19. Bergson, World Politics, p. 600.
20. Compare Feiwel, "Poland's Industrialization Policy," p. 47.
21. U.S., Congress, Joint Economic Committee, Economic Developments, pp. 456-58.
22. United Nations, Economic Survey of Europe in 1970, Part 2 (New York, 1971), p. 72.
23. Ibid., p. 118.
24. Compare B. Balassa and T. J. Bertrand, AER, May 1970, p. 315; Tinbergen et al., Optimum Social Welfare, p. 58ff.
25. Compare U.S., Congress, Joint Economic Committee, Economic Developments, pp. 271, 443.
26. United Nations, Statistical Yearbook 1972 (New York, 1973), pp. 398-99.
27. Handel zagraniczny a wzrost krajów RWPG (Warsaw, 1969), p. 62.
28. A Zauberman, Industrial Progress in Poland, Czechoslovakia and East Germany (London, 1964), pp. 275-82.

29. H. Chenery, AER 4 (1960): 624-55.
30. F. T. Pryor, The Communist Foreign Trade System (Cambridge, Mass., 1963), pp. 27, 275-79.
31. P. J. D. Wiles, Communist International Economics (New York, 1969), pp. 419-53.
32. A. Brzeski, "Remarks on East European Trade," mimeographed (Milan, 1971).
33. Compare Zauberman, Industrial Progress, p. 279; Pryor, Communist Foreign Trade System, pp. 100ff.
34. Rocznik statystyczny handlu zagranicznego 1968 (Warsaw, 1968), pp. 78-79.
35. For a discussion of the factors reducing and increasing the level of trade, see A. Brown and E. Neuberger, eds., International Trade and Central Planning (Berkeley, Calif., 1968), pp. 283ff.
36. Handel zagraniczny, p. 62.
37. United Nations, Economic Survey . . . 1970, Part 2, pp. 92, 139. Compare Przemysl w Polsce i wybranych krajach 1950-1968 (Warsaw, 1970), pp. 2-9, 36-39.
38. Przemysl w Polsce, pp. 36-51.
39. Brown and Neuberger, International Trade, pp. 130-73.
40. U.S., Congress, Joint Economic Committee, Economic Developments, p. 255.
41. Compare M. Kolanda, Podnikova Organizace (Prague) 12 (1966): 551-56; IM 9 (1974): 3.
42. United Nations, Economic Survey . . . 1971, Part 1, p. 55.
43. Ibid.
44. Ibid.
45. U.S., Congress, Joint Economic Committee, Economic Developments, pp. 551-54.
46. U.S., Congress, Joint Economic Committee, Reorientation and Commercial Relations of the Economies of Eastern Europe (Washington, D.C., 1974), p. 82.
47. U.S., Congress, Joint Economic Committee, Economic Developments, p. 554. Compare Uneshnaya Torgovlya SSSR za 1971 god (Moscow, 1972), pp. 184-92.
48. On problems of integration, divergent interests, and disputes among CMEA partners, see M. Kaser, Comecon (London, 1967); U.S., Congress, Joint Economic Committee, Reorientation, pp. 79-134. On the Rumanian position on integration, see J. M. Montias, Economic Development in Communist Rumania (Cambridge, Mass., 1967), Chapter 4. Kaser suggested that political factors plus Bulgaria's particularly favorable terms of trade and recognition that it can do better inside than outside the CMEA account for Bulgaria's apparent "love affair" with the CMEA: Kaser, Comecon, pp. 105-06. Moreover, the high dependence on the USSR reduces the "unplannable"

character of external trade and bilateral agreements are beneficial to the c.p. as they reduce the risks of plan underfulfillment.

49. B. Reutt, GP 9 (1968): 35.
50. Z. Knyziak, GP 2 (1970): 1.
51. K. Botos, Kulgazdasag (Budapest) 12 (1974): 923.
52. United Nations, Economic Survey of Europe in 1973 (New York, 1974), pp. 153-54.
53. For an interesting hypotheses and testing of the factors influencing CMEA imports of industrial goods from the West, see U.S., Congress, Joint Economic Committee, Reorientation, pp. 672-75. According to Montias in that study, Bulgaria displayed the greatest loyalty to the CMEA, for it imported a smaller share of Western manufactured goods than it could afford to pay for from earnings of its sales to the West of surplus raw materials and agricultural produce, tourist earnings, and Western credits. On the useful distinction between goods hard to sell in the West and those that are not, and for data on surpluses and deficits in each category, see Brown and Neuberger, International Trade, pp. 132-38.
54. Compare J. Hardt, Tariff, Legal and Credit Constraints on East-West Trade Commercial Relations (Ottawa, 1975), pp. 27ff.
55. Compare J. Wilczynski, The Economics and Politics of East-West Trade (New York, 1969).
56. U.S., Congress, Joint Economic Committee, Reorientation, pp. 693-718. For an estimate of the Eurocurrency position and hard-currency indebtedness in Eastern Europe, see Hardt, Tariff, Legal and Credit Constraints, pp. 11-13.
57. United Nations, Economic Survey of Europe in 1975, Part 1, Chapter 2, prepublication text, ECE (XXXI)/1 Add. 1, Table 5.6.
58. Ibid., Table 5.1; United Nations, Economic Survey of Europe in 1971, Part 2 (New York, 1972), p. 94; United Nations, Economic Survey . . . 1973, p. 155.
59. United Nations, Economic Survey . . . 1975, Table 5.3.
60. Ibid., Table 5.5.
61. Y. Genovski and A. Marukyan, VT 5 (1975): 10-13.
62. Statistical Yearbook of Bulgaria 1971 (Sofia, 1971), p. 195.
63. Ibid., pp. 198-226.
64. Tass, December 17, 1975.
65. Statistical Yearbook 1971, pp. 198-205.
66. P. Somlev, ZZ, May 28, 1975, p. 1.
67. Statistical Yearbook 1971, pp. 206-16.
68. Ibid., pp. 198-226.
69. A. Kadiyski, PS 9 (1972): 44.
70. U.S., Congress, Joint Economic Committee, Reorientation, p. 669.

71. United Nations, Economic Survey . . . 1973, p. 155; Economic Survey . . . 1975, Table 5.1.
72. Tass, November 27, 1975; December 1, 1975.
73. A. Bikov, IZ, June 4, 1975, p. 9; A. Gitsov, IZ, February 12, 1975, p. 9.
74. T. Velev, PS 7 (1975): 17; Ivanova, IM 1 (1974): 38; I. Bozhkov, IZ, February 19, 1975, p. 10.
75. Handel zagraniczny, p. 282; Statisticheski Godishnik na Narodna Republika Bulgaria 1967 (Sofia, 1967), p. 298; Statistical Yearbook 1971, p. 186; Statisticheski Godishnik na Narodna Republika Bulgaria 1973 (Sofia, 1973), p. 327.
76. L. Ciamaga, Od wspolpracy do integracji (Warsaw, 1965), p. 39.
77. United Nations, Economic Survey of Europe in 1960 (Geneva, 1961), Chapter 6, pp. 36-37.
78. R. Hristozov, RD, March 17, 1958. Compare M. Nikolev, RD, June 18, 1959.
79. G. Popisakov, NV 6 (1959).
80. K. Nestorov, OF, September 20, 1959.
81. G. R. Feiwel, New Economic Patterns in Czechoslovakia (New York, 1968), pp. 68-71.
82. T. Zhivkov, Problems of the Construction of an Advanced Socialist Society in Bulgaria (Sofia, 1969), p. 34.
83. D. Davidov, IM 6 (1965): 6.
84. E. Mateev, NV 8 (1961); Mateev, NV 5 (1963): 75-76.
85. Mateev, NV 5, p. 80.
86. Zhivkov, Problems, p. 45.
87. Davidov, IM 6, pp. 4-5.
88. Zhivkov, Problems, p. 87.
89. Ibid.
90. G. Petrov and P. Stanoev, FK 7 (1963): 48.
91. N. Gesheva, PSS 5 (1961): 64-76; D. Markov, IM 3 (1962): 10.
92. Zhivkov, Problems, p. 44.
93. Petrov and Stanoev, FK, p. 48.
94. Compare N. Yankov, PSS 9 (1965): 61-63; B. Purvulov, IM 6 (1965): 24-25.
95. Yankov, PSS, p. 63.
96. Zhivkov, Problems, p. 19.
97. Ibid., p. 50.
98. Ibid., pp. 19-20.
99. A. Dobrev, TT 1 (1964): 58.
100. Compare Davidov, IM 6, pp. 19-20.
101. G. Kostov, TT 4 (1964): 35.
102. Zhivkov, Problems, p. 43. Compare I. Mironov, TT 10 (1963): 32.

103. Zhivkov, Problems, p. 47.
104. Mironov, TT, p. 32.
105. V. Rankov and R. Angelov, Izvestiya na Visshiya Institut za Narodno Stopanstvo "Dimitur Blagoev," 1 (1965), p. 85.
106. Petrov and Stanoev, FK, pp. 48, 55-57.
107. Zhivkov, Problems, p. 35. Compare Mironov, TT, p. 33.
108. Zhivkov, Problems, p. 8.
109. Ibid., p. 25.

CHAPTER

4

ECONOMIC REFORM: TRANSITIONAL PERIOD

Since the early 1960s, particularly between 1963 and 1966, the usually prosaic Bulgarian economic literature was enlivened by various reform proposals. These proposals were more or less "advanced" within the context of the existing system, and more or less adaptable. The discussion flourished after <u>official</u> recognition of past abuses and calls for reform were sounded. Most of the proposals were along the lines officially propounded and were in a similar "quasi-profit-oriented" family. Concurrently a series of experiments was initiated along the lines that were later given official sanction in the "Draft Theses of the Politbureau of the Bulgarian Communist Party on the New System of Planning and Management of the National Economy" (hereafter referred to as the Theses). The <u>a tatonnement</u> process of reform implementation was fraught with a number of snags, resulting in numerous amendments. Within four years of the initial experiments the general direction of Bulgarian reform was reversed at the July 1968 Plenum of the Central Committee, Bulgarian Communist Party (CC BCP)—the real clue to what the Bulgarian reform had become.

THESES ON THE NEW SYSTEM OF PLANNING AND MANAGEMENT

The initial reform blueprint was promised on several occasions in 1965 and was finally published at the end of the year. It was approved in April 1966, also after several delays.[1] The system outlined in the Theses was perhaps closest to that tried at that time in Czechoslovakia, particularly as regards the three-tier price construct (which is also a feature of the Hungarian reform) and the concept of gross income, borrowed and adapted from Yugoslavia.[2] In the mid-1960s, from the vantage-point of STE reform blueprints, the Bulgarian one could probably be classified with the more advanced (Czechoslovakia), as contrasted with the more orthodox solutions in Poland, the GDR, and the USSR. However, it was still far behind the much bolder solution in Hungary and ostensibly less reformist than the Yugoslav system.

The Theses attempted at restricting the scope of direct administrative orders down the vertical pyramid of the industrial system, and at enlarging the scope of economic calculation. More specifically, it aimed at enhancing the role of profit at tying wages in closer with performance, and at extending horizontal relations. The blueprint involved controlled market relations, but conditions for their implementation were not provided.

The Theses reinforced the principle of branch management, by confirming the association as the immediate and powerful superior. The association would have much to say about the scope of freedom left at the enterprise. It would function on <u>khozraschet</u> (business calculation) and be self-supporting like its wards. Obviously, the grouping of enterprises producing a similar line of goods under the aegis of a powerful unit entails grave monopolistic dangers (quite well understood by the Hungarians). Such dangers were minimized in the Theses. A sharp contrast was drawn between monopolies under capitalism and those under socialism. The socialist state, it was claimed, possessed the following tools to counteract monopolistic practices: (1) price controls, (2) assignment of specific tasks or inducements for their fulfillment, (3) control over investment, and (4) control over production quality and assortments.

But the initial experience with associations was not overly impressive. Their authority was sometimes bypassed with direct orders being given to enterprises. The associations also showed a penchant for administration of enterprises by fiat. Both their enlarged powers and retained traditional management methods enabled many associations to exercise their monopolistic powers in decisions regarding production volume, composition, quality, and prices. Even despite state price setting, "cases [were] not infrequent when in the name of closely-knit interests, profits [were] increased."[3]

The Theses reaffirmed the power of the c.p. The central administrative organs would have to implement the party's economic development policy especially in the areas of foreign trade, investments, wages, prices, finance, and instruments of plan execution. The Theses foresaw a certain abrogation of ministerial powers in favor of associations. It prescribed the following duties and rights for ministries:

1. To ensure execution of state policy and to coordinate the work of their associations in this respect

TRANSITIONAL PERIOD

2. To outline the branch's development and to participate in drawing up its annual and long-term plans
3. To manage intra- and interbranch specialization and to participate in foreign-trade specialization
4. To assign plan directives to associations (together with the SPC) and to control their fulfillment
5. To pursue a unified personnel-training policy
6. To direct research and development (R&D) work
7. To control observance of state discipline.

The peremptory administrative acts of ministries (encroaching on lower units' funds, shifting and redistributing them and so forth) would no longer be tolerated.

Planning

The Theses explicitly stated that a most important function of planning was to create a framework and instruments to ensure the fulfillment of preset goals of development. The shift pertained to the mode of implementation—that is, relying to a much greater degree on indirect economic instruments rather than on direct orders. The number of directive indexes would be reduced to a minimum.[4] Some of the former directives, such as gross output, construction and assembly, total wage fund, and production costs, would be changed into indexes set by the enterprises themselves. The approved Theses retained the following as compulsory directive indexes: (1) output volume in physical terms, (2) investment limit, (3) commissioning of capacities, (4) procurement limits, (5) foreign currency earnings, and (6) foreign currency limit for imports.

The enterprise's production plan was to be drawn up on the basis of state orders and the "on-file" and anticipated orders of trading partners. The state orders would be issued for the basic producer and consumer goods and for all goods that were "in short supply." The latter would presumably be dropped from the state orders as their individual production satisfied requirements. The list of goods ordered by the state would "be constantly changing so as to correspond to the concrete conditions of development." The state would also have to guarantee the producer the inputs to fulfill these orders. But a producer was to be allowed to enter into contract with a buyer only insofar as his capacity and ability to secure inputs permitted.

The Theses affirmed "planning from below"—in essence, the aggregation of plans prepared by lower units on the basis of a limited number of state directives and limits (control figures). The burden of comprehensive planning was to be shifted to the enterprises and associations so as to render the plan more realistic.[5] But the center would continue to map out proportions and the production profile of the lower units.

Contrary to the requirements of the planned economy, the time horizon of actors traditionally tends to be short. To overcome this drawback the reform naturally stressed long-term normatives that would increasingly replace outright orders. Of particular importance were the long-term stable norms regulating the relations between enterprises and the budget and those for determining the enterprise's special purpose funds (s.p.f.).

The Financial System

The cornerstone of the new financial system was the principle of self-financing, together with credit financing of investments to encourage parsimony and effective use of capital.[6] The enterprise's income was to be distributed in such a manner that after the collection of the progressive income tax, the capital charge, and so forth and the allocation of funds into the enterprise's individual s.p.f., the residue would serve to pay fixed (or guaranteed) wages and premiums (variable wages). Briefly, the enterprise's s.p.f. would be formed and disbursed as follows:

1. The Development Fund (DF) was to be determined as a percentage of fixed capital differentiated according to branches and would include a part of depreciation. The DF could be used both for investment and repairs and for repaying investment and working capital loans.
2. The New Products Fund (NPF) would be formed as a percentage of production costs and serve to finance the extra costs of developing new products.
3. The Welfare Fund (WF)—to be used for housing, cultural, and other welfare activities—was to be determined at a centrally set rate.
4. The Reserve Wage Fund (RWF) would be a source of guaranteed wage payments whenever the regular funds would fall short. It would be determined as a percentage of the wage fund, centrally differentiated by branches.

The entire system of distribution hinged on the stability of norms, presumably for an FYP. In the tradition of priority planning, the norms were to be differentiated by branches, allowing the c.p. to discriminate in the development of the various branches.

TRANSITIONAL PERIOD

Not surprisingly the blueprint provided for fairly narrow decentralization in investment. It involved financing, rather than decision making—shifted the financing to the DF and credits, which presumably would instill in the investor greater awareness of the efficiency of a given project. Enterprises could decide freely only on investments of quick return. Budget subsidy for the investments of certain branches (mainly heavy industry) was still upheld.

Monopoly of foreign trade was to be retained, with modification in management not explicitly outlined. Some organizations, still controlled by the Ministry of Foreign Trade (MFT), could be "permitted to come into direct contact with foreign enterprises." The prices received or paid by the foreign-trade enterprises (f.t.e.) would be aligned with prices that determined the revenue or costs of domestic producers or users.

The most novel scheme in Bulgaria, again similar to the Czechoslovak system, was that of the progressive income tax, which was instituted as a tool for regulating both enterprise profitability and the wage fund. The experiments showed that it was totally wanting in the latter function, but it performed somewhat better in the former. (It is claimed that in Hungary the use of taxes to control excess wage increases has proved fairly effective.) But as yet the Bulgarian system's designers were wary of returning to central regulation of the wage fund. They decided to regulate "the ratio of wages of the various groups of workers employed in enterprises."

The Theses condemned wage leveling. Whereas efficiency dictates relating income to performance, political and other considerations (even stronger during reforms) restrict maneuverability and impose strong limitations on the variable share of earnings. In Bulgaria it was stipulated that enterprises should guarantee full wage payments to blue-collar workers and 80 percent wage payments to white-collar workers. The Theses contended with the problem of moral versus material incentives by acknowledging both and claiming that the material incentives—"a powerful factor in socialist development"—would support the moral ones. Here again, the Bulgarian stand (even in its most progressive form) could be contrasted with the Hungarian "new economic mechanism," which overwhelmingly stresses material incentives. It is not altogether clear whether the Bulgarian insistence on moral incentives (though vague in form) is, as some political commentators suggest, an indication of a pro-Chinese faction in the Bulgarian leadership.

Prices

Despite the awareness that, without an appropriate price reform, the new system would flounder, there was considerable uncertainty and confusion about the type of price reform required. The Theses reiterated that prices should reflect better the "socially necessary" costs, promote technical progress, and that "various factors" should be taken into account. Thus, domestic prices should be related to world market prices. Planning requires certain price rigidity, but more flexible prices should reflect better changing conditions. The system's designers continued to advocate the basing of prices on average costs plus a mark-up and central price fixing or control. All this was sufficiently vague to lend itself to various interpretations. Such general pronouncements are not very enlightening about the specific direction of price reform. No matter what the proclaimed basis for price formation, the exceptions are so omnifarious, large, and frequent that it is difficult to discern between them and the rule. In addition, specific rules for the deviations were not advanced.

The Theses advocated the establishment of a three-tier price system, featuring stable state-fixed prices for basic producer and consumer goods, prices that would be allowed to move between state-established maximum and minimum limits for goods of secondary importance, and free prices for a restricted number of goods of local nature. The price revision was not to raise the retail prices of consumer goods.

EXPERIENCE WITH REFORM INTRODUCTION

In Bulgaria economic experimentation began on April 1, 1964, in 30 industrial enterprises. It was introduced by a resolution of the Politbureau and the CM, along the lines outlined by Todor Zhivkov in his address to the January 23-24, 1964, joint session of the CC BCP and the CM.[7] Many enterprises producing consumer goods were included.[8] By July 21, 1964, the CM decreed that by 1965 experimentation would expand to include a number of branches and subbranches: all enterprises of light and food processing industry, all state and cooperative farms, the entire trading apparatus, all rubber and chemical-pharmaceutical enterprises, all lumber and furniture producers, enterprises producing machinery for cultural and household uses and those producing tools and metal goods, all automotive and other transport repair shops, several f.t.e., regional and state construction firms, and some branches of the Central Administration for Material Technical Supply (such as the branch for by-products and the branch for petroleum products). By the end of 1965 experimenting enterprises produced 43 percent of total industrial production,[9] and by 1966, 65 percent.[10] During 1967, reportedly 70 percent of industry, all of agriculture, water transport, and some enterprises of domestic and foreign trade were switched over to the new system.[11]

It is relatively easy to set down desiderata on paper, but implementing them in practice is fraught with many dangers. Some of the problems encountered were due to the initial vagueness of pronouncements. Others could be traced to partial implementation (for example, new prices had not been set; the system of financing was not yet fully operational; and so on). Predictably, one of the most nagging problems was the deficiency of existing prices and the obstacles in changing them. The reports seem vague about the extent to which enterprises themselves were allowed to fix prices. Nachev suggests that experimenting enterprises were allowed to fix prices of second- and third-grade fashion and luxury goods for which the first-grade price was centrally fixed. But he also points to a "certain fear of fixing prices" felt by enterprises. New models awaited pricing interminably. Moreover, when prices of new goods were fixed they often included a very minimal rate of profit (2 percent) or were a priori fixed below costs. Those factors discouraged production of high-quality new goods.[12] For example, one of the problems encountered by the "reformed enterprises" was the conflict between fulfilling the plans for profit and for turnover tax. It was reported that in 1966, while the industrial output plan was fulfilled by 104.6 percent, that for turnover-tax payments was executed only by 99.3 percent. In view of the existing price system, neither the fulfillment of the profit nor turnover-tax plans indicates that market demand had been satisfied. But the "reformed enterprises" strove to produce the most profitable goods. However, the financial organs, responsible for the turnover-tax plan, applied pressures on enterprises to produce goods with high turnover tax.[13]

On the whole, only partial corrections were made in factory prices. Formally, only profitable enterprises were switched over. Subsidies were used for the so-called nonprofitable goods to raise profitability.[14] Apparently, about 40 percent of the growth in profits was attributed to so-called external factors, largely manipulations. In most cases enterprises felt themselves free to produce goods that were not actually in demand, but that brought in high profitability. For example, as a result of assortment manipulations in the first nine months of 1965, profits were increased by more than 12.6 million leva over those of the same period in 1964.[15]

The experimenters apparently suffered in their contacts with the "outside world." Some of the experiments' shortcomings were attributed to the poor supply situation and the disorganized transportation system.[16] It was reported that several experimenters in Sofia encountered difficulties in concluding agreements both with their suppliers and buyers for 1965, and as the main culprit they cited the traditional planning system that made both their customers and suppliers wait for orders from above.[17]

The Bulgarian variant, like most other STE reforms depended considerably on stable normatives to crystallize long-range incentives and to extend the planning horizon. However, either such normatives were not introduced or they were not sufficiently adhered to during the course of experiments.[18]

The experiments were also found wanting in the area of technical progress, partly in view of the lopsided endowment with capital. Investment that did not make for substantial employment reduction (but merely, say, for upgrading quality, especially, as was often the case, without rising prices) was in disfavor.[19] On the whole, the experimenters were reluctant to invest. This was at first explained by their inexperience. But it was increasingly attributed to inertia, unwillingness to take risks, and the traditional fear of disrupting regular procedure and everyday activities.[20]

It was difficult to detect any definite pattern in overall income distribution. Some tendencies toward wage equalization were noted, with the highest increases going to the lowest-paid employees or to the proverbially underpaid, considering their skills and education, such as the white-collar worker. In certain cases there were pronounced discrepancies in wage distribution. For example, at the Central Administration for Material and Technical Procurement the wages of workers rose by 18 percent, while those of the engineering staff went up by 41.1 percent. At the Pharmaceuticals Association the wages of workers rose by 12.3 percent, while those of the engineering staff rose by 25.6 percent. The nonindustrial personnel of the Ministry of Food Industry received an average 21.5 percent wage increase, while the blue-collar workers got only 5.7 percent.[21] The incentives to raise productivity were weak, and enterprises were still induced to keep more workers than they needed, and even to hire additional workers in low-paid categories.[22]

The practice of superiors "imposing tasks from above without economic justification and in contradiction with the interests of individual enterprises" ran counter to many experiment undertakings.[23] The experimenters continued to suffer all sorts of damages due to sheer negligence, obvious mismanagement, theft, and so on. This was particularly obvious in experimenting associations of the food industry.[24] In many cases it was beneficial for the producer to manufacture goods other than those ordered and even to pay the penalty (assuming that it was exacted) rather than to fulfill contractual obligations.[25] Some assortment manipulations occurred not only in the area of plan overfulfillment, where they are usually prevalent, but also in basic assortments of producer and consumer goods.[26] Concurrently, standards and quality specifications were disregarded. Lower-grade output was sold at higher-grade prices.[27]

TRANSITIONAL PERIOD 99

Toward the end of 1967 an attempt was made to revise somewhat and concretize the scope of the reform. The November 1967 decree pinpointed the pressing problems, among which were the ever-present issues of qualitative performance, labor discipline, and turnover.[28] It indicated that insufficient attention was being paid to technical progress, to new organizational forms, to appropriate schemes of remuneration, to the implementation of the financial and credit rules (by 1967 only 10 percent of investment was financed by bank credit), to the adaptation of bookkeeping and of the entire information system to the new requirements, and so on. It was obvious that the reform was introduced without sufficient preparation, that trained personnel was lacking, that the necessary parameters were not fully worked out, and that planner's pressures continued to exert their nefarious effects.

The decree was particularly emphatic about the need to overcome the conservatism of the "reform executants." Obviously, there were a number of bureaucrats who felt threatened by the new methods and were afraid of the infringement on their arbitrary rule. Nor were they prepared to go through a learning process—they preferred to adapt the new rules to the old ways of doing things. Yet there must have been those who (like Evgenii Mateev) realized that the so-called new system was only a half-baked one, and it was at cross-purposes not only with many legacies of the traditional system but also with the strategy of development that the leadership so relentlessly pursued.

Neither the reform nor its amendments did away with bank control of the wage fund, which was supposed to center mainly on the enterprise's observance of planning, formation, and expenditure of the wage fund as stipulated by the state. The bank was also to control the enterprise's fulfillment of its plan, especially for high-grade consumer goods, and supervise the internal distribution of the wage fund to prevent wage leveling—a tall order, indeed. Particular stress was placed on bank control of the relationship between growth of value added and of the wage fund, for it was found that the reformed enterprises tended to have their wage funds grow faster than value added.[29]

From the earliest stages of experimentation the setting and maintenance of stable norms was a vexed question. In 1965 in at least 200 enterprises the initial norms were immediately corrected in order to enable them to pay higher wages. This was done on the pretext of structural changes. The scheme of profitability regulation was highly inequitable, for it allowed some enterprises "unjustifiably high incomes" dependent on the volume of sales, rather than on that of profit.[30] Moreover, initially, the progressive income tax had been so constructed that the enterprises with a larger number of workers found themselves in a lower tax bracket. Enterprises then attempted to hire more workers. In addition, seasonal enterprises hired extra workers at below-average wage rates that served to compute their tax assessment. This assessment was computed on planned profit. In fulfilling their plans enterprises switched their assortments to more profitable goods. A new method of computing the income tax was introduced on June 1, 1966, relating the tax rates to fixed and working capital.[31]

The rates and norms issued with this changed computation were again unstable. By May 1, 1967, the norms of 341 enterprises were corrected. Characteristically, calls for changing norms were only issued when the new norms cut into the income for distribution at the enterprise. These changes in computation involved dozens of pages of additional instructions. They also confused considerably the accounting process, so that costs and profits had to be computed differently for different purposes.[32]

The price revision was again postponed until mid-1968. The decree was characteristically vague in this area. It did underline, however, that both factory and retail prices should vary to a larger extent with the quality and novelty of goods with a more extended scope of trade markups for highly effective modern goods. Price changes were not to affect the retail price level, although speculation was rife (and subsequently justified) that retail prices were going to rise.

Apparently, world prices were being used more extensively in preparing the price revision. This created certain problems of gross income distribution among branches. But it provided better criteria for judging the effectiveness of production and rate of return in various branches.[33] By the end of 1967 the first deputy chairman of the Price Commission wrote that the reform of factory prices was being conducted so as to achieve "to the highest possible degree a proximity between them and the price ratios established on the world market." The new prices would reward quality. The price ratios of substitutables would be considerably differentiated to take account of use values. He reaffirmed that a degree of price flexibility would be introduced by giving wider price-setting prerogatives to regional authorities, ministries, and associations, and more importantly through the three-tier price system. Subsidies were to be contained and considered only as a "temporary necessity."[34]

THE 1968 REFORM IN REVERSE

The July 1968 Plenum was a turning point for the reform.[35] Although the rhetoric of Zhivkov's speech and the subsequent resolution did not sound retrogressive, the reform's three most "progressive" features—that is, planning from below, the lifting of controls over the wage fund, and the three-tier price system—were abandoned. Considerable stress was laid on catching up with world technical standards and on more effective administration. The list of directive indexes was expanded, but the Theses' skeleton of the financial system was preserved, although its meaning was largely lost. In addition, there was serious consideration of improving the management of the economy by computerization. All these foreshadowed the technocratic approach to management evinced in the 1971 modifications of the traditional system.

It would be oversimplifying to state that the Bulgarian recentralization (at least on paper, for essentially decentralization had not been put into practice) was prompted solely by the example of Czechoslovakia, from fear of deteriorating relations with the USSR. Doubtless, the factors that played a significant role included impatience with the piecemeal reform implementation and its disappointing results, difficulties in price and norm setting, unwillingness to push further the half-measure initial concept, and conciliation of the bureaucracy and the forces of status quo.

The list of what had so far gone wrong with reform implementation indicated the way to consistent reform, but instead of attacking these ills more forcefully, the redesigned concept adapted itself in more traditional ways to the ills. Zhivkov admitted that the expected results could not have materialized, for many of the reform's features were neither worked out nor implemented. For example, the planning system remained vague; new prices had not been introduced; annual rather than long-term norms were being used; the supply system had not been changed; in general, superiors continued to lead their wards by the hand and to attach little importance to profitability as a performance criterion.

Zhivkov also perceived certain major economic conditions for successful reform implementation: (1) a buyer's market, (2) larger material, financial, and foreign currency reserves, (3) competition among enterprises and branches, with investments channeled to higher returns activities, and (4) greater openness of the system, including larger import of consumer goods, as a stimulus for improving the performance of domestic producers. This was the rhetoric, but how far was he prepared to go and what were the concrete steps proposed?

The reversal of planning from below was indubitable. Zhivkov pointed out that the "<u>national economic proportions and the structure of the national economy should be centrally determined</u>," and, therefore, "the hitherto prevailing concept that the state plan should be established mainly by starting from below" would be "substantially modified."

The new list of directive indexes read as follows: (1) physical production volume of basic goods, (2) basic subcontracting deliveries, (3) limit of investment, (4) key construction projects and commissioning deadlines, (5) procurement limits, (6) foreign-currency earnings, (7) foreign-currency limit for imports, (8) maximum wage fund per 100 leva of gross income (or some other indicator), and (9) main technical progress tasks of the branch. A comparison of this list with that provided in the Theses shows that the number of directives had almost doubled. The key additions pertained to subcontracting deliveries, a wage-fund limit, and technical progress. The number of directive indexes is misleading. It is a minimum only and each of these directives can be, and usually is, split into many specific directives, depending on the attitude of superiors. The number of directives can also be expanded by imposing restrictive "back door" conditions. On the whole, while the Theses promoted "indirect centralism," the Plenum was an overt reversal in favor of predominantly administrative methods.

Zhivkov noted that at that point the few new performance criteria introduced had not set in, for evaluation continued to be made according to traditional criteria to which the units had to submit. He did not advocate profit as the only performance criterion; rather he offered a whole gamut, including labor productivity, profitability, technical parameters representing world standards, quality of output, and gains in foreign currency. Moreover, performance would continue to be judged in terms of plan fulfillment.

Competition between enterprises was to be encouraged by (1) eliminating monopolistic tendencies, (2) establishing interassociation competition so as to promote technical progress and reduce prices, and (3) setting up reserve funds (for example, in foreign currency) at the association's level. But Zhivkov did not explain how this was to be accomplished; nor did he perceive that associations, especially with their increased powers vis-a-vis enterprises, would be a potential hotbed of monopolization.

Although the economic functions of government and state organs were to be more clearly defined, the Plenary decisions seem to have added to duplication and layering of responsibility by superimposing new organs and demanding greater state control. In addition to the National Assembly's economic functions, a new body, the Council of the People's Republic of Bulgaria, elected by the National Assembly, was proposed to "combine the taking of decisions with the implementation." Another supraadministrative body created was the Committee for Economic Coordination (CEC); this was to be an organic part of the CM, with the right to decide on plan changes, on financial questions, on deliveries, supply of materials and subcontracting, and so forth. The CEC would control and coordinate the functioning of ministries; its primary task was to ensure plan fulfillment. With that assignment, many of its functions seemed to duplicate those performed at the SPC. The setting up of the CEC was indicative of the resurgent importance of plan fulfillment, of the c.p.'s dissatisfaction with the sluggish and lax plan execution, and of the malfunctioning of the existing administrative organs.

The creation of the Ministry of Supplies and State Reserves (MSSR), to coordinate, allocate, and control the distribution of material inputs, bore witness to the importance of the problem, to the increasing difficulties, and especially to the continuing centralized distribution of materials. The increasing importance attached to strong and comprehensive price control and central price setting was also evident from the organization of the CP, responsible to the CM, to replace the Price Committee attached to the SPC.

NOTES

1. RD, December 4, 1965; RD, April 29, 1966; T. Zhivkov, The New System of Economic Management (Sofia, 1966).
2. Compare G. R. Feiwel, New Economic Patterns in Czechoslovakia (New York, 1968), Chapters 3-5.
3. L. Radulov, PSS 7 (1966): 27, 32-33.
4. A. Pashev, PSS 5 (1966): 3.
5. Compare ibid.
6. Ibid.
7. Compare S. Radev, TT 9 (1964): 36.
8. KhP 2 (1965): 1; KhP 3 (1965): 8.
9. A. Pashev, IZ 2 (1966): 1.
10. I. Vulkov, NV 12 (1967): 18.
11. Compare I. Paliyski, FK 2 (1967): 11.
12. D. Nachev, PZ 13 (1964).
13. D. Bazhdarov, FK 2 (1967): 27.
14. Ts. Petrov, PSS 1 (1966): 41.
15. S. Vulchanov, PSS 1 (1966): 34-35.
16. S. Bonev and V. Vakanov, IM 7 (1966): 112.
17. N. Yankov, PSS 9 (1965): 63-64.
18. Compare Vulkov, NV, pp. 22-23.
19. Ibid., p. 23.
20. Y. Khazan, FK 5 (1965): 37.
21. Bonev and Vakanov, IM, p. 116.
22. Vulkov, NV, p. 24.
23. Ibid., pp. 25-26.
24. D. Karagazov, KhP 4 (1966): 7.
25. Vulkov, NV, p. 25.
26. Khazan, FK, pp. 37-38.
27. Compare M. Tanev, Pogled 4 (1967): 2-3.
28. DV 88 (November 10, 1967).

29. S. Drumev, FK 2 (1966): 40-47; D. Davidov, IZ 1 (1966).
30. K. F. Shopov, FK 5 (1967): 4.
31. Y. Yanev, NA 20 (1966): 14-22.
32. Shopov, FK, pp. 4-5.
33. S. Bonev, IM 10 (1967): 95.
34. I. V. Videnov, RD, November 15, 1967.
35. T. Zhivkov, RD (July 25, 1968), translated and reprinted as Fundamental Trends in the Further Development of the System of Public Administration in Our Society (Sofia, 1968); RD, July 27, 1968.

CHAPTER

5

PRICES AND DECISIONS

Economic efficiency can almost never be governed by technological considerations alone. Effective choices require reliable information. The role of prices to convey such information was aptly stressed by Tjaling Koopmans:

> Price is regarded as a label, a signal, a piece of information that is attached to the good or service traded. This information expresses simultaneously the ultimate usefulness to consumers of this good, and the foregone usefulness of other goods or services that could have been produced alternatively from the resources absorbed in making this good. Choices about methods of production and about amounts to be produced are based on this information; if these choices are to be good choices, the information used has to be accurate.[1]

This is not to say that there are no other signals for transmitting information about scarcities; nor is it to argue that prices provide sufficient information for decision making. Scarcity prices are not necessarily efficient prices, and they should not under all circumstances be used as guides for resource allocation.[2]

Shifting attention from what prices ought to be to the dismal realities of prices in STEs provides some insight into the "dynamics of perpetuation." Some of the relevant questions are the following: What are the consequences of wrong prices, and what could be considered an improvement? How much do different prices really matter, and what are the factors restricting their sphere of influence? Are prices consistent with other elements of the functioning system, and, if not, is it enough to provide effective prices, or do they have to be accompanied by a whole network of conditions? Would prices discharge their functions satisfactorily and promptly, even if the major constraints on decision making by the periphery were removed? Are the alternative ways of generating prices equally efficient? Can static price theory be applied to dynamic conditions? What are the limitations of general equilibrium theory and neoclassical prescription?[3] In this and subsequent chapters we shall explore some of these questions, though not so exhaustively as they merit.

With his usual expository skill Oskar Lange pointed to the consequences of using prices that fail to reflect conditions of supply and demand:

> If the price of a scarce productive resource is planned too low and the price of an abundant resource is planned too high, the cost of production shown in the bookkeeping of the plants can be reduced by substituting the scarce resource for the abundant one. Considered from the point of view of the economy as a whole, such a substitution is a waste of resources and the reduction in the bookkeeping costs of the plants represents a decrease, not an increase, of their economic efficiency. The discrepancy between the cost accounting of the plants and their true economic efficiency can be avoided only by pricing the productive resources according to their scarcity relative to demand.[4]

In the economist's world of free and unimpeded mechanism of the world market the forces of competition should equate the country's internal and world market relative price structure. Thus, in principle, domestic prices should reflect the proportions and scarcity relations on the world market. The socialist economy provides a sweeping scope for price arrangements, and obviously the existing one is only one of the alternatives. In traditional STEs domestic prices are significantly divorced from correspondence to world market price ratios, and producers' (industrial wholesale) prices are separated from consumers' (retail) prices. To the extent that production decisions are influenced by official actual prices (as contrasted with shadow or black-market prices and physical allocations and commands) this obfuscates economic calculation and distorts comparative costs and benefits to be derived from indirect production. For prices on Western markets, as imperfect as they are, reflect more or less marginal opportunity costs for the country's effective export and import substitution, and tend to stimulate diffusion of world technical and organizational progress. The dissociation of domestic prices from those prevailing on Western markets provides certain benefits, including almost complete shielding from imported cost inflation. But the costs are admittedly high in terms of static and dynamic efficiency. Thus, most STE reform blueprints provided for internal prices to be brought into some correspondence with relative prices on Western markets, but, with the notable exception of Yugoslavia, only a limited attempt was made to align domestic with world-market prices. The half-measure reforms for the most part preserved the split between internal and external price patterns.

The STE price system provides an opportunity to insulate changes in some prices from others, to have different prices for one product depending on the sphere of circulation, to use different standards in price formation, and, in general, to manipulate the "budget constraint" on industrial management, income distribution, real purchasing power of households, and the size of the economic surplus to be extracted (capital accumulation). For example, retail prices of consumer goods can be insulated from changes in their factory prices, or vice versa. Production decisions do not have to be justified by the test of realization at market prices, and the producer is not subjected to the "discipline of the market." However, in practice the total insulation has not been carried out. But the application of divergent principles of price formation for producers' and consumers' prices has involved widely differentiated sales taxes and subsidies.

At first sight such a price system seems "irrational," but one has to keep in mind how and for what purposes it was developed. The ever-present conflict between the requisites of efficiency and the dictates of expediency provides a partial explanation.[5] Whether the system suited well the requisites of "extensive growth" is another matter. But even if the historical explanation is not questioned, it does not provide justifications for continuing along the same lines.

As in other STEs, the most serious stumbling block of the Bulgarian reform was fallacious, irrational (even from the standpoint of planner's preferences), and rigid price system. Bulgarian economists and official reformers recognized that prices would have to be considerably altered if they were to serve as effective choice coefficients; they had significant differences in opinion, however, as to the nature and scope of this alteration. The price reform (which actually ended up in a revision) was many years in the making. It is difficult to judge from the available information to what extent the authorities were really prepared to remodel the price system and whether the various postponements were an expression of the leadership's vacillation toward economic reform and of the winds of recentralization.

There are good indications that revised prices for all of industry were not introduced until 1971, almost a decade after the previous revision. As a background to this revision we shall outline broadly the dual-price system that took root in Bulgaria, its fallacies and shortcomings, and some of the proposals to reform it. For a number of reasons the traditional price system has shown a remarkable resistance to fundamental remodeling. Thus, its basic flaws are not only of theoretical and historical importance. Although the successive price revisions have endeavored to bridge the gap between the levels of producers' and consumers' prices, by increasing the level of the former, their goal was not to eradicate the dual-price system, and to this day the principles of price formation for producer and consumer goods vary considerably. After a brief sketch of the 1956 and 1962 price revisions, we shall discuss the history of the most recent revision, and glance cursorily at retail, agricultural, and foreign-trade prices.

THE DUAL-PRICE SYSTEM

As in other STEs, the dual-price system adopted in Bulgaria meant not only a divergent and separately moving price level of producer (group A) and consumer (group B) goods with the bulk of surplus collected in the form of turnover tax in the consumers' prices, but also different principles of price formation. The prices of producer goods were, as a rule, computed on the basis of the average attributable branch-production costs (with the notorious failure to account adequately for costs of capital employed, rent, depreciation and obsolescence, meager returns for entrepreneurship and various skills) that were generally cumulatively understated, plus a small (about 2 percent) "profit" mark-up on costs to impose a budget constraint, but not to empty the system of incentives. In setting consumers' prices, an attempt was made to find the level and relative structure of prices that equates demand with available supply for each commodity ("neoclassical prices"). But prices were strongly conditioned by political and distributional considerations, restricting the use of the pricing mechanism. Thus, in practice the relative prices of consumer goods frequently seem to be nowhere near their microeconomic equilibria in (partial) markets. With prices below equilibrium, there is excess demand: Buyers compete among themselves for the sellers' favors, and black-market profiteers appear. The budget collects a whopping share in the form of turnover tax: the difference between the average production costs plus an insignificant "profit" mark-up (the factory price) and the price at which the product is sold to the domestic trade organ. Upon sale of the product, the bank automatically extracts from the industrial enterprise the turnover tax, which is transferred to the state budget. Thus the firm's revenue is generated excluding turnover tax.

Basically, prices of producer and consumer goods fall into the two categories of wholesale and retail prices although there is some blurring of the lines when some producer goods are sold at retail prices to the population and then include turnover tax. First, there are wholesale prices, which include purchase and procurement prices, factory prices, import prices, prices of construction and installation jobs, fares and rates, and so on; these apply in transfers among socialized enterprises. Second, there are retail prices, which include rates of municipal services, transportation fares, admission tickets, and so on, applied to the sale of goods and services to the public.[6]

A Bulgarian critic of the dual-price system, Georgi Petrov, writes: "Consumer goods and a small share of producer goods have two types of prices: selling (retail) and factory (settlement) prices. There is no direct link between them." The difference between them is determined on "the basis of different principles" of price setting, often performed "by different authorities and at different times," and is expressed, after deducting trade mark-ups, in the form of turnover tax collected by the state budget. Thus, "turnover tax rates are differentiated not only by branches and sub-branches, but also for each individual product." This system features differences between the selling and factory prices for the same product, depending on the user. For example, sugar, while having a single factory price, has a multiplicity of different selling prices depending on the purchaser. There is a uniform retail price per kilogram of chocolate, but the factory settlement price varies considerably, according to individual plants' production costs. Thus, the turnover tax per kilogram of chocolate is different for each producer.[7]

Under the dual-price system

> [there are] no common prices for the participants in the commodity exchange. For the producer the only relevant price is the factory (settlement) price that determines his revenue. The buyer is interested exclusively in the retail (selling) price which determines his real income . . . whereas the factory price is completely irrelevant to him. Thus the producer is interested in the highest factory price and does not care about the retail price. In such a case the customary procedure is to reduce the difference between the two prices, i.e., the state's proceeds from turnover tax. . . . Conversely, if a lower retail price is set under the pressures of the buyers, it is again the turnover tax that is usually lowered and not the factory price.[8]

But the producer's pressures to obtain a higher factory price and the buyer's proclivity for a lower price are "countered by the administrative responsibility of the financial organs of the state not to allow a reduction in budgetary receipts."[9]

Such a system tends to indicate spurious relative contributions of various sectors, branches, and activities to national production; it distorts distributive shares and obfuscates the efficiency calculus. Instead of promoting technical progress, the artificially low prices of producer goods reinforce the propensity to waste materials and capital goods.

With due caution as to the refractory nature of the statistical material and factory prices used, the data in Table 5.1 indicate that whereas both groups A and B shared almost evenly in total industrial output, the factors used (labor plus capital) in the production of group A far exceeded those in group B (in terms of the stock of capital more than three times as much, and in terms of invesments within four years almost five times as much), while the financial accumulation (profit plus turnover tax) in group B was more than three times as large as that in group A. During this period the bulk of "surplus product" was being realized in the prices of consumer goods and centrally redistributed to finance development of the producer-goods sector.

The dual-price system gives rise to considerable subsidies that, for reasons of planning and financial expediency, tend to be concentrated at the early stages of the production process (mining) and on fewer items in basic industries. But such prices of basic inputs artificially lowered production costs (and hence prices) at the successive processing stages. One of the principal aims of price revisions has been to reduce subsidies (which, with rigid prices, grew because wage increases outpaced productivity gains that were proceeding at

TABLE 5.1

Relative Share of A and B in Industrial Output, Inputs, and Financial Accumulation, 1957-61

	Group A	Group B
Total industrial output in 1961 at 1962 prices	49.5	50.5
Wage fund in 1961	62.4	37.6
Stock of capital in 1961	77.2	22.8
Investments in 1957-61	83.3	16.7
Financial accumulation in 1961*	23.3	76.7

*Rough estimates.

Source: Y. Shekerov, TT 3 (1964): 69.

PRICES AND DECISIONS

different rates in various activities, deteriorating geological conditions, and rising marginal costs). With deficits or low rates of financial accumulation, the producer-goods branches not only had to receive subsidies for current operations, but the financing of development was wholly undertaken by the budget, which redistributed funds for capital formation.

Clearly, costs are neither confined to nor necessarily reflected in the amounts paid and recorded (explicit historic costs). The c.p. can determine administratively what constitutes a cost element and assign appropriate values. But to determine at what economic—as distinct from accounting—cost the production of a good is secured, one needs to take into account implicit (opportunity) costs. But the state budget carries some of the costs of inputs, and relative price stability is ensured primarily through the redistribution of funds and the subsidizing of priority activities, resulting in the failure to account fully for costs of inputs employed. Thus, prices calculated on the basis of average (and sometimes individual) production costs do not give proper weight to marginal opportunity costs of all inputs. Even those costs that are explicitly accounted for are based on prices of inputs that are more or less arbitrary. Moreover, since management is interested in having the highest possible costs allowed, calls for strengthening administrative discipline over pricing are largely ineffectual.

This raises the cardinal question of the nature of costs, prices, and profit. If price is a derivative of costs, its constituents are cost elements that are themselves not independent variables. The cost of a given commodity cannot be said to be price determining since the constituent cost elements are determined by prices. Planners allocate inputs and then set prices. Effective prices cannot be set unless inputs were beforehand allocated efficiently to various uses.[10]

But in many cases the cost basis is considered merely a "point of departure" for price determination. Many deviations from costs are allowed to take into account the relative technical "use values," to enhance or curtail usage in view of relative shortages or surpluses or to alleviate balance of payments and like matters. This raises questions about the real nature of the rules and the extent to which prices are arbitrary.

Among others, Petrov challenges the prevalent view that the "real efficiency of production is expressed relatively correctly by the profit of enterprises calculated as the difference between factory (settlement) price and production cost." He argues that it is more likely that "the enterprise's profit reflects a fictitious and not actual efficiency of production."[11] "Wrong prices" (in Petrov's interpretation persistent and significant deviations of selling prices from scarcity prices) contribute to ineffective structure of production and consumption: "The greater or smaller profit (net income) realized by the individual enterprises cannot serve to evaluate their performance as better or worse." Costs tend to be arbitrary: "Higher or lower producer costs in the various enterprises and branches may be due to the use of producer goods with fallacious prices."[12] Thus, with fallacious choice coefficients, profit cannot be used as a measure of efficiency of current production and returns on investment and foreign trade.

Export and import efficiency calculations are distorted by applying fallacious prices, for it may appear that "Bulgaria would benefit from developing one type of product and not another, although realistically the opposite could be the right direction. The incorrect placement of Bulgaria in the international division of labour leads to great losses for society and limits the possibilities for the development of production and the raising of living standards."[13]

If a given product were produced in larger quantities than those demanded, the factory price would not be changed: "The producer will not be interested to comply with market conditions and correspondingly to reduce his output. Contrariwise, if the factory price assures him a relatively higher profitability, he will strive to expand the output of a given product and sell it." In practice, when the quantity demanded exceeds the quantity supplied, the factory price usually remains unchanged. If it should assure lower than average or even average profitability, the producer will not be interested to increase output to alleviate the excess demand. On the contrary he may continue to curtail production and thus exercise pressure to increase the price.[14]

Backwardness is perpetuated by the practice of setting prices of new products on the basis of planned production costs plus a 2 to 5 percent "profit" mark-up. Thus, the disincentives affecting the producers are very much strengthened and assume a dominating and determining significance. These prices remain unchanged for years. The longer the same product is produced at a growing rate, the greater the cost reduction and the higher the profit. Under the existing system the enterprise is thus vitally interested in continuing production of obsolete (and therefore highly profitable) products for as long as it possibly can, and in not adopting—or in delaying the adoption of—new (and therefore less profitable) products. One should also keep in mind that because factory prices of new goods are determined on the basis of their planned production costs, and in some cases also in accordance with some norms specifying the weight and size of products, even if enterprises should adopt new products, they would not be interested in subsequently improving the quality of those products and in reducing their planned production costs. Their interest dictates exactly the opposite: the adoption of new products with inferior technical parameters and lower quality (because they are easier and faster to produce) and submission of highest possible cost estimates. By its very nature the factory price is designed to cover planned production costs, irrespective of real "use values" of the products, which should be roughly reflected in the retail price. These are some of the determinants of the technical retardation of certain branches, and of the continuous production of obsolete, ineffective, and expensive producer and consumer goods.[15]

TURNOVER TAX

Turnover tax separates the price that the consumer pays (retail) from the price that the producer receives (factory). Turnover tax is computed as the difference between the retail price (less distribution costs and mark-ups) and the factory price. It is collected at the producer's when the product is sold to the trade organ; hence, the consumer is unaware of its size. If the turnover tax were undifferentiated and levied as a uniform percentage rate on all goods, the relative price ratios would be unaffected. As it is, the process of computing and levying turnover tax is complex, cumbersome, and quite prone to manipulations. To ensure inflow of revenue to the state purse, the financial authorities are likely to be swayed by expediency and control considerations. Some of the shortcomings of the turnover tax structure could be classified as follows:

1. In many cases the turnover tax is very large and highly irregular in its incidence. For example, on mattresses there are some 70 various turnover taxes, 60 taxes for bedsheets, and 25 for pillows and pillowcases.[16] According to Ivanov this wide and arbitrary differentiation impedes economic calculation for the following reasons: (a) the turnover taxes and profitability of various goods do not vary proportionately, so prices of some goods include a high profitability and relatively low turnover tax, and those of others include a low profitability and high turnover tax (obviously, enterprises lean toward producing the former); (b) quality differentiation is often not reflected in profitability ratios but rather in turnover tax variances, and the producers remain unaffected by producing low-quality output; (c) it is the turnover tax, rather than profit, that is linked to demand, but since the producer remains unaffected by its magnitude, there is a rift between him and the market for his goods, disorienting him, so that stocks of certain goods build up and others are perpetually in short supply.[17]

2. The amount of turnover tax collected depends on the volume of production and the product mix. Audits have indicated that the discrepancies in payment of turnover tax are usually to the detriment of the budget. In the course of transactions, both producers and buyers verify the prices they pay or receive, but there is no interest in assuring the budget's share.[18] The producers are interested in selling their goods at the highest possible factory prices, while purchasers aim at buying them at the lowest possible retail prices. Both tendencies naturally encroach on the amount of turnover tax collected as the difference between them. Such a collection of turnover tax creates a buffer zone between producers' and consumers' prices, rendering the producer largely insensitive to the changes on consumers' markets. Due to the variation in turnover tax and the lack of interest in enterprises in producing goods with higher turnover tax, sales may increase rapidly but turnover tax does not keep pace or sometimes even falls off.[19]

3. Imposition of high turnover taxes to discourage quantity demanded of some consumer goods might not be prudent if such products could be obtained with relatively small investments, as is often the case in light industry.

4. The complexity of turnover-tax payments requires a growing body of clerks to compute these payments. Because it is open to all sorts of manipulations and even inadvertent errors, turnover-tax collection requires stringent state control and a large contingent of controllers.[20]

One of the problems of the STEs, and to varying degrees in other countries, is that of ensuring that the population consumes only a part of what it produces. To ensure appropriate capital accumulation and macroeconomic stability, the c.p. should choose a certain effective aggregate price-wage ratio (mark-up). Thus, the price level of consumer goods should help to yield the needed surplus over the wage fund expended. The problem is one of finding an effective device for extracting the surplus. Turnover tax is such a device. The nature of turnover tax, and even its very name, is a controversial issue in socialist economic literature. It is rather immaterial whether it is classified as a tax, for essentially it does perform the function of a sales tax in siphoning off purchasing power and thus reducing real income. It appears that for political and incentive reasons (primarily to maintain the gross-wage illusion), the original designers of the Soviet system preferred indirect reduction of spending power (turnover tax) to direct (income) taxes or lower nominal wages.

Reformers operate within the framework inherited from the traditional system. Even should they wish to lower the existing wage level, they cannot do so (and strong pressures are exerted to satisfy the expected "normal" increases). Similarly, consequential direct taxes to reduce disposable income are political dynamite. Consequently, reduction of purchasing power via the highly differentiated turnover taxes still appears to be the more expedient course.

Until recently, turnover tax was called upon to furnish the bulk of budgetary revenue.[21] The problem was not only that the tax level was high, but that the average conceals highly differentiated group or individual magnitudes (violation of welfare rules).[22] In some instances the gradation of the tax is related to and increases with the classification of a product as a "nonnecessity," "luxury," or "socially harmful," but the exceptions are so numerous that the rule is not dependable. Obviously, the problem is not confined to macroeconomic equilibrium of the system, for the microeconomic components of the aggregate sales-wage ratio have important welfare, incentive, and microdistribution implications.

In the reformed system the fiscal role of turnover tax was to be considerably reduced and the new taxes (profits tax, capital charge, and so forth) were to constitute the core of the tax system. Even in the 1960s turnover tax was receding as a source of budgetary revenue, with considerable shift toward the budget's share in profits of industrial enterprises. However, by 1968 turnover tax still constituted more than 35 percent of the budget's revenue, and its decline was disappointing.[23] But the relative shares of turnover tax and profit tax as sources of budgetary revenue are not indicative of decentralization. They are simply alternative methods of collecting "financial accumulation." With the increased level of producers' prices (including larger profit mark-ups), the budget siphons off larger amounts of profit from enterprises producing producer goods. Concurrently with quasi-stable consumers' prices and increasing producers' prices, the production costs of consumer goods rise, and the turnover tax makes up the difference. Although the weights of the sources of budgetary revenue have little economic significance, they matter to financial authorities, for turnover tax collection is a "safer" method in that it is less open to manipulation by lower units than is the collection of the profit tax. Whatever the merits of the new fiscal structure, it is noteworthy that the advent of economic reform has not affected turnover-tax computation and collection, simply because such changes would be directly related to alterations in the price formation principles. As one Bulgarian economist bluntly put it: "[These] have not been made."[24]

ON REFORM PROPOSALS

Bulgarian economists argued that the new system makes it necessary to improve price formation because of

(1) the adoption of profit as a basic performance criterion of enterprise activity and of direct linking between enterprises' financial results and the possibilities of self-financing and labour's rewards requiring a precise determination of the level of prices by branches and sub-branches; and (2) greater enterprise autonomy making the production profile dependent on market relations and not on the nomenclature of output imposed by the plan.[25]

Effective prices are required in order to reduce to a minimum "cases of 'advantageous' or 'disadvantageous' production and to protect equally both the interests of producers and consumers."[26] In general, economists viewed the principal shortcoming of wholesale prices as a considerable neglect of market conditions: "Consequently wholesale prices have the property of a financial accounting concept and to a large extent they have been deprived of economic character and content."[27]

Officially, average production costs were still viewed as the correct basis for prices. However, the criterion of "socially necessary expenditures" was construed as approximating world-market prices, but with such flexible interpretation that it was difficult to fathom what was actually meant.* The price system should induce technical progress and the growth of the "progressive branches of industry." It should discriminate against poor-quality goods. The different natural production conditions of mining and agricultural enterprises should be taken into account by means of differential rent, regional prices, transport and delivery charges, and so on. The price system should reflect more fully the terms of foreign trade. The current retail-price relations should be maintained. The state should continue to fix prices of basic producer goods and consumer necessities. Prices should be changed promptly and flexibly to reflect current production and sale conditions. The sale prices could be adjusted between producer and buyer so long as retail prices and the budget (in terms of turnover tax) were not affected.[29]

In the course of the reform discussion a number of proposals on price formation were submitted. Many of these proposals fell within the Marxist framework of price formation.[30] Some were along the lines of the "optimal solution" propounded by the Soviet mathematical school.[31] A few notable voices advocated market socialism.

---

*In preparation for reform a very general study of tying in domestic to foreign prices was prepared by the Economic Research Institute of the SPC.[28]

Most economists trod on "secure grounds" and naturally advocated that prices reflect "socially necessary costs." On a more pragmatic plane, they propounded that a certain flexibility was mandatory if the system was promptly to absorb technical progress and changes in quality of output. This presupposed some price-formation leeway at lower levels. But for consumer necessities and other goods of basic importance the state should retain exclusive price-setting prerogatives.[32] State price regulation could take various forms, such as setting maximum and minimum limits, approving prices agreed upon by buyers and sellers, and giving permission to amend certain prices.[33] The scope of price fixing by enterprises was to be enlarged. They were to be able to negotiate prices of certain goods that were not in short supply. Their participation in setting other prices could only be beneficial for the latter's accuracy.[34]

Granted state price fixing for the majority of goods, what basis was to be used? So far the method of adding profit mark-up to the average cost had proven unsatisfactory. Some economists advocated "marginal-type" cost pricing: basing prices on the costs of the least-efficient enterprise or group of enterprises.[35] This was objected to on the grounds that it would relax the budget constraint on enterprises and that it was not "justifiable" except in the extractive industry and agriculture.[36] Others proposed that the supply of consumer goods could be regulated satisfactorily by varying the rate of turnover tax.[37] A good number of economists backed the idea of a price of production.[38]* Arguments for and against the various types of price formulae were aired, with many authors favoring some variation or an eclectic or mixed formula; they referred to the mixed-type or multichannel-type of formula adopted in the GDR and Hungary.[41]

---

*If the price of production were adopted, the redistribution of surplus product would favor capital-intensive branches, such as power, ferrous metallurgy, and transportation, whose prices would increase considerably. Rough estimates, based on the input-output table for 1964 suggested that the production prices in electric power would be higher than the prevailing ones by 180 percent; those in ferrous metallurgy, by 97 percent; those in transportation, by 60 percent. In view of the repercussions on the users it was feared that the cumulative effect of such a rise in the price level would be intolerable.[39] With value-type prices, extractive branches would have relatively higher profits, since the relative labor intensity there is higher than the average for industry. Ipso facto, more capital-intensive (mechanized) branches would obtain smaller profits. The Economic Research Institute of the SPC has done considerable preparatory work for price reform. It appears that most of it was centered on price recomputations within the Marxist framework of price formation.[40]

But the problem is not confined to redesigning the price structure to reflect more closely relative costs. For prices cannot be governed by costs alone and have to correspond also to scarcities relative to demand.[42] In addition, the establishment of a mechanism ensuring that changes in underlying conditions are registered in new prices seems to be even more important than the basis on which the original price is set.[43] For prices have to be permitted to respond more readily to changing conditions of production and demand.

At the other end of the spectrum were a few economists who advocated controlled market-price formation; their leading representative was Petrov.[44] He advocated a single market (scarcity) price, common to sellers and buyers alike, reflecting conditions of both production and demand. The profit of the enterprise would represent the difference between the market price, reduced by the turnover tax (calculated as a uniform percentage rate of the sales price), and more realistically determined production costs (including all other taxes, levied as percentage rates uniform throughout the system). Petrov inordinately stressed market determination of prices and was not explicit about removal of market distortions. He was vague about the role of the center. He asserted that even the state organs equipped with modern mathematical models and computer technology are "incapable of determining economically sound price," but "such a price can result from contracting among the parties to the exchange," for the "buyers (enterprise or private individuals) are more competent to derive and justify a given price." Above all, "under normal circumstances they are in a better position to force the producer to accept a right price." A key condition for the derivation of such a price is the bargaining power of the participants in the exchange: "As long as the final consumers are millions who compete for cheaper and better products, they will be on an equal footing with the producers who will also be competing among themselves." Petrov realized that this assumes absence of monopolistic producers—a very unrealistic proposition, indeed. However, he asserted the superiority of a socialist system in its ability to limit monopolies and to check monopolistic behavior. He also assumed, without elaborating on it, that the state would ensure realistic "material, labour, financial, and foreign exchange balances and reserves."[45]

The market price derived by negotiation was considered by Petrov of special significance for stimulating technical progress, for it "reflects relatively most correctly the quality of individual products." Negotiations will ensure that contract prices of obsolete or low-quality products will be significantly lower than prices of high-quality products. The stimuli of the market on quality improvement is undoubtedly great, but Petrov seems to have overstated it. Again, he paid little attention to the conditions that must be fulfilled in order

PRICES AND DECISIONS

to exploit this mechanism more fully. The market price "will exert a powerful stimulus on the producers to adopt continuously and rapidly new products with high technical parameters and to discontinue output of obsolete ones."[46]

The prices of coal, ores, fuels, and other extracted minerals should be determined on the basis of production costs of marginal (high-cost) producers. Differential rent should be siphoned off in the form of land tax to the budget from supramarginal producers. To the extent that imports play a significant role for some products in heavy industry, the selling price would be "some average between the domestic price and the world market price." In such a case domestic producers of that product would receive subsidies (to be deducted from the turnover tax or land tax paid to the budget). The subsidies would signal the branches of the extractive industries with high costs. The high costs and prices of many domestic producers of basic raw materials and fuels would adversely affect competitive position of the remaining branches on foreign markets. The losses from exports would be subsidized by the state.[47]

As an addendum to this discussion, mention should be made of a proposal for "restricted" reform along different lines, advanced by Evgeni Mateev, one of Bulgaria's leading economists.*

---

*He suggested that in order to improve the system of functioning, material incentives be related to actual progress, that is, improvement over the previous period. The premiums of managerial personnel should increase in proportion to the growth of output over the preceding period. But due to divergent possibilities for output expansion, owing primarily to different quantities and qualities of capital employed, the relationship between growth of output and increase of premiums (quotas) should be differentiated. These quotas should be stable for a "rather lengthy period." Furthermore, if the enterprise produced the same volume of output as the preceding year, the same level of premiums would be retained. However, two important conditions would have to be met: The qualitative indexes would have to be raised in comparison with the previous year, and output would have to be sold. Mateev stipulated four qualitative premium conditions (nonfulfillment of one would not preclude receipt of premium for another): (1) raising labor productivity, (2) increasing quality and technical level of output, (3) enhancing the share of export goods in total output, and (4) augmenting the volume of output per 100 leva of capital employed. To safeguard the realism of the plan he proposed that the enterprise be induced to reveal all its potential for growth at the drafting stage. But he weakened the inducement by stipulating that the enterprise be punished for underfulfilling the plan. Mateev also

Having a good grasp of the realities of planning, he considered application of a "quasi-profit-oriented" reform inconsistent with the preservation of other components of the system (such as planning by material balances, high-pressure economy, and rationing of goods in short supply). With a continuation of the industrialization rush and a proclivity to limit reforms to half-measures, price reforms are likely to be partial administrative rearrangements. With continuing faulty prices and taut planning, the "quasi-profit-oriented" reform would fail, and the periphery would again be deluged by direct orders. Whatever the limitations of Mateev's proposals, with the benefit of hindsight, one is inclined to speculate whether, given the growth policy pursued, the Bulgarian economy would not have been better off had similar, but more coherent, partial improvements been attempted, rather than the course adopted. This might not have led to the strong recentralization movement with stress on the association that was the end result.

RETROSPECT ON PRICE REVISIONS

The system is never completely protected and sufficiently isolated from changing conditions in a dynamic world. With rigid prices, distortions manifest themselves with increasing intensity, and removal of some of them becomes unavoidable. Periodic, one-shot, price revisions are required, among other things, to mitigate the general and cumulative price distortions that initially appeared to be only faulty particular microeconomic cases. If for no other reasons, fiscal expediency and control require strict limits on the periphery's liquidity (the budget constraint) and elimination or greater cohesion of subsidies, or at least their concentration on fewer activities. In turn, financial rearrangements require modified prices. The increased stress on the self-supporting economic unit requires a change in the mark-up and a gradual shift from the Stalinist notion of budget constraint to self-financing. But such a shift usually creates the problem of larger enterprise liquidity, which makes it less susceptible to control. Still there is a world of difference between the periodic price revisions, which mainly aim at removing the subventions in the sphere of producer goods production and at bringing producers' prices

---

suggested that the volume of output be computed in terms of value added, that is, by subtracting from the volume of output, in comparable prices, the value of materials in these same prices. This index not only counteracts the proverbial shortcomings of gross value of output, but also promotes material savings—a very important problem on the Bulgarian scene.[48]

closer to costs, and the kind of price reform required to support an effective economic reform.

Thus far, the Bulgarian price system has featured periodic revisions of factory prices every decade or so, with the aim of eliminating some evident accumulated inconsistencies and streamlining industrial prices. During the relatively long intervals between price revisions, some individual prices did change; new (or spuriously new) products were introduced; the relative price structure shifted. Of course, technical and economic conditions changed; productivity and wages grew at varied rates, and so forth. Prior to the latest (1971) revision, there were two others in 1956 and 1962.

The early price revisions were mainly intended to bring producers' prices closer to costs (as traditionally conceived). As a result, within a few years profits in certain branches rose considerably, as indicated in Table 5.2. The 1956 profit plan was set at 2.551 billion leva, and by 1960 the plan rose to 7.113 billion leva. By 1960 calls were sounded for restricting "excessive" profits at enterprises.[49] By 1962 another set of revised producers' prices went into effect and on the whole profitability was pared down. The main aim then was to cut down the profitability of certain branches (chemical and rubber, cellulose and paper, electric power, railroad and automotive trans-

TABLE 5.2

Changes in Profitability, as Percentage of Costs,
Between 1956 and 1961

Branch	1956	1961
Electric power	3	37.79
Fuels	5	3.29
Ferrous metallurgy	3-5	-22.18
Nonferrous metallurgy	3-5	-1.51
Machine building and metal working	3	20.17
Chemical and rubber	2-4	21.16
Lumber and woodworking	3-5	9.81
Cellulose and paper	3	22.28
Glass and porcelain	3	11.94
Food	3	4.66
Other	3-5	9.67

Note: 1956 was the first year of revised prices, and 1961 the last year that these prices were in effect.

Source: S. Obreshkov, TT 10 (1964): 29.

port) and to increase the prices of others (lumber, coal, ferrous metals, footwear, and so on).[50] It was not difficult for the processing industries to report increasing profitability so long as the basic inputs were still priced considerably below costs. Although the 1962 revision attempted to remedy this situation, as shown in Table 5.3, ferrous metallurgy continued to be deficit, and the profit rates included in the prices of fuels, nonferrous metals, and building materials were much lower than those in other industries. After the April 1, 1956, revision, the only planned-loss branch was ferrous metallurgy. By 1961 it was joined by nonferrous metallurgy. After the January 1, 1962, price revision, the losses in ferrous metallurgy were reduced, and glassware and porcelain broke even.

The 1962 profitability situation at Bulgarian enterprises is illustrated in Table 5.4. Although the majority of enterprises (57.6 percent) reported a profit from 1 to 10 percent, there were still 161 deficit enterprises (14.2 percent of the total), and 37 percent of enterprises reported a profit of less than 1 percent. Only 209 enterprises (18.7 percent of the total) reported a profit of from 10 to 20 percent.

By 1964 the ferrous metallurgy and cellulose-paper industries again reported a deficit. The number of deficit enterprises rose to 186 in 1963 and fell to 170 in 1964. In 1963 the largest number of planned-loss enterprises could be found in timber and woodworking, building materials, machine building, metal processing, electric power, fuels, nonferrous metallurgy, and chemical and rubber industries. Some branches and enterprises that worked at planned-loss reported high turnover-tax payments. For example, glassware and porcelain paid 9.5 million leva in turnover tax, while it broke even on its profit-and-loss statement. The cellulose and paper industry paid 6 million leva in turnover tax and reported a loss of 1.4 million leva. The 1962 prices were so computed that even the largest and technically well-equipped enterprises had difficulties reporting a profit.[51]

THE 1967-71 PRICE REVISION

The price reform (which actually ended up as a revision) that was supposed to coincide with the introduction of the reform was a very protracted process. In 1966-67 prices of some goods were changed, mostly to increase profitability so that the experimenting enterprises could meet the payments to the budget and contribute to the s.p.f. But the partial price revisions had little impact on inducing enterprises to adopt their production to the pattern of demand. They also failed to stimulate greater use of substitutes in relatively abundant supply. They seemed to have created havoc in the profita-

TABLE 5.3

Profitability, as Percentage of Costs, of Industrial Branches, 1959-62

Branch	1959	1960	1961	1962
Electric power	18.82	35.98	37.79	10.25
Fuels	8.99	7.34	3.29	3.87
Ferrous metallurgy (and ore mining)	-17.96	-22.97	-22.18	-0.94
Nonferrous metallurgy (and ore mining)	2.31	2.45	1.51	2.54
Machine building and metal working	14.99	18.64	20.17	8.29
Chemicals and rubber	14.39	18.58	21.16	4.43
Building materials	7.58	4.16	3.61	1.21
Lumber and woodworking	11.33	10.67	9.81	6.43
Cellulose-paper	17.30	19.79	22.28	3.27
Glass and porcelain	15.55	16.13	12.56	0.00
Textile	12.53	12.05	11.94	7.51
Garment	3.64	2.87	5.24	6.52
Leather and footwear	23.35	8.67	8.84	8.22
Printing	13.11	9.96	8.33	12.20
Food	5.49	3.56	4.66	5.43
Other	12.87	9.82	9.67	7.18
Total	9.46	8.98	9.21	5.87

Source: Y. Bozhilov, TT 2 (1965): 33.

TABLE 5.4

Industrial Enterprises According to Profitability,
as Percentage of Costs, 1962

Profitability Group	Number of Enterprises	Percent of Total Number
Unprofitable	161	14.2
Up to 1 percent	37	3.3
1.1 to 5 percent	296	26.6
5.1 to 10 percent	347	31.0
10.1 to 15 percent	151	13.5
15.1 to 20 percent	58	5.2
20 percent and up	70	6.2
Total	1,120	100.0

Sources: Finansova statistika 1963 g. (Sofia, 1964), p. 50; Y. Bozhilov, TT 2 (1965): 34.

bility of branches, with profitability rising above average in machine building, chemicals, and timber industries, and dipping dangerously in light, food, and power industries, with such industries as metallurgy, coal mining, and some food processing becoming unprofitable.[52]

There were a number of pronouncements at various stages of the reform about the function and setting of producers' prices. At the early stage a three-tier price construct, similar to that in Czechoslovakia and Hungary, was officially espoused. However, there were no specifications about the size of each tier—that is, the number of goods or the percentage of turnover that would fall into the category of centrally set prices, the number that would have a lower and/or upper price limit, and the number that would be "freely" set by the parties to the contract. However, it was mentioned that only goods of local importance would fall into the third category. Not too great a store should be set by the three-tier price construct anyway, because, among other things, the grouping of prices into the three categories can be extremely restrictive and distorting. The second group (limited prices) might include only the goods of nonrepetitive production, produced under contract on a cost-plus basis, whose prices, even in the most restrictive traditional system, are difficult to control. In a seller's market the prices in this group would tend to hover at the upper limit (as was the experience in Czechoslovakia and

Hungary); thus, for all intents and purposes there is no flexibility, and the upper limit becomes the rigidly fixed price. Moreover, the third category (free prices) may apply only to few and very insignificant goods that affect neither the consumer's budget nor the enterprise's balance sheet.

To recall, the July 1968 Plenum abandoned the concept of the three-tier price construct. As the reform entered its recentralization stage, the guidelines for price revision were as follows:

1. The main emphasis was on revising prices so as to draw them closer to the price pattern on the world market (interpreted in Bulgaria as primarily the CMEA market).

2. Price ratios should be set so as to reflect technical parameters and quality and the "use values" of substitutes.

3. The mutually contradictory tenets of price stability and flexibility were reiterated. Price stability was to be maintained in order not to provoke unrest on the consumer market and for computational reasons, and price flexibility was advocated to take cognizance of changing conditions of production and sale. Price flexibility would involve increasing price negotiation between producer and user, under supervision of price agencies.

4. Prices should be made consistent with new financial arrangements; that is, they should provide for payment of the capital charge, repayment of credits and interest charges, contributions to the enterprise's and association's s.p.f., and payments to the budget. Generally, depending on the particular branch, the mark-up on "value added" was 8, 10, or 15 percent. Such a percentage was established on the basis of the following calculatory needs: 5 percent for the Additional Material Incentive Fund (AMIF), 3 percent for the DF, and 2 percent for the WF. In view of the wide discrepancy in the "profitability" of different enterprises, a uniform progressive rate of participation was established so that, for example, for 1 percent profitability the share of the enterprise is 0.25; that of the association, 0.25; and that of the budget, 0.50. The respective figures for 2 percent profitability are 0.30, 0.30, and 1.40.[53]

One may infer from the various sources that a price revision had been in preparation since 1967. However, writing in mid-1969, one observer pointed out that it was still in progress and had not been implemented. Presumably, the revised prices introduced at the beginning of 1971 were based on the principles outlined for the revision in preparation.

Of note is the stipulation that the revision broke away somewhat from the cost-plus principle, and a limited attempt was made to relate domestic wholesale prices to world market prices, at least for goods subject to foreign trade. However, it has been implied in personal communication that this was applied to a restricted number of key products. But, in view of the considerable import of key raw materials, intermediate goods, and machinery, the influence of their prices on those of the rest of the products (whether or not subject to foreign trade cannot be disregarded, and apparently the present domestic prices are "more realistic." Costs continued to be used in setting prices of goods that were not "permanent" objects of foreign trade.[54]

The official indication was that the new wholesale prices would be "connected with and dependent on" standard production costs.[55] Hitherto, costs had been planned and measured in relation to past performance, thus "legitimizing" backwardness, slackness, and inefficiency. Domestic costs were not being compared with those of other countries whose competition Bulgaria encounters on the foreign market. It was suggested that the main criterion in the cost sphere should be the achievements of the industrially developed economies. Concurrently, one of the means for objective cost measurement should be a well-developed standard cost base.[56]

It was reported that in computing prices in relation to world-market prices considerable changes in price levels for the various branches were noted, together with wider discrepancies between these levels. The maximum increase of prices took place in the lumber and woodworking industry (+28.5 percent), and the maximum reduction in the textile industry (-22.3 percent). In adopting world market prices as a basis for price setting, attention centered on the choice of the relevant world market prices and on cleansing them from distortions. Several distorting effects were taken into account, such as restrictive trade practices, monopoly prices, short-term fluctuations, and terms of trade. Allowances were made for some of these factors, and correction coefficients were applied. Obviously, the "correctives" were partly dictated by self-interest and were not themselves free of distortions. In principle, the new prices are supposed to reflect international price relationships. However, where these relationships do not correspond to the "use values" of products of similar technical characteristics, these disparities are eliminated. A feature of such a price-setting system is that it departs from the procedure whereby the profit norm is established, after which a price is calculated. The new procedure starts out with a given price from which the cost is deducted, and profit constitutes the residue. The size of the profit is an indication of the comparative advantages in producing a given product. But no matter what its other shortcomings, the procedure is applied to a limited number of goods. Moreover, the persistence of the traditional attitude to pricing was revealing. At one stage of the price revision it was found that prices for individual goods were widely differentiated. In order to forestall enterprises

PRICES AND DECISIONS

from manipulating product mix, price differences among individual goods and groups of goods were toned down, blunting the variations in profitability.

The financial arrangements were again handled in the old familiar way. A number of enterprises, producing low-profit or deficit goods, "which are at the same time essential for society" would find themselves in difficulties. An attempt should be made to raise the prices of these goods to a profitable level, but, if this were not advisable, for certain products a system of subsidies would be evolved to cover the financial requirements of these enterprises. The subsidies would be no larger than what is considered the "necessary profitability."

The price-setting system is intended to encourage producers to produce technically advanced, high-quality, new goods by compensating them for expenditures in connection with design and development and by affording them larger profits on these goods. Here, again, world market prices should be used as a basis. In the case of an improved product, the vague stipulation was that the price should be set in relation to that of the existing product, but reflect the improvement made.[57] Profit mark-ups were to be differentiated according to quality: A superior-quality good could carry an extra 10 percent mark-up on the set price, and the price of an inferior quality good would be marked down by 20 percent.[58] However, in practice there are many complaints that the quality of the product is not adequately reflected in its price. For example, the quality incentive on a man's suit amounts to only 5 leva, which is distributed among trade (50 percent), the budget (20 percent), and the producer (30 percent). The producer receives the supplement only after the product has been sold to the final consumer (this seems to be a move to reflect consumer's approval of the increased quality). However, the producers complain that the delay has a disincentive effect on production. Furthermore, this "incentive supplement" does not apply to the factory prices, but to the retail price, which means that it has no impact on the size of the AMIF and does not affect the enterprise's self-interest.[59]

With regard to fixing prices of new products, it was decided to refer to either of two methods. First, the comparative method is used in all cases where international or domestic prices exist for similar goods. The prices are established by comparing technical and quality indexes. Alternatively, the calculatory method is used where comparative bases are unavailable or inexpedient, and prices are brought into line with production costs and are adjusted to quality standards and sales conditions.[60]

A precondition for the success of a meaningful reform is to integrate price planning with the plan construction and execution process. Although much lip-service was paid to price planning, there is good indication that this planning is not being implemented. The law on the Sixth FYP stressed the need to announce factory price changes in advance to facilitate financial planning at the production level. At the same time, the emphasis on the need to differentiate prices so as to stimulate efficiency was another sign that, despite much rhetoric on the subject, in practice prices failed to reflect technical progress, substitutability, and the like.[61] Price planning remains one of the most underdeveloped activities.

Without a change in the very basis of price formation it is doubtful that significant improvement can be expected from some measures to give greater cognizance to world-market prices, quality differentials, modernity, and so on in setting factory prices. Although he offered only a fragmentary explanation, Petrov, writing in 1966, still sounds true today:

> No significant improvement has yet been made in the determination of factory prices. . . . Under the existing price setting and with dual prices which are not directly interconnected, factory prices cannot be formed in any other way but on the basis of proposals submitted by the producers themselves. They cannot be flexibly changed in accordance with changes in value, demand, and supply, and quality, for the price fixing organs are incapable of expressing these changes quantitatively and of continuously recalculating the impact of price changes in some items on the production costs and prices of the remaining products. Thus, the existing dual price system cannot be improved but must be rejected.[62]

Administrative price fixing tends to be very comprehensive and ludicrously detailed, not only because of the logic of the bureaucratic system but also because of price interdependencies, which tend to make the price-fixing process very comprehensive. The locus of decisions tends to shift upward along the vertical lines. The main price-setting functions are performed by the CP, with certain key prices subject to confirmation by the CM. The ministries set factory prices only for products transferred exclusively within their wards, and, similarly, associations' price-setting functions are confined to products transferred among their enterprises. Peoples' councils are allowed to set factory prices for inconsequential goods transferred within their district.

Thus, officially the micro levels have very restricted price-setting functions. But their sphere of influence is de facto much wider than usually conceived. By providing the cost information on which the authorities base prices and via "new products" (only slightly different from those listed in price catalogs) the enterprises do influence prices appreciably. Costs are usually overstated so that even if the percentage of profit included in the price is low, almost immediately after a price revision the reported profit is considerably higher (especially in the processing industries). Moreover, in the case of nonrepetitive single-batch output, produced on specifications, the producer usually sets the price in agreement with the buyer, on the basis of planned costs, plus a low profit mark-up. However, in this case as well, the costs are considerably inflated, and such production (especially in machine building) is a consequential source of profits.*

In a centralized system one should not confuse the locus of price fixing with that of price making. The further removed from actual production and sale the price fixer, the more dependent he becomes on the information generated by the lower levels. In fact, the central price authorities more or less "rubber-stamp" prices for a good share of products. Their control and bargaining power is limited, for they lack the specific knowledge of the particulars possessed by the periphery.

A NOTE ON AGRICULTURAL AND RETAIL PRICES

One of the key problems of price policy is setting effective terms of trade between agriculture and industry and distributing the burden of industrialization rush between the industrial and agricultural workers. Excessive extraction of "surplus product" from agriculture is counterproductive. Although the state can impose an unfair distribution of the burden, it undermines the "will to produce" and cannot compel the peasant to produce effectively. Since the mid-1950s increasing attempts were made to redress the situation, involving, among other things, raising procurement prices of agricultural products to ensure the farmers not only coverage of rising costs but also a certain profit margin. The "scissors" are reexamined, and periodic adjustments of the terms of trade take place.

The deficiencies of agricultural prices are beyond the scope of this study, but the very general points need to be made, for such prices affect eventual changes in producers' and consumers' prices, costing and subsidies, and the contribution of the agricultural sector.

---

*For example, it was reported that in 1960-61 cost-plus pricing was instrumental in helping two combines in Stara Zagora and Yambol realize considerable "illegal" income that was spent on higher salaries for some key personnel, compensations for quality control inspectors for letting shoddy goods through, and so on.[63]

The latter is of particular importance as it affects the living standards of the industrial worker, and consequently his will to produce, and thus his performance. Moreover, the agricultural dilemma is indicative of a larger problem, for what matters is not so much the quantity and quality of resources that are made available to that sector as it is the mismanagement in use and lack of palpable incentives to produce more cheaply and better. Perhaps the resources poured into agriculture could be employed more effectively in expanding the food-processing industry. But here again there is a lack of prices that reflect and reward high-quality processed output.

To recall, the periodic general price revisions are supposed to be confined to producers' prices and affect the turnover among enterprises of the state sector. The level, relative proportions, and scope of revision of consumers' prices is considered a separate problem. The planner faces the problem of finding the overall level and relative pattern of prices that will achieve equilibrium between supply and demand for consumer goods and services on a global scale and that matches demand with available supply of each commodity. The quantity demanded is, of course, a function of the price of the product in question, the relative prices of substitutes and complementary goods, consumers' preferences, spending power, the availability of goods and expectations as to future availabilities and price movements, the stock and quality of goods on hand, and so forth.

One of the important features of the dual-price system as it developed is the population's incredulity (with good reasons) toward the official position that a revision of producers' prices or agricultural purchase prices will not permeate the retail-price level. Both the 1962 and 1967-71 price revisions involved increases in retail prices; there were some compensatory cuts or subventions to lower-income groups, but they were insufficient to absorb the increases. In 1962 the largest increases were recorded in foodstuffs, especially vegetables and meat products, which account for the largest part of the family budget of lower-income groups. Increases in procurement prices were followed in 1968 by retail-price hikes. Prices of mostly slow-moving or unwanted industrial goods were reduced. The increases in the purchase prices of meat, milk, and various types of fruit and vegetables were explained as incentives for the development of animal husbandry and to ensure necessary output of foodstuffs for the population.[64]

In mid-1973 the prices of children's clothing, knitwear, socks, and stockings were cut by about 13 to 35 percent; those of medicines, by 24 percent; and those of television sets, by about 20 percent. At the same time, higher prices were imposed on high-proof alcoholic beverages, luxury products, and hand-made goods or those produced from costly raw materials, such as carpets and fur coats.[65] These price changes were explained to the population by stressing the commitment to a "policy of stable prices." In the past five years the price index registered a slight downward trend. The policy was oriented to lowering living costs of the young and large-size families and to reducing the costs of medical treatment at home.[66]

The main problem with the policy of price reduction is that it tends to aggravate, rather than alleviate, shortages. This policy has a long tradition in STEs, particularly in the USSR and Bulgaria.* On welfare grounds it would be preferable to increase living standards through wage increases, rather than through price reductions. But this task is complicated by many noneconomic questions, and even in Hungary difficulties were encountered in this area.[67]

The process of price inflation takes many forms. The price level increases as cheaper goods are displaced by the more expensive varieties, with no concomitant changes in quality; Zhivkov noted "a trend towards increasing the average level of prices of a number of commodity groups even though the individual retail prices of some assortments remain unchanged."[68] The latent inflation takes place at the expense of the consumer. The center is quite aware of the situation. Condemning the prevailing practice, Zhivkov preached: "The economic organizations must clearly understand that raising the average price level is justified only when prices in fact reflect the quality of the goods and when the market is supplied with an adequate quantity and variety of inexpensive consumer goods. In no case should we allow the sale of low-quality goods at prices set for higher grade products."[69]

DOMESTIC AND FOREIGN TRADE PRICES

At all stages of reform implementation in Bulgaria there was a wide gap between what needed to be done in pricing and what was being done. Even assuming that the Bulgarian leadership (like their Hungarian counterpart) had been committed to market-type reforms, it is not at all certain that the pricing reform would have been implemented. Conditions differ and comparisons cannot be carried too far, but the Hungarian experience is of particular interest in this respect.** Hungarian reformers were fully aware that a sweeping

---

*Still, in China the level of money wages remains pretty much unchanged, and workers are rewarded for gains in productivity primarily by lowering prices of wage goods.

**We are not concerned here with the details of the "grim reality," but merely summarily with some causes of the phenomenon, for in judging Bulgaria's aborted reform we cannot abstract from this perspective.[70]

price reform was necessary, and the area received priority treatment. The blueprint provided for gradual transformation of centrally set prices into controlled market prices, whose first step was the introduction of the four-tier price construction: (1) centrally fixed, (2) maximum, (3) upper and lower limits, and (4) free. During implementation, the share of goods for which flexible prices were allowed was disappointing, and "free" price setting was shackled by many restrictions. Under the conditions of a seller's market, which persisted for many goods, prices reached immediately and remained at the upper limit, so that they were de facto indistinguishable from centrally fixed. With the overwhelming bulk of prices in the rigid or nearly rigid category, the movement of free prices was distorting. To maintain many of the fixed prices, a number of activities had to be heavily subsidized, and tax reductions had to be allowed. In fact, during the period after the reform, the amount of subsidies rose, and tax rates were manipulated for reasons of financial expediency. The dual-price system was not abolished, and there was little success in linking domestic to world market prices. Consumers' prices continued to be separated from producers' prices by a cumbersome network of highly differentiated (but reduced in number) turnover taxes and subsidies. Much was said about the conflict between price stability and the requisites of efficiency. But in practice the system continued to uphold stability. At the outset the state guaranteed maintenance of retail prices of basic necessities and stipulated that the real income of the majority of the population could not be lowered. The scope for introducing the necessary changes was severely circumscribed. Thus, improvement in relative structure of consumers' prices had not "come about," and in many cases new distortions were introduced.[71]

In an open system an improvement in efficiency can be gauged by exposing enterprises to international competition and adopting the measures of profitability reckoned in prices that more or less reflect conditions in the world market. The creation of a strong link between domestic and world market prices has a long literature in STEs.[72] World market prices were allowed to penetrate domestic price structure at varied rates in different countries, and substantial differences could be observed.[73] Despite the increasing "consideration" of world market prices in the process of price determination and revision, there is considerable inconsistency in linking domestic to foreign trade prices. The 1965 reform and subsequent measures in Yugoslavia had a greater impact on the domestic set of prices than they had in any other East European country. Domestic prices of goods that were, or could become, subject to foreign trade were administratively fixed on the basis of foreign exchange prices, converted at uniform and reasonably realistic exchange rate. This resulted in a significant change in the relative price structure.

PRICES AND DECISIONS

Despite some progress made, the Hungarian reformers did not succeed in their limited attempt at establishing a realistic congruence between domestic and foreign trade prices.[74] We cannot do more here than merely indicate some of the problems encountered. The identification of the world market price is not a simple matter. World market prices can really be ascertained only for raw materials, typical semifabricates, and more or less homogeneous products. This raises the question of setting prices for such products as complex machinery and equipment, ships, and other items for which no actual world market price exists. Thus, it appears that the relative "power to negotiate" and noneconomic considerations play a much greater role than the firm basis or rules in determining the price. The identification of the world market price is not merely a difficult technical question; in many cases there is a wide set from which one can choose. What matters then are the politics of selection and the various corrections and adjustments the price undergoes.

With some oversimplification, a CMEA country faces two kinds of foreign markets. The first is the CMEA or "ruble trade market"; there, the dollar can be used in intra-CMEA trade, but ruble balances are nonconvertible in trade with the West, and there is only fairly limited convertibility of ruble balances among CMEA partners, with the rule of bilateral arrangements. The second is the Western (and Third World) market, where the generally acceptable currency is the dollar.[75] Generally, dollar price proportions differ considerably from ruble price proportions. World market prices are translated into the domestic price structure via exchange coefficients, and the latter assume considerable importance. To promote more realistic foreign trade decisions, the Hungarians have established separate foreign exchange conversion coefficients for hard and soft currency areas to reflect different conditions on each of the markets (indicating that the ruble is overvalued in relation to the dollar).

But the problem is not confined to the rationale of the relative shadow rates set, for the link between domestic and foreign prices is often more tenuous than is claimed. Prices are distorted by widespread and differential subsidies and preferential treatment. Export-promoting schemes, protective tariffs, and all kinds of restrictions inhibit import activity.[76] Import-using enterprises pay in domestic currency the equivalent of the actual price paid for imports abroad. Subject to numerous exceptions, changes in foreign prices of imports should affect the domestic prices paid by import users. But domestic and import prices should generally be the same for those commodities whose domestic prices are fixed or limited (for example, most of the raw materials imported for the chemical industry).[77]

The world market prices chosen are not the current Western market prices; frequently, they are some averages of previous years.

The adoption of the base depends on the movement of prices and on the preponderant interpretation of what the "representative" prices are. One of the major problems is the rise in world market prices, particularly of raw materials and fuels. Domestic prices of those inputs tend to be fixed to neutralize the effect of increases on domestic prices, and on the consumer price level in particular. If the government wishes to insulate the domestic price structure from this impact, the state budget has to cover the difference. The alternative is to allow the impact to affect the domestic price structure with an almost certain jump in the prices of major goods affecting living standards. But neutralization of the impact of world inflation can be achieved only by various types of budgetary subsidies (which weaken and distort the allocative function of prices).

In common with other STEs, Bulgaria suffers from the arbitrariness of foreign exchange rates. The failure of such rates to reflect anything distantly approaching supply-and-demand clearing levels is a major problem.[78]

The proposal advanced in Bulgaria for relating domestic prices of imports and exports to those paid for or received on foreign markets was similar to that adopted in Hungary. As in Hungary, two foreign exchange conversion coefficients were advocated, one for dollars and the other for rubles, uniform for exportables and imports. In addition, certain protective measures (tariffs, direct and indirect subsidies, and taxes) were envisaged. In practice, in Bulgaria the accent is on the "discriminatory approach" and various adjustments in calculating domestic prices of various commodities. Foreign exchange prices are multiplied by the appropriate official rates of exchange for the various currencies and then modified by vastly differentiated coefficients for different products or groups of products.[79] It is difficult to tell to what extent the complexity is a "roundabout" way of determining meaningful shadow rates or merely a reflection of artibrariness.

There are multifarious problems of price setting in intra-CMEA trade; we can only allude to them here.[80] Again, the base of the world market price is subject to flexible interpretation. Naturally, national interests dictate that prices exceeding world level should be sought for exports and prices below that level for imports.*[81]

---

*When an independent CMEA price system was being discussed, Bulgaria opposed the "world market basis" and favored man-hours instead. Since, in terms of relative labor productivity in the Communist bloc Bulgaria was on the bottom, prices favoring labor intensity would bring relatively high returns for Bulgarian exports and enable it to pay relatively low prices for capital-intensive imports.[82]

Whatever else needs to be said about CMEA prices, the problem of the "tie-in" arrangements in bilateral agreements (the relative terms of exchange) and the quantitative, rather than price, preferences cannot be ignored.[83]

Obviously, the importance of linking domestic price formation to foreign trade prices depends on the sensitivity of the economy to foreign trade, the flexibility allowed in shifting the geographical and commodity composition of foreign trade, and the periphery's real sphere of decision making. In this, as in other respects, Bulgaria is in a different position from that of Hungary, for the pressure to undertake major steps to improve foreign trade is not so imperative in Bulgaria, and perhaps the scope of maneuverability is much narrower. Parenthetically, it appears that various kinds of formal and informal import and export restrictions imposed on Hungarian firms make the problem of alignment of domestic with world-market prices less urgent on pragmatic grounds.

There is an unassailable argument for realistic prices in any economic system. But their importance depends largely on the extent and weight of economic decisions made on such prices. Surely, such prices are not enough, and the periphery must be allowed to make consequential decisions. When the scope of decision making at the perhiphery is narrow (which is still the prevalent tendency in STEs), prices matter less. Moreover, the c.p. relies primarily on output adjustments, rather than on price adjustments. The half-measure reforms are aimed at tightening up the system, rather than at introducing basic changes. But a minimal requirement for the success of even such reforms was the introduction of prices that would support what the c.p. was trying to accomplish to insure coherence between the various elements of the system of functioning. Effective production and distribution require effective coefficients of economic choice. An authoritative system may dispense with such coefficients and rely largely on physical targets and allocations. The economic question is not so much the feasibility as the relative costs of alternative arrangements for resource allocation.

NOTES

1. T. C. Koopmans, Uses of Prices (Chicago, 1954), p. 2.
2. The literature on the subject is extensive. See, among others, the following (and the references therein): K. J. Arrow, "Information and Economic Analysis," Harvard Project on Efficiency, Technical Report No. 14 (1973); K. J. Arrow, AER, March 1974; A. Bergson, Essays in Normative Economics (Cambridge, Mass., 1966); B. Csikos-Nagy, Socialist Economic Policy (Budapest, 1969); G. R.

Feiwel, The Soviet Quest for Economic Efficiency (New York, 1972); G. R. Feiwel, JEL, December 1974; M. Kalecki, Theory of Economic Dynamics (London, 1965); Koopmans, Uses of Prices; J. Robinson, Collected Economic Papers, vol. 2 (Oxford, 1964); P. J. D. Wiles, Communist International Economics (New York, 1969); A. Zauberman, Aspects of Planometrics (New Haven, Com., 1967); J. G. Zielinski, Rachunek ekonomiczny w socjalizmie (Warsaw, 1967).

    3. Compare Feiwel, JEL.
    4. O. Lange, The Working Principles of the Soviet Economy (New York, 1944), p. 14.
    5. Compare Feiwel, The Soviet Quest.
    6. DV 31 (April 19, 1968): 3.
    7. G. Petrov, PSS 1 (1966): 17.
    8. Ibid., p. 18.
    9. Ibid., p. 19.
    10. Compare G. R. Feiwel, The Economics of a Socialist Enterprise (New York, 1965), Chapter 2.
    11. G. Petrov, PSS, p. 22. Petrov offered the following example to illustrate what he meant. Prices of natural and artificial fibers were set so as to ensure equal profitability; in fact, though, the profitability of artificial fibers is considerably greater than that of natural ones. This, however, was not reflected in the profit of enterprises as it was expressed in the higher turnover tax on synthetic materials.
    12. Ibid., pp. 20-21.
    13. Ibid., p. 22.
    14. Ibid., p. 20.
    15. Ibid., p. 22.
    16. K. Ivanov, SK 7 (1968): 18.
    17. K. Ivanov, PS 8 (1968): 45; K. Ivanov, FK 7 (1968): 43-44.
    18. Ivanov, SK, p. 17.
    19. Ibid., pp. 21, 25-26.
    20. Ivanov, FK, p. 44.
    21. Compare F. D. Holzman, Soviet Taxation (Cambridge, Mass., 1955); R. A. Musgrave, Financial Systems (New Haven, Conn., 1969), pp. 48-51.
    22. Compare Bergson, Essays.
    23. Ivanov, FK, p. 40.
    24. Ibid.
    25. S. Kalinov and I. Yordanov, PSS 5 (1966): 9.
    26. Ibid.
    27. I. Videnov, IM 4 (1968): 5.
    28. T. Kutiev, Vzmozhnosti za preminavane km sobstvena baza na tsenite na osnavata na diyestvuvashchite v stranite-chlenki na SIV tseni na edro (Sofia, 1969).

29. T. Zhivkov, <u>Problems of the Construction of an Advanced Socialist Society in Bulgaria</u> (Sofia, 1969), pp. 52-57, 111-18.
30. For a discussion of the various price proposals within the Marxist framework as it developed in the USSR, and on which most of the Bulgarian proposals drew, see Feiwel, <u>The Soviet Quest</u>, pp. 185-97.
31. Compare ibid., pp. 197-216.
32. D. Davidov, <u>IM</u> 6 (1965): 11.
33. D. Yanchev, <u>IM</u> 4 (1965): 88.
34. A. Milvshevski, <u>NV</u> 1 (1965): 52.
35. For a discussion of this method, see Feiwel, <u>Economics of a Socialist Enterprise</u>, pp. 66-68. Such a basis for the "point of departure" price was advocated in Poland during the debate on prices (late 1950s) by W. Brus.
36. I. Mironov, <u>TT</u> 2 (1965): 23-24.
37. B. Dinova, <u>IM</u> 10 (1964): 22-24.
38. Yanchev, <u>IM</u>, pp. 90-93.
39. Kalinov and Yordanov, <u>PSS</u>, p. 11.
40. For the interesting results, see G. Kotsev, P. Skachkova, and I. Radylova, <u>Vliyane na podgotviyanata nova sistema na tseni na edro vrkv razkhadite za proizvodstvoto na promishlenosta</u> (Sofia, 1969).
41. These problems are discussed in among others, Kalinov and Yordanov, <u>PSS</u>; Videnov, <u>IM</u>; S. Obreshkov, <u>TT</u> 10 (1964): 30-37.
42. Compare Robinson, <u>Collected Economic Papers</u>, pp. 27-48.
43. Compare Kalecki, <u>Theory</u>, pp. 62-63.
44. Compare G. Petrov, <u>IM</u> 9 (1964).
45. G. Petrov, <u>PSS</u>, p. 23.
46. Ibid., p. 25.
47. Ibid., p. 26.
48. E. Mateev, <u>NV</u> 5 (1963): 74-89.
49. G. Chalukov, <u>FK</u> 8 (1961): 22.
50. Y. Shekerov, <u>TT</u> 3 (1964): 72.
51. Ts. Petrov, <u>PSS</u> 1 (1966): 57.
52. <u>Finanse</u> (Warsaw) 4 (1968): 44.
53. <u>DV</u> 31, p. 2.
54. Compare Z. Stanev, <u>PT</u> 7 (1969): 3-5.
55. T. Zhivkov, <u>RD</u>, December 14, 1972, p. 13.
56. K. Kostov, <u>NA</u> 6 (1970): 17, 21.
57. Stanev, <u>PT</u>, pp. 3-13.
58. <u>VE</u> 8 (1969): 60.
59. D. Svilenov, <u>IZ</u>, October 3, 1973, p. 2.
60. <u>DV</u> 31, p. 2.
61. <u>RD</u>, December 17, 1971.
62. G. Petrov, <u>PSS</u>, pp. 22-23.
63. D. Markov, <u>IM</u> 3 (1962): 11.

64. P. Takov, RD, January 2, 1968.
65. BTA, June 1, 1973.
66. I. Videnov, RD, June 2, 1973.
67. I. Friss, EEE 3 (1973).
68. Zhivkov, RD, p. 5.
69. Ibid.
70. Compare Friss, EEE, p. 10; O. Gado, ed., Reform of the Economic Mechanism in Hungary (Budapest, 1972).
71. Friss, EEE, p. 16.
72. Compare M. Kalecki and S. Polaczek, GP 4 (1957): 18-22; M. Kalecki and S. Polaczek, GP 9 (1957): 66-68; J. Zachariasz, Finanse (Warsaw) 6 (1956); F. Pryor, The Communist Foreign Trade System (Cambridge, Mass., 1963), p. 24.
73. Compare M. Kaser, Comecon (London, 1967); F. Pryor, Property and Industrial Organization in Communist and Capitalist Nations (Bloomington, Ind., 1973).
74. Compare Figyelo (Budapest), March 18, 1970, p. 3; T. Nagy, AER, May 1970, p. 431; R. Portes, SST, April 1972, p. 644.
75. Compare Wiles, Communist International Economics.
76. Compare B. Balassa, "The Firm in the New Economic Mechanism in Hungary," in Plan and Market, M. Bornstein, ed. (New Haven, Conn., 1973), p. 368.
77. B. Csikos-Nagy in Reform of the Economic Mechanism in Hungary, I. Friss, ed. (Budapest, 1969), p. 153.
78. Compare Wiles, Communist International Economics, Chapters 6, 7; Pryor, Communist Foreign Trade System, pp. 100ff.
79. W. Sztyber, Ekonomista (Warsaw) 6 (1969): 1381. Compare N. Mitrofanova, PKh 9 (1973).
80. Compare Kaser, Comecon; Wiles, Communist International Economics; Pryor, Communist Foreign Trade System.
81. Compare Wiles, Communist International Economics, p. 240.
82. Compare Kaser, Comecon, pp. 196-97.
83. Compare Wiles, Communist International Economics, p. 243; A. Brown and E. Neuberger, eds., International Trade and Central Planning (Berkeley, Calif., 1968); F. D. Holzman, Foreign Trade under Central Planning (Cambridge, Mass., 1974).

CHAPTER

6

**REDESIGN OF
THE SYSTEM**

Following the July 1968 plenum the modus operandi was again redesigned. Many of the partial decentralization measures of the Theses were reversed, but the system remained in flux, and only in the 1970s did it begin to crystallize. The period 1969-70 could be seen as an interregnum that preserved some of the features of the original "market-oriented" reform of the mid-1960s (although these were largely bereft of their essence as they were increasingly inconsistent with the changing philosophy of reform) and contained the seeds of the "technocratically oriented" modification of the traditional system of the 1970s.

ADMINISTRATIVE STRUCTURE

Since 1971 industry has been organized on a four-level pyramid headed by the c.p.

First, there are seven industrial ministries, which are the direct link between the center and production. They represent the c.p. vis-a-vis the lower levels. They are concerned with elaborating forecasts for the development of their industries, taking measures to remove development bottlenecks, and generally implementing state development policy within their associations. But the ministries are not supposed to participate actively in operative planning. In this scheme of things one should underline the increased role of the Ministry of Finance (MF) and the bank.

At the second level is the association (primarily a horizontal organization), which has become the basic unit, endowed with a number of essential decision-making prerogatives. At the end of 1970 the CM decreed a drastic combination of associations (from 120 to

64), presumably to provide for intensified concentration, improved synchronization of investments, and tighter links between R&D and production.[1] The association integrally incorporates production enterprises, with all their assets and liabilities, and the related R&D institutes. It now includes units performing production, R&D, design and construction, and foreign-trade functions. The existing MFT performs mainly nonoperational and control functions and incorporates directly only a few f.t.e., especially those dealing with raw materials. As a rule, the association does not incorporate domestic trade enterprises, which have been localized under the Ministry of Domestic Trade. Aside from its functional departments, the association includes the two forms of production units: subsidiaries and subdivisions.

At the third level are the subsidiaries (multiplant enterprises, large factories, combines, mines, and so on); they are part of the unified organization but enjoy relative independence, which allows them to conclude contracts, have independent bank accounts, and dispose of incentive and other noninvestment funds. However, they are no longer legal entities. These subsidiaries are responsible for (1) timely production of a given volume and product-mix of output, (2) adherence to standard production costs, (3) achievement of stipulated technical and quality levels, and (4) full utilization of materials, capacity, and labor allocated to them. To fulfill these tasks the subsidiaries are allotted certain financial and material resources. The extent of horizontal contacts between subsidiaries is delineated by their supervising association.

At the fourth level are the subdivisions (factories and small enterprises), which are fully dependent and either incorporated into the subsidiary or directly subordinated to the association.

THE ASSOCIATION

The Bulgarian experience features the transformation of the association from an intermediary link in the chain of command into a fully integrated "economic-management" organ. The association becomes the basic unit that not only groups enterprises of a given branch or subbranch but also encompasses the entire production process of given products. It also becomes the basic unit on khozraschet, with enterprises functioning on a form of internal khozraschet.

As the legal entity, the association disposes of all fixed assets and inventories and redistributes them gratis among its wards. Investments are undertaken only by the association, which may, however, assign to the subsidiary certain executing functions. The association is in direct planning and reporting contact with the SPC, the MF, the budget, and the bank. It is legally and financially responsible for import and export of its products.

The association is managed on the basis of "one-man responsibility" by the general director, who is appointed by the CEC at the minister's suggestion. He is assisted by the economic council and the operative executive bureau, which, aside from the director and his deputies, also includes, among others, some enterprise directors, heads of association's departments, and political and regional representatives. The association's three directorates (production, economic, and R&D), headed by the deputy directors, have become increasingly important. The first is responsible for current production activity, for fulfillment of the production plan, for raising output quality, and so on. The second prepares the association's economic policy and manages trade and investment activity. The third oversees the R&D institutes, design offices, and so on.

To invigorate the association's role all plan indexes, normatives, financing, and crediting are addressed directly to the association, which is fully responsible for plan fulfillment and settlement with the budget. The enterprises in turn receive all assignments from the association and are accountable only to it. The association not only maps out development and undertakes investment but also decides on volume and product mix of output, supply, labor, and so forth. The enterprise is mainly responsible for organizing and executing production, becoming a plant on internal khozraschet.

Some of the claimed advantages of such a system, which have probably swayed its designers, are the following:

1. In addition to other "economies of scale," as a large unit the association is in a better position to make use of computer technology, and above all it is also better suited to control from the center.[2] The limited number of associations simplifies the information flow, thus permitting a smoother computerized central direction.[3]

2. The association is presumably familiar with the conditions at each enterprise, so it can determine the "objective and subjective factors" in the enterprise's performance and judge which indexes the enterprise is able to improve. The association can use individual settlement prices allowing for divergent quantity and quality of capital. It can provide its enterprises with an "equal start" by differentiating normatives for the distribution of profits.[4]

3. The incorporation of R&D institutes should draw a closer link between R&D and production.[5] In the past, R&D was remote from production and not attuned to the arising snags and future practical requirements. The association can provide the R&D institute with better facilities and a broader overview of the problems of the given branch than the enterprise can provide. An institute within the association is much closer to production than it would be as a "budgetary unit" outside the production process.

4. The process of concentration is better served by the association than by combination of several existing enterprises. Specialization can be easily developed under the tutelage of the association, which can compensate enterprises for temporary disadvantages, shift assets among enterprises, and ensure better coordination of production flows.

5. The association is better equipped than either the ministry or the enterprise for investment decision making. It possesses an overview that the enterprise lacks, yet it is not so far removed as the ministry and is able to perceive certain interrelations from a closer vantage point. In addition, its interest in profitability should prompt it to decide on more effective investments.

6. The supply system could be streamlined when the supply organs deal with only a restricted number of associations. To ensure smoother production flows the association would perform better than the traditional supply organs the function of distributing supplies among its wards.

7. As the sole producer of a line of goods, the association is better equipped to engage in domestic and foreign market studies.

8. The association can employ skilled manpower more effectively. It is in a position to retrain redundant workers and make use of them in other areas.[6]

Although the "ugly word" of monopoly was not uttered, the specter of monopolistic practices was raised. The solution was neither imaginative nor practical: Given adequate foreign trade, imports could play the competitive role.[7] However, in practice, due to the balance-of-payments problems, imports tend to be restricted to the most necessary materials. Moreover, trade with the West—the source of high-quality goods—is quite limited, and the continuing stress on economic integration with the USSR and trade with the CMEA does not augur well. In the final analysis it is the consumer who suffers from the monopolistic practices.

The potential advantages and drawbacks of associations notwithstanding, in practice much is attenuated by the fact that these units are staffed by old-time bureaucrats used to getting orders from above and distributing them peremptorily and often arbitrarily among their wards. Despite the transformation of associations into "economic units," there is a lack of economic intelligence and managerial ability at that level. The associations' work-style is bureaucratic and sluggish, and they show little initiative in solving the problems of their wards.[8] In general, the poor economic performance in the first years of the Sixth FYP gave rise to rather severe public criticism of high-ranking management (ministers, deputy ministers, directors of associations, department heads, and so on).[9]

PLANNING

The administrative structure is closely related to the planning process, characterized by the following:

1. The basic planning unit is the association and not the enterprise.
2. The association receives its tasks (targets, limits, normatives) directly from the SPC and <u>not</u> via the ministry.
3. Only some of the indexes received from the SPC are supposed to be of directive nature; the rest are "guides" for the drafting of the association's plan.
4. The basic plan, and point of departure for setting annual plans, is supposed to be the FYP, elaborated at the ministerial and association levels; the annual plans are not subject to approval and can deviate from their respective segments of the FYP if necessitated by changed conditions.

But in an overheated economy the exception often becomes the rule. Too great significance should not be attached to the differentiation between guiding and directive indexes: In many cases guiding indexes are transformed into directive ones, and the distinction between the two is often blurred and spurious.

The plan draft is prepared by the association on the basis of the following binding indexes received from the SPC: (1) volume of output in physical units, (2) allocation of materials in short supply in physical units, (3) employment limits, (4) the wage fund normative (related to gross income), (5) the normative of export budget subsidies, (6) limit of investment expenditures, (7) turnover tax, (8) s.p.f. normatives. Item 7 has been added to the usual list, presumably because, guided by the profit indexes, the producer paid little attention to the assortments with high turnover-tax content, and the budget suffered in the process. In addition, item 3 appeared for the first time since reform implementation--an indication that the propensity for overemployment persisted. The association, in turn, distributes these indexes (except for the investment limit), and any number of others, as it sees fit.

The enterprise drafts its financial plan on the basis of indexes approved by the association, such as size of turnover tax; standard costs for products or product groups, with the specification of material and labor costs, together with standards of overhead, such as costs of repairs; and the normative of contributions to the WF related to the wage fund. The directive specification of standard costs is directly tied into the new incentive system. The last plan version prepared by the enterprise is the so-called counterplan, which, in the interest of the staff, should incorporate a given improvement in cost reduction in relation to the approved standard.

PROFIT DISTRIBUTION

Profit distribution was changed to fit the reorganization. The change was also explained in the light of the alleged malfunctioning of the reformed system, which featured a proclivity to distribute profits in favor of consumption, rather than investment, and a dispersal of funds among enterprises, rather than their concentration in the budget. In addition, enterprises attempted to increase their AMIF by inflating factory prices, by eliminating from production low-profit items, and by delaying commissioning of new capacities so as not to have to pay increased capital charges.[10]

Profit remains a performance criterion only at the association. At the subsidiary the success indicators are rather numerous, covering the old standby production-plan fulfillment in volume, assortment, technical, and quality terms; the standard costs of production; the normative wage per unit of output; utilization of equipment; and so on. Increase of the association's profit should be promoted by the standard production costs, which should be so designed as to ensure the most efficient production and to link incentives to stimulate fulfillment of plan targets, increase of labor productivity and output quality, strengthening of internal <u>khozraschet</u> within the association, and "mobilization" of the entire staff toward exploiting all reserves.

The total profit of an association is the sum of profits from domestic and foreign sales, increased by per-unit-of-output budget subsidies. Although realized through the intermediary of subsidiaries, only the consolidated profit of the association is subject to distribution. The subsidiaries contribute to their own AMIF and WF and set aside the amounts for the capital charge, the interest charges on working-capital loans, and social security contributions. The remainder of profit is then transferred to the association. Payments from profit to the budget are made exclusively at the association.

The association's profit is apportioned among the following funds, according to approved normatives: WF, AMIF, and Economic Assistance Fund (EAF). After the above contributions are made, the remainder of profit is subject to a profits tax that gravitates at about 80 percent in industry.[11] However, the profitability differences among various branches, due, among other things, to the continuing irrational price system and the past and continuing discriminatory investments, make the application of a uniform tax rate very tenuous. Many associations find themselves without the means for self-support; others accumulate funds in various bank accounts.[12] When the reported profit is insufficient to make the necessary contributions and payments, the CM may grant a given tax relief, approved in the form of a limit and deducted from the computed profit tax. The profit tax is collected before the repayment of investment and working-capi-

REDESIGN OF SYSTEM   145

tal credits. This was done apparently to prevent repayment of credit at the expense of budget payments. The data gathered for the previous periods indicated that with the considerable increase in investment crediting, the funds flowing into the budget declined.[13] Previously, the bulk of investment was financed from the budget and the revenues had to be relatively large, but nonbudgetary financing of investment should cut into these revenues. However, in the increasingly centralized Bulgarian system credit financing of investment continues to be promoted at the same time as increased centralization of funds in the budget.[14]

After the tax payment to the budget, the remainder is the so-called profit for distribution allocated to payment of interest charges on working-capital loans, social security contributions, Increased Foreign Trade Efficiency Fund (IFTEF), specific funds of a given branch (usually investment funds, and Maintenance of Association Fund (MAF). If there still remains a certain amount of profit it is earmarked for the DF. Other s.p.f. are formed from cost surcharges.

In practice one continues to hear of financial manipulations in profit distribution. For example, early in 1973 the MF appealed to enterprises and associations to become more familiar with the scheme of profit distribution, to compute accurately the profits tax and remit it promptly, and to accumulate and spend s.p.f. lawfully; it appealed to financial organs to control this activity more closely.[15] This was an indication of laxity and manipulations in financial computations and also to some extent of the continuously changing norms for computations that caused such confusion and played into the hands of those who were willing to manipulate.

FINANCING OF INVESTMENTS

The present system has upheld self-financing in considerably altered form, since self-financing no longer applies to the indivudal enterprise, but only to the association as a whole. As such it has been implemented with greater vigor than in the past, since it is easier for the c.p. to control the revenue and expenditures of some 60 associations than those of hundreds of enterprises. This has changed considerably the structure of revenue and expenditures of the state budget, with funds for "self-financing" increasing annually. For example, according to the 1972 budget, financial resources left at the disposal of associations were to be 17 percent above those in 1971.[16]

According to the 1972 budget the estimated revenue from industry was 4.6 billion leva, out of which more than 50 percent was to be paid into the budget (2.718 billion), and 1.866 billion leva were to remain for self-financing and other purposes.[17] In 1973, out of an aggregate investment outlay of 4.268 billion leva, 3.242 billion leva were to be self-financed (investors' funds and bank loans), and 1.026 billion leva were to be provided from the state budget. The state purse would be dipped into mainly for expansion and modernization of facilities in the service sector and housing.[18]

Capital repairs are financed from the enterprise's Capital Repairs and Improvements Fund (CRIF), which derives from depreciation. The size of the fund is determined for each enterprise by the association, in accordance with needs for capital repairs, with a floor of 30 percent of the part of depreciation earmarked for this purpose, and the possibility of subsidy from the DF. The association is also entitled to redistribute the CRIF among its enterprises.

From the standpoint of financial arrangements the main investor is the association, with all funds for financing such activity created only at that level. The basic fund is the DF, whose main source is depreciation, consisting of the entire part earmarked for replacement and also the part for capital repairs that has not been left with the CRIF. Previously, depreciation could only be used for financing investment, but at present it can be used for all the purposes for which the DF is used. These purposes include repayment of investment loans and interest charges; R&D in investment activity; payments to the budget for the industrial use of agricultural land, in the amount of 40,000 leva per hectare (a part of investment costs); subsidies for enterprises' CRIF; personnel training and salaries of foreign advisers; contributions to a retraining fund (accumulated at the Labour Ministry) in the amount of 0.3 percent of the wage fund; financing the increase of working capital; financing investments; membership dues in scientific organizations; and aid to combines. The DF should be spent first for purposes for which credits cannot be granted, and only in the last instance can it be spent for financing investments.

The association may seek investment loans, which are granted by the bank when the claimed pay-off corresponds to the normatives of pay-off differentiated by branches, with heavy industry still getting the best terms, irrespective of the underinvestment of light industry (the average repayment period for heavy industry is ten years, and for light industry six years).

The budget usually finances nonproductive investments or those of long pay-off periods, such as harbors, railroads, water-protection projects, and so forth. It also takes part in financing investments in metallurgy in view of the large size of outlays earmarked for this branch.

The shift in the sources of investment financing in the last decade has been from the budget to bank credit, with the share of investors' own funds remaining relatively stable. For example, in 1964 investments were financed as follows: 36 percent from investors' funds, 6 percent from bank credits, and 58 percent from the budget. By 1973 the three sources of financing shared almost equally (30, 34, and 36 percent, respectively) with the budget still having a slight edge over bank credits.

Interest on investment credits is differentiated as follows, according to investments in different areas: raw materials and intermediate goods, 1 percent; domestic and foreign trade, 3 percent; others, 4 percent. Interest rates may be increased as follows when the period of repayment has been exceeded: up to six months, 6 percent; up to nine months, 7 percent; 12 months and more, 8 percent. The participants in the investment process are rewarded for not exceeding the completion deadline and the cost estimate. These premiums are set at 0.1 to 1 percent of the project's cost estimate. If the execution period is shortened, the premium is increased proportionately to the time saved. The premium is paid from investment funds. The figures illustrating the shifts in methods of financing should not be taken as a measure of decentralization. They have little economic content, as distinct from financial and accounting content.

Investment outlays continue to be strictly limited, with nonlimited decentralized investments playing a very minor role. The FYP lists investment limits for individual associations. The annual plan further details these limits. Projects costing more than 2 million leva are listed by name—that is, decided upon by the c.p. The others are decided upon by the association within the framework of the remainder of the association's limit. Investments in utilities and communal services are centrally decided (and listed) if they exceed 300,000 leva. The plan is supposed to include an investment outlays reserve at the disposal of the SPC and the bank, earmarked for widening bottlenecks and distributed among associations by "competitive bidding," supposedly depending on "objective" efficiency calculations of the projects submitted. But the problem is not so much one of securing financing (the information submitted can be doctored) as it is one of securing real resources.

In 1971, in an attempt to curtail the frittering away of investment funds and resources, a ceiling was imposed on total nonlimited investments, which in principle should not be specified in the plan. For example, whereas total investment funds were earmarked at 3.293 billion leva, nonlimited investments could not exceed 150 million leva—considerably less than was spent for these purposes in the past.[19] This clamp-down on the only decentralized investments was to release resources for the large projects decided upon centrally. The move indicates that within the past few years decentralized investments probably competed successfully for real resources, impinging on construction of large centralized projects. Furthermore decentralized investment activity is very strictly limited by the availability of funds and the size of projects allowed. The association can undertake non-limited investments up to a limit of 2 percent of the DF's share in above-plan profits. The projects in this category cannot cost more than 20,000 leva each.[20]

As in other STEs, the Bulgarian experience shows that even restrictive and minor decentralization of investment activity creates formidable problems for the planners and appears to be incompatible with the logic of the system. The conservative financial apparatus seems to be right on this point. Increasing the share of decentralized investments, whatever the arguments on efficiency grounds, is not enough.

By 1973 all the well-known shortcomings in the investment process continued to plague the Bulgarian economy: Construction periods were drawn out; costs were high; there was continued preference for construction, rather than purchase of equipment; new projects still flagrantly outranked modernization; and so on.[21]

SPECIAL PURPOSE FUNDS FOR FINANCING TECHNICAL PROGRESS

The IFTEF is created in associations conducting export activity for their own account. The sources of the fund are contributions from profit according to norms approved by the CM, export premiums, and special premiums for exceeding the exports to the West. The fund is earmarked for financing investments for export purposes (by means of transfers to the DF), developing new product lines for export (by means of transfers to the NPF), premiums for executing export tasks, advertisement, and so on.

The Inventions and Rationalization Fund (IRF) is created from 30 percent of reported profit on the application of an invention or rationalization, and earmarked for financing such activity. Six percent of this fund is transferred to a centralized fund at the Institute of Inventions and Rationalization.

Created by means of differentiated norms (approved by the CM) applied to gross output in wholesale prices and incorporated in costs, the NPF is used for R&D, experimental projects, and improvement of existing ones. It is formed at the level of the enterprises, but expenditures are usually made at the association's level.

The Automation of Production and Administration Fund (APAF) derives from reported savings on the introduction of automation. The norms of contributions are set annually in accordance with the effects achieved and the needs for this fund. The APAF finances the designing the automation process, the wage fund of the personnel employed in the "automation" and "organization" departments, and research and design work on automated systems and purchase of foreign or domestic software.

The Foreign Currency Fund (FCF) is formed from a part of above-plan foreign currency revenue from exports, from savings on imports, and from export premiums. It finances primarily the import of machinery and equipment, spare parts, and electronic calculators. But it can also be spent on the import of raw materials for the production of highly profitable export goods; for repaying foreign currency loans; for purchasing patents, licenses, and prototypes; for advertising; and for premiums for executing export tasks. The FCF may be spent through other funds, such as the DF and AMIF.

FINANCING OF WORKING CAPITAL

Working capital continues to be financed by the unit's own funds (about 70 percent of the normative) and bank credits (about 30 percent). The former are supplemented from the DF. This apportionment is not rigid; it is subject to negotiation between the association and the bank. The planned normative of working capital is set by the association itself.

Credit for working capital, within the normative, bears an interest rate of 4 percent. Above-norm inventories can be credited at a higher interest rate, depending on the period of repayment: up to six months, 6 percent; up to nine months, 7 percent; and up to 12 months, 8 percent. Increased interest charges on working capital credits (included in production costs) detract from the financial result, the wage fund, and premiums. This is expected to induce dishoarding and to discourage above-plan inventories. The c.p. still deludes himself that financial measures are able to combat inefficiencies, whose roots are much deeper. Since the supply system has not been improved and the overheated economy still involves a seller's market, hoarding of inventories remains in the association's interests. Thus, associations tend to use their own funds for accumulating above-plan inventories and to resort to bank loans for other purposes.

THE COUNTERPLAN

The counterplan campaign began in 1972. It was created as a device to overcome what the c.p. considered slack in the early years of the Sixth FYP. The point of the exercise is for units to undertake and fulfill obligations in addition to their regular plan. This could also be termed as a "plan of plan overfulfillment." It is supposed to reveal additional reserve capacity and augment tautness.[22] But the "material-technical and financial coordination of the counterplans must take place without additional investment, additional foreign exchange, or additional materials and manpower."[23] It was stressed that fulfillment of those counterplans was necessary not only to implement the living standards program but also for capital formation and structural changes of the economy.[24]

Despite the many references to the counterplan as democratization and increased participation of workers in the planning process,[25] it is nothing but another attempt at "mobilizing" planning. Obviously, the plan supplements do not really originate from below. Whatever the political advantages, the results are often disruptive and abortive, with predictable upsets in coordination, balances, and internal consistency. Overfulfillment is not always desirable (for example, when goods are produced from imported raw materials, are of low quality, and cannot find buyers) and nonuniform plan fulfillment in various activities might be conducive to waste. Predictably, the original plan is understated by the extent of the counterplan. That both still reflect understated capacity is obvious from reports at the end of 1972, boasting that in many enterprises not only the original plans but also the counterplans were overfulfilled. Management has learned to play the game, but this involves even stronger incentives to understate reserves that are to be revealed later. The unit is supposed to prepare its counterplan independently; however, in fact, it is often told what the counterplan should be. For example, a report revealed that a commission was touring Pleven district to assess the possibilities for counterplans at various enterprises and to recommend targets to be adopted by the enterprises.[26]

In many cases counterplans are fulfilled and overfulfilled at the expense of other indexes. For example, in the important Balkankar Association—producing various types of electric and motor cars—counterplans were supposed to have resulted in 1972 in additional output of about 8.5 million leva. But, like other branches of machine building, it failed to fulfill its assortment profile—despite the fact that the output of this association is being closely watched by the SPC and CM. Generally, the failure to fulfill the plan in kind tends to be "compensated" by fulfillment of the plan in value terms—a situation characterized by the press as "alarming." The association also failed to fulfill the profit, profit-tax, and gross-income plans, and wages stagnated.[27]

Countrywide overfulfillment of the counterplans (by 2.8 percent for the first half of 1973) was reflected in a rise of profits. The share of profits in the total income of industrial organizations rose from 49.6 percent during the first half of 1972 to 51.1 percent during the same period in 1973. However, investigations disclosed that in industry many units overshot costs, and those costs were not likely to be made up in the second half of 1973, since in the last months of the year, when the lagging plan implementation is "caught up with," higher outlays are allowed, mainly for materials. Despite the campaigns for material savings, in the first half of 1973 increased material intensity of production was recorded and overexpenditures of raw and other materials were allowed.[28]

A party conference on counterplans for 1974 and 1975 focused attention on the fact that "some enterprises are seriously lagging in the fulfilment of their 1973 counterplan," and on "certain negative results" of counterplan experience (in 1971, 1972, and 1973.)[29] Qualitative indexes, such as gross income, profit, and cost reduction, were not fulfilled. Material input norms would have to be revised with special attention to save on imports—a direct admission of waste of materials. The growth of industrial output attributed to higher labor productivity declined in 1972, and decelerated even further in the first eight months of 1973. These weaknesses were blamed on such "subjective" factors as inept management. But there were even stronger arguments mustered against the counterplans themselves, which induced further overheating, with increased disregard for qualitative indexes. Supply difficulties were acknowledged as one of the main factors responsible for underutilization of capacity and for the deleteriously fluctuating rate of output (especially at such key factories as the Kremikovtsi Metallurgical Combine, the Sixth September Electric Cars Plant, and the Telegraph and Telephone Equipment Plant).

Obviously, much of the chaos resulted from lack of specification, from concretization, and from the dovetailing of counterplans with remaining plan indexes. The c.p. is usually prompt to blame the executants for stressing "extensive" growth criteria, without realizing that his "mobilization" tools are the villains. Thus, "some managers and specialists continue to focus their attention mainly on the implementation of the volume indicators; and others, instead of creating an atmosphere of full mobilization of the collectives, try to convince the superior organs to amend their control figures."[30] Apparently, in formulating quality indicators, the units tend to ignore changes in product-mix. This is supposed to create real difficulties in the coordination of physical indicators and makes for unrealistic counterplans of cost reduction, profits, and gross income.

Moreover, some key construction projects are seriously behind in completion, although their output was included in the plan.[31] In addition, shortages of skilled personnel persist, especially for newly commissioned capacities. "The question of ensuring the production process of the necessary import of materials in 1974 is quite alarming." Furthermore, the persisting shortages of spare parts account for many stoppages and underutilization of capacity.[32]

The production units' management complains that the state plan norms are unrealistic.[33] Pricing was also blamed for hindering plan fulfillment. The director of the Sixth September Electric Car Plant, Ivan Mindov, pointed out that the prices of electric cars (the factory's final product) were lowered without amending the prices of subcontracted components; this depressed the marketable output, profits, and gross-income indexes.[34] Moreover, the periphery as a whole blamed the malfunctioning of the supply network for "seriously disturbing the rhythmical flow of production and for becoming a permanent negative factor in the fulfilment of the counterplan." This was particularly the case of subcontracted deliveries.

Trade unions were under attack for their failure to exhort the units to implement the counterplans. The mid-1973 trade union plenum was called to "mobilize" economic activity. Kostadin Gyarov, the chairman of the Central Council, Trade Unions (CC TU), responded by pointing to the instability of plan indexes and the unions' inability to ensure that planning authorities provide stable plan indexes. This is one of the main "demobilizing" factors and a source of distrust of employees. The diminishing marginal productivity of campaigns was also increasingly recognized, even by the top party officials.[35]

WAGES AND PREMIUMS

The so-called material incentive system is composed of four separate funds based on given normatives that are supposed to induce the units to exploit production reserves and mainly to lower costs.

The basic fund is the wage fund, related to gross income in order to stimulate increase of the latter, especially since wage-fund savings (the difference between the wage fund computed according to the normative and that spent) flow into the AMIF and the EAF. But the association determines the wage funds of its various wards on the basis of the volume of output and product-mix. For this purpose, the association approves various norms of wage-fund expenditures per unit of output (or group of products). The wage fund of the association's headquarters and of the f.t.e. is determined by limits. In these units, the savings of the wage fund are used not for premiums but for the EAF; this seems to be largely counterproductive to the well-advertised aim of cutting down on administrative personnel. When the wage fund is exceeded, only management and responsible specialists suffer cuts of up to 20 percent of their basic salaries.[36]

The AMIF is related to cost reduction, derived from the total profit of the association but created at the level of enterprises. The plan of the AMIF takes into account: 100 percent of wage-fund savings between the wage fund initially set and that computed according to the counterplan, 40 percent of the planned savings of material costs, and 40 percent of the planned improvement of product quality. The AMIF also benefits from overfulfillment of the counterplan as follows: the above-plan share of top-quality output plus 50 percent of above-plan savings of material and labor costs. The AMIF is decreased when the share of the lowest-quality goods is overfulfilled and the costs in the counterplan are exceeded. This is supposed to counteract unrealistic counterplan drafting. However, it would seem that counterplan overfulfillment is still sufficiently attractive to induce enterprises to preserve a considerable safety margin, for the cost-benefit analysis might show that, in view of the uncertain supply system and other factors, it might be more beneficial to overfulfill a lower counterplan than to risk underfulfilling a tauter one.

When the approved wage fund is exceeded, and such excess is not covered by a wage fund reserve or the EAF, it is deducted from the AMIF. However, compensation for excesses of material and labor costs can be gotten from increased profits from larger output of higher quality. Despite strict restrictions, some associations continue to exceed the wage fund. In 1972 these overexpenditures reached 52 million leva.[37]

The previous system made the size of the premium dependent on profit. The new solution, which relates premiums to cost reduction, was applied, among other reasons, because of the use of internal <u>khozraschet</u> within an association and of the very restricted influence of enterprises on the structure of output and sales.

The EAF is created at the associations. In a way it is the association's reserve fund and is earmarked for assistance to wards and for premiums for particularly difficult tasks set by the association. This fund is created from the association's profit, equivalent to the difference between the wage fund computed according to the approved normative and the wage fund (plus the AMIF) actually paid out. It is planned at least at 2 percent of the approved wage fund. In addition, the EAF benefits from budgetary premiums for above-plan production of raw materials and other goods, according to a list approved by the CM, 55 percent of the premium for the production of highest-quality goods, and savings in the maintenance costs of the association's headquarters.

Individual premiums continue to be distributed on the basis of two or three indicators and one or two conditions. The management of the association is paid premiums when the association as a whole fulfills its profit and labor-productivity indexes, and when one chosen premium condition (such as technical level, output quality, standard cost, or export profitability) is met. The management of subsidiaries is paid premiums for the fulfillment of the production program, according to its main product lines, drawn up for the association, for meeting the standard costs of output, and for achieving given technical parameters.

If some indicators are not fulfilled or if the wage fund is exceeded, the culprits are punished by withholding the premiums for other indicators and, if that is insufficient, by cutting basic salaries for management (by 20 percent) and other employees (by 10 percent). If the stipulated levels of output quality, exports, and subcontracted deliveries have not been achieved, management is deprived of premiums. Upper limits (related to basic wages) for premiums are set for different groups of employees in different associations. The Ministry of Labor classified associations and their subsidiaries into three groups according to the size of output, complexity of production and management, and the priority accorded to the branch of industry. The upper limits of premiums in associations are 60, 40, and 20 percent of basic salaries, and the corresponding limits in subsidiaries are 50, 40, and 30 percent.[38]

In practice, wages are being used for attracting and keeping manpower when shortages arise. This is done extensively by reclassifying workers into higher grades, providing slacker norms that can be easier overfulfilled, padding the results, and so forth. However, wage differentiation is not systematically and officially used. Thus, wages are not increased to overcome labor shortages in the textile industry and in services. There is a shortage of economists, yet their salaries are lower than those of engineers. There is also a shortage of well-trained and experienced engineers in all fields, yet their salaries are on the same level as those of skilled workers. The wage level is not even used as a means of releasing redundant manpower.[39]

As a rule, it is the variable share of the take-home pay (the premium) that has come to play the part of wage differentiator, and as such it can only be effective if it is sufficiently large. But then it is claimed that centralized control over wages is weakened. Such a system is considered particularly suitable when the production program is determined by "horizontal ties," and performance can influence overall results substantially. Contrariwise, with comprehensive production indexes, vertically determined, premiums can be only a minor share of total pay. Thus, it was urged that the wage system be streamlined by reducing the weight of premiums and making the system more dependent on basic wages.[40]

Increasing disapproval is voiced against the alleged maldistribution of premiums—a subject that was widely discussed in the much-publicized reports of the Committee of State Control on the violations of discipline.[41] Apparently, premiums were paid mainly for plan fulfillment, without due regard for diversity of "objective conditions." Premiums were used as mere supplements to wages to dissuade

REDESIGN OF SYSTEM 155

workers from leaving. As in Hungary, the technostructure tends to grab and allocate among its members the lion's share of distributable income. This emphasis on maldistribution of premiums, at the expense of workers and lower management levels, and its demoralizing effects, has its political underpinnings. In addition, it seems to indicate a desire on the part of the c.p. to clamp down on the size and concentration of this source of spending power in order to contain the intensified difficulties on the consumer market.

An effective reform requires that a significant part of take-home pay be in the form of premiums, and the technostructure tends to appropriate them as a reward for its contribution. There are also well-known difficulties in relating premiums to individual performance. The premium becomes an uncontrollable variable that the c.p. attempts to contain at periods of intensified industrialization and resultant disproportions, as a measure to restrict the growth of spending power. Political problems of income distribution are not easy to solve. The resentment of workers definitely circumscribes any strong measures to reward good managerial performance. While acknowledging the dissatisfaction of workers with their share, the c.p. might wish to channel it away from the roots of the problem and to incite their traditional animosity against management.

FOREIGN TRADE MANAGEMENT

The traditional STEs featured almost complete organizational separation of production from trade with formidable problems of coordination and lack of flexibility and adaptability of the producer to the foreign market—at which the Japanese excel. Similarly, in Bulgaria there was little contact between the f.t.e., the producer, and the user. Prices received and paid in foreign markets were of no interest to export-producing or import-using enterprises. This became a growing handicap due to the increasing interest in selling industrial, rather than agricultural, products and to waste of imported goods. When the Bulgarian reform centered on the enterprise, the only organizational changes in foreign trade were on an experimental basis. But when the spotlight shifted on the association, the c.p. probably considered it "safe" to allow a partial decentralization of foreign trade and to expose the easily controllable association to foreign trade pressures. The half-measures in foreign trade aimed at closer coordination between f.t.e. and production activity during plan construction and implementation, organizational integration of foreign trade and production, and reflection of foreign trade profit or loss in the balance sheet of the industrial association.

The MFT continues as the stronghold of state foreign trade monopoly. Its main functions are active participation in plan drafting, surveillance over foreign trade plan fulfillment, coordination and control of foreign trade activities at associations and their f.t.e., and control over the balance of payments.

With the exception of a few f.t.e. (dealing primarily in raw materials) that have remained under the direct aegis of the MFT, most f.t.e. have become subsidiaries of associations, but unlike the production subsidiaries, they have remained legal entities.

The association's account is credited with payments for exports. The revenue is based on foreign trade prices converted at unrealistic foreign exchange coefficients, discriminated according to country of sales. In turn, the association settles with the producing enterprise on the basis of factory prices or approved standard costs of production. Thus, enterprise revenue is still calculated at prices set for sale to the domestic market (sometimes adjusted for cost differentials for better quality, packaging, and so forth). Special price lists were issued for imported products. Domestic prices of imported goods are set by adding differentiated percentage mark-up to the "foreign exchange value" (arrived at by converting the foreign currency price plus costs of acquisition into leva at stipulated exchange rates). Prices of imported goods, similar to those whose bulk is produced domestically, are set corresponding to domestic prices.

Profit or loss on exports is in principle retained by the association, while the profit or loss from imports is transferred directly to the state budget. For exports at prices below the domestic ones, the association may obtain a per-unit standard subsidy. The subsidy may be cut or withdrawn by the MFT if the export plan is not met. An entire network of export subsidies is to be used by the c.p. to regulate, stimulate, and constrain foreign trade activity.[42]

By design, the association as a whole is much more sensitive to the good performance of its f.t.e. than are the enterprises themselves, for the association stands to benefit or lose most. In case of deviations from the profit normative the f.t.e. only gains or loses 5 percent; the association absorbs the rest. The producer is not penalized for exports that do not measure up to domestic quality ratings. But the association is penalized if it sells goods abroad at less than their average rate of return.[43]

The MFT distributes premiums to the f.t.e. for above-plan foreign currency earnings and the like. These funds are planned in proportions of 30 percent of the AMIF, 35 percent of the DF, and 35 percent of profit or loss. The associations are granted premiums from the budget for fulfilling the export plan to the West, amounting to 5 percent of the factory prices if a rate of return of 5 percent is reported (the index of the rate of return is computed on the basis of production costs and the shadow exchange rate for the dollar).[44] Special indicators are used to relate individual wages to the financial results achieved by the f.t.e. The penalties for nonfulfillment of premium indexes cannot exceed 20 percent, and premiums cannot exceed 60 percent of the basic wage.[45]

Although formed at the association, accounts of the FCF are maintained for the production and foreign trade subsidiaries. The creation of the FCF is supposed to stimulate (1) efforts of the f.t.e. to search for new customers, getting better terms and the like, and the efforts of the production units to produce new goods, improve quality, and lower costs and (2) better selection and reduction of imports by savings of imported raw materials, their replacement by domestic production, and replacement of imports from the West by those from the CMEA. This reflects the antiimport policy and the forced shift to imports from the CMEA, while exports to the West are being encouraged by a discriminatory export schedule.[46] The FCF is distributed as follows: 50 percent for the association, 30 percent for the producer of export goods, 15 percent for the local peoples' council, and 5 percent for the f.t.e. This is a weak incentive for the commercial activity of the f.t.e., unless the system's designers banked on the association to exert sufficient direct pressure on the f.t.e. But here again, the tinkering with the incentive system is likely to have only a limited impact.

The foreign currency bonus is 60 percent of the savings of imported materials from the West and their replacement by domestic goods (while maintaining quality standards), distributed as follows: 60 percent to the FCF and 40 percent exchanged at the bank at 1.85 leva to the U.S. dollar. The FCF so derived is distributed as follows: 90 percent to the association that produced the substitute (with 50 percent granted to the specific production subsidiary), 5 percent to the supply unit, and 5 percent to the f.t.e. In the case of such import savings from the CMEA, the foreign trade bonus is only 15 percent.[47]

Despite constant "administrative control," substantial violations of the "existing foreign exchange regime have been discovered," and cases of administrative incompetence in foreign trade were disclosed, with reported foreign exchange losses due to improper selection of trading partners and widespread "inertness, bureaucracy, and crude mistakes due to negligence" of the trade apparatus. Moreover, applicants for export or import were accused of supplying deliberately false information on the compulsory five quotations. Often, unfavorable supply or sale prices are chosen, and "very few foreign trade functionaries watch systematically the movement of world market prices." The f.t.e. seldom cooperates; as a result, there are losses from lack of synchronization of various export and import transactions.

The applications for licenses are not accompanied by supporting materials; thus, the controlling organs have no way of assessing the proposed transactions. The return "is calculated on the basis of presumption" without examining prospects for the product for which licenses are purchased. Often, such products are obsolete, and purchased licenses are merely shelved.[48]

CONCLUDING REMARKS

There were no fundamental changes in the methodology of plan construction. A key feature of the latest planning and financial changes is that the SPC prepares the plan by associations, and directives are sent directly from the SPC to the association, bypassing the ministry and thus doing away with one intermediary level of aggregation in the information system. In this context the ministries are outside the process of current planning and management, at least in the design of the system. Whether they actually continue to interfere in the process is another question, more in the domain of strong roots of old habits and power structure.

The reform started out with the idea of enterprise-oriented decentralization. Bulgarian economists claim that although this experiment had some positive aspects, it showed that such a system led to "dispersal of investments, underutilization of capital and labour, and suffocation of the scientific and technical revolution and progress."[49] Apparently, the experiment indicated that enterprises were unable or uninterested on their own to solve such problems as adaptation of output to the user's needs, introduction of technical progress, and personnel training. According to the redesigners of the system, the multifarious problems of concentration, specialization, subcontracting, mechanization, and automation could not be successfully tackled within the existing framework of legally independent enterprises. These were issues to be approached by large integrated and powerful units. Moreover, the c.p. expected that the divergence of interests of the state and of individual enterprises would somehow be reconciled by the association.[50]

By far the most important managerial role is assigned to the association. It is difficult to see why such an association would eliminate the "contradiction" in interests between the state and the enterprise. On the contrary, being a more powerful unit, the association would be more forceful in arguing its case with the c.p. Its direct contact with the c.p. eliminates intermediary "costs" of bargaining and probably institutionalizes it as a "countervailing power" to the ministry, while allowing the c.p. more leeway in controlling it.

REDESIGN OF SYSTEM 159

Nolens volens the designers of the system have created a set of predesigned monopolies. The retort, that monopolistic behavior is attenuated by the very subsidiary role played by the market, is largely misleading. Although the association has no price-making prerogatives to bamboozle the consumer, it does have a relatively wide gamut of other possibilities, including: manipulation of product-mix in favor of plan-satisfying output (restricting output of low pay-off), deterioration of output quality, neglect of servicing and information about products, and decisions on delivery dates. The state-created pressures for implementation of technical progress notwithstanding, there is neither strong compulsion on the association to do so nor sufficient incentive to generate innovation from below. This is primarily because the "revealed planner's preferences" indicate continuous stress on volume of output and intolerance for qualitative improvements that might infringe on the short-term growth rate of output. Plan fulfillment of primarily quantitative production continues to be the key performance criterion. With the existing seller's market, which is not about to be replaced by a buyer's market under the high-pressure economy, the system's designers have not created any compelling economic pressures for the producers to adapt to buyers' demand. On the contrary, by creating a powerful network of monopolies, the designers have inured the producer even more to the influence of the consumer than he was in the traditional system.

The latest version of the functioning system is definitely of a directive nature. The well-known drawbacks of such a system are supposed to be offset by the system of counterplans intended to do away with plan bargaining and to induce the lower units to reveal reserves. However, in practice the counterplan has failed. Instead of inducing more "mobilizing" planning, it mobilizes the economic units' planners to bargain for lower initial targets and higher inputs, so as to be able to offer a more "mobilizing" counterplan at the final stage. The frequent substantial overfulfillment of counterplans is an indication that the system has not reduced the propensity for "planning with a precautionary margin" at the production level.

Despite the leadership's proclaimed endorsement of high-quality output, the changes in the system of functioning did not solve this very important problem. The differentiation in profit rates of factory prices was supposed to be a step in that direction. Even assuming its success, such a stimulant would be weakened by the stress on cost reduction, rather than on increase of profits. In fact, the insistence on cost reduction is often at cross purposes with increased output quality, for in the last analysis quality deterioration is probably the easiest road, and the one most often resorted to, to cost reduction.

The c.p. determines not only the residue of profit to be left at the unit's disposal but also the distribution of this profit for consumption and production purposes, and within the latter the distribution among the various s.p.f. Ceteris paribus, the higher the level of the s.p.f., as internal sources of financing (self-financing), the lower the share of financial accumulation extracted by taxation and redistributed through the state budget. Shifting the responsibility of financing investment from the state budget to the unit (whether from its own funds or from bank loans) should correspond to lower budget receipts. To some extent this is merely a shift in fiscal procedures, with innovations restricted to accounting treatment of the financial flows. In many respects the distinction between the various forms of siphoning off financial accumulation is administrative rather than economic. While meaningful decentralization requires self-financing, the latter can exist without the former. Investments and the production volume and profile can be decided at the center while financed at the lower units from their own revenue and bank credits.

The system features a good many s.p.f. with an attempt at restraining managerial maneuverability to substitute inputs (director of outlays). In terms of the c.p.'s traditional pattern of behavior, the number of s.p.f. can be used as one of the indicators in gauging the degree of centralization versus decentralization of a system. By establishing the approximate size of a s.p.f. (to vary within a given limit so as to have an incentive effect) and by prescribing strictly for what purposes the fund may be used, the c.p. maintains direct control over the expenditures of units. Hence, the larger the number of s.p.f., and the more restricted the unit's latitude in shifting resources among funds, the more centralized is the system. This is so, at least on paper. However, in practice, in view of the existence of a large number of s.p.f. and the often contradictory regulations governing their formation and expenditure, there is considerable scope for manipulation, which affords the manager greater practical flexibility.

The system relies heavily on cost reduction as a performance criterion and on standard costs as yardsticks of comparison. The notion of standard costs, well entrenched in Western accounting, seems very appealing. But it is technocratic rather than economic and reflects the industrial-engineering approach. The variance between the standard and actual costs should be a good indication of the quality of performance. This immediately focuses attention on the frame of reference (quantities and prices of inputs) and the meaning of standards—that is, how they are determined.

Standards are usually set with reference to experience, and—in common with prices based on costs—they suffer from legitimizing inefficiency. In a certain sense, although they undergo some adjustments, such standards tend to be a projection of past errors into the future. The setting of standards suffers from a divergence of inter-

REDESIGN OF SYSTEM 161

ests between those who confirm the standards and those who supply the cost information. While tight standards are in the interest of the c.p., cost information is supplied so as to slacken the standards. These are then the result of the tug-of-war between the conflicting forces, and can only be as good as the setter's knowledge of the actual state of affairs at the producer and his ability to forecast the future.

If the standards pertain to specific inputs, one has to remember that savings in one input can be reported by overspending of others. A favorable variance might be a result of inefficiency. Neither is cost reduction of attributed total costs a good indication, among other things because it might disregard some implicit costs. In a dynamic system (or due to economies of scale) some cost reduction could be expected. One of the problems is to distinguish to what extent cost savings are dependent on or independent of managerial action. Above all, the methods of cost manipulations are well developed, intricate, and difficult to detect. The power of the periphery in this area should not be underrated.

The entire construct of the system is heavily dependent on the accuracy of norm setting and the stability of norms. The law of the Sixth FYP stressed that it was mandatory to improve both the organization and methods of elaborating the various norms, and that norms should be set in advance of a FYP, so that the production units can take them into account when elaborating their plans.[51] Ivan Iliev, then head of the economic planning department of the CC BCP, and at the end of 1972 appointed chairman of the SPC, also stressed the need for stable five-year and annual plans so as to provide the units with a sense of security that the employees would be adequately rewarded for executing their tasks. A resolution of a plenary session of the CC BCP in November 1971 that discussed the Sixth FYP called for final elaboration and application of the stable indexes, limits, and norms that were to be the basis for planning, supply organization, and labor rewards.[52] Hence, once again, norm setting was lagging behind, and the planners were not drafting their plans on the basis of such norms. This, together with other veiled complaints, was a good indication that all was not well in the domain of norm setting, that the process of altering "stable" norms persisted, and that the Bulgarians had not succeeded in elaborating stable norms, on which so much of the reformed system depends.[53]

As in the case of other "insurmountable" difficulties of reform (notably price setting), one wonders whether these difficulties are not as much caused by "objective" reasons as by the confining effect of such stable norms on the c.p. A high-pressure economy is incompatible not only with an extended time horizon but even with abiding by the plan for a limited period (the operational plan). Thus, whereas the problem of planning is to reduce the uncertainty component in economic decision making, the high-pressure economy tends to increase it. Given that the plan is unrealistic (with built-in imbalances and intolerable pressures), the proclaimed norm stability is foreign to the system. Such norms would constrain the c.p.'s maneuverability in shifting the plan and redressing priorities to alleviate accumulated pressures. Because of the difficulties in affecting the "last-minute" changes in other parts of the plan so as to preserve its coherence, frequent changes often tend to aggrevate imbalances. Adjustments tend to be ad hoc, and by eliminating some dislocations others are created. This does not imply that calculation of "stable" norms is simple, rather that the difficulties encountered are manifold and that it is not particularly relevant which carries more weight. It would only become of greater importance when and if the c.p. would decide to shift the economy to a lower gear and to provide the necessary slack for meaningful reform implementation. But such a solution does not seem in the offing.

Finally, the numerous reorganizations of the last decade have not proceeded smoothly. They were usually not fully thought out, and even while they were being introduced, their shortcomings were quite evident. Subsequently, they were discreetly reversed, making for a layering and overlapping of jurisdiction, without clear-cut definition of responsibility. In the meantime, the bureaucratic apparatus proliferated and flourished.

NOTES

1. Council of Ministers (CM), "Decree No. 27, Dated November 20, 1970," DV, January 19, 1971, pp. 1-3.
2. Compare B. Ilev, PP 2 (1970): 75.
3. Stat. 6 (1971): 10.
4. Ilev, PP, pp. 76-77.
5. Interview with M. Tringov, TG, December 22, 1970, p. 3.
6. Ilev, PP, pp. 74-75.
7. Ibid., p. 75.
8. I. Iliev, NA 24 (1971): 15.
9. Compare RD, December 23, 1972; December 27, 1972; March 22, 1973.
10. Compare G. Zhelev, PS 3 (1971).
11. D. Popov, FK 1 (1973): 5.
12. Compare Zhelev, PS.
13. Ibid.
14. Compare D. Popov, RD, December 19, 1972.
15. Popov, FK, p. 6.
16. D. Popov, SK 1 (1972): 4.

17. D. Popov, RD, December 15, 1971, p. 4.
18. Popov, RD, December 19, 1972, p. 2.
19. S. Dulbokov, NS 1 (1971): 4.
20. Zh. Zhivkov, IM 1 (1973): 4.
21. Popov, FK, p. 7.
22. Compare RD, January 15, 1972.
23. G. Filipov, IZ, October 3, 1973.
24. B. Belchev, IZ, June 20, 1973.
25. Zh. Zhivkov, IM, p. 4; M. Dakov, PS 9 (1972): 17.
26. RD, January 15, 1972.
27. A. Dimov, IZ, August 22, 1973, p. 5.
28. D. Fratev, IZ, September 19, 1973, p. 6.
29. IZ, October 3, 1973, pp. 1-3.
30. Ibid., p. 1.
31. Ibid.
32. D. Svilenov, IZ, October 3, 1973, p. 2.
33. D. Marinchev, IZ, October 3, 1973, p. 2.
34. I. Mindov, IZ, October 3, 1973.
35. Compare D. Dimitrov, RD, August 14, 1973; RD, January 2, 1974.
36. Zhelev, PS.
37. Belchev, IZ.
38. B. Zielinska, GP 4 (1972): 245.
39. A. Dobrev, PT 1 (1970): 79.
40. I. Mironov, NV 9 (1972): 30-42.
41. RD, May 29, 1974.
42. Compare Zhelev, PS.
43. A. Kadiyski, PS 9 (1972): 49.
44. Ibid., p. 48.
45. T. Gonchev, FK 5 (1972): 11.
46. Kadiyski, PS, p. 44.
47. Ibid., pp. 46-47.
48. L. Buchkov, SK 6 (1973): 9-15.
49. Zhelev, PS.
50. Ibid.
51. RD, December 17, 1971.
52. RD, November 16, 1971.
53. Dakov, PS, p. 11.

CHAPTER 7

CONTINUITY AND CHANGE

Despite the tinkering with the system of functioning and the partial remedies, as the Bulgarian economy entered the Sixth FYP (1971-75), it was increasingly beset by problems. Some of them were system-made, reminiscent of those that originally prompted the leadership to consider reform, but with time and the increasing complexity of the economy they had become more acute. Others were definitely attributable to the growth policy whose basis had not undergone fundamental changes through the years. These problems involved primarily the system's dynamic and static efficiency.

Faced with the growing intensity of dilemmas, the Bulgarian leadership was forced to take some action. As we have seen, by the end of the 1960s it had abandoned the marked-oriented reform and returned to centralization, with some rationalization. Borrowing and dissemination of technical progress on a large scale was seen as the possible solution, if not to all at least to a good number of the most grievous problems. However, the attempted solutions primarily involved increased pressure from above for implementing technical progress, without sufficiently powerful and motivating incentives for the lower levels to take the technical-progress initiative. Increased rationalization was expected from concentration, automation, and computerization. Despite the growing pressures for "more scientific" central planning and use of mathematical models, sectoral balances, forecasts, and so on, once again the 1971-75 FYP proved unrealistic.

SKETCH OF ECONOMIC PERFORMANCE

In the first half of the 1970s evidence of deteriorating performance was mounting. Something had to be done to cope with the rankling situation. The intensity of the problem is not easy to discern, but it appears that the taut Sixth FYP had to be shifted midway, due to the usual accumulated disproportions, growth barriers, and ceilings, in an attempt to shift priorities to redress the dislocations. The manifest problems and the c.p.'s pattern of response were not much different from those in the past (and from those in other STEs). The seriousness of the problems is indisputable. The differences of opinions center primarily on attributing the proximate causes, diagnosing the ills, and prescribing the cures. The point is that a wrong or tendentious diagnosis will be followed by the "wrong" cures; that is, if the c.p. does not attribute the shortfalls in performance primarily to the overheated economy, he will merely resort to ad hoc responses to bottlenecks and half-measure organizational improvements to attempt to put into motion some "intensive-growth" factors, rather than revert to a resolute slowdown of the growth momentum, which would enable these intensive factors to take root and flourish, and in the end would result in higher and more sustained growth.

Naturally the state of the economy in the 1970s is affected not only by the decisions made for the Sixth FYP and by the exogenous or semiexogenous factors but also by the past pattern of development. Past experience should have taught the c.p. considerable caution in postulating growth rates so as to avoid the deleterious dislocations. But more often than not history teaches us only that it teaches us nothing. The sources of the c.p.'s optimism are many and varied. Temporary success whets his appetite and reassures him that he can overcome difficulties. The c.p. is a victim of his own wishful thinking. He assumes that the proposed reforms will improve efficiency and relies increasingly on intensive growth factors, without providing either a consistent set of reform measures or the necessary conditions (slack) for their implementation. But, more important, goals are partly imposed on the leadership by the realities of the political process. The "short-term" growth-rate performance has to keep up with the past and should not fall back in comparison with other CMEA countries. The institutionalized growth mania is difficult to extirpate. The leadership is judged by those well-entrenched standards—by which, after all, it does perform best.

The c.p.'s nervousness halfway through the Sixth FYP was not caused so much by the economy's failure to measure up to the "new" qualitative performance indicators, as by the deteriorating performance in terms of traditional quantitative criteria. Economic performance was disappointing not only in comparison to promulgated targets but also in comparison to the past growth momentum and to growth dynamics within the CMEA. It matters not that economic performance cannot be evaluated exclusively on the basis of growth dynamics and that the official statistics do not fully represent the velocity of indexes but that the c.p. relies on these variables and that they affect his behavior and shifts in policy.

The reader will recall that the aggregate statistics do not fully portray the dimensions of the problems and that average rates of growth conceal intraperiod fluctuations. Growth rates depend on the level achieved, and unsatisfactory performance depresses the base. Aggregates really have to be broken down into components to perceive the imbalances and bottlenecks that proliferated from one branch to another. Nor do the statistics fully reflect the benefits and costs of growth. In addition, variances between planned and reported data are not overly meaningful, since the final plan is often tailored to the report, and vice versa. Moreover, short-term statistics do not reflect lags. And, above all, qualitative performance is not fully depicted in quantitative indexes.

The 1966-70 Five-Year Plan

As a background to the problems encountered in 1971-75, a brief retrospect on the preceding FYP is in order. Some of the features of the 1966-70 plan and its fulfillment were as follows:

1. In comparison to 1961-65 FYP, performance in terms of average annual growth rate of national income improved and was still better than in any other CMEA country, but it deteriorated in terms of the growth rate of industrial output and was outdone by Rumania. The accumulation rate rose from 27.5 percent in 1960 to 28.3 percent in 1965, jumped to 34.2 percent in 1966, and remained above 30 percent until 1970. The consumption fund remained stagnant in 1968 and declined in 1970. The rate of growth of fixed investment jumped from 8.7 percent in 1961-65 to 12.5 percent in 1966-70—only exceeded by that in 1956-60: 18.3 percent. (See Tables B.18, B.20, B.24, and B.76).

2. The industrial investment share grew from 44.8 percent in 1965 to 47.8 percent in 1969—the highest thus far. Within that, the primary beneficiaries were the chemical, fuels, machinery, and metallurgical industries (see Tables B.27-28).

3. From 1965 to 1971 there was a considerable restructuring of industry, with the share of the fuels industry increasing from 3.8 to 6.1 percent; that of ferrous metallurgy, from 2.2 to 3.6 percent; that of machine building, from 16.5 to 22.2 percent; and that of chemicals and rubber, from 4.7 to 8.4 percent.[1] The largest growth rates of output were registered by petroleum, ferrous metallurgy, machine building, and chemicals and rubber (see Table B.38).

CONTINUITY AND CHANGE   167

   4. The growth of real incomes throughout that period was largely
due to a considerable increase of nominal wages in primary industries
that went into effect on October 1, 1966, and at the beginning of 1967—
hence, at the early stages of the period. In electric power, mining,
ferrous and nonferrous metallurgy, timber, and so forth the average
increase for workers was 14.5 percent, and that for managerial and
engineering personnel, 20 percent. The increase for construction
workers averaged about 21.3 percent.[2] From 1965 to 1969 the average annual nominal wage increased from 1,109 leva to 1,402 leva, or
by 26.6 percent. The average annual rate of increase of nominal wages
was 6 percent, and that of real wages, 5.4 percent in 1966-70.[3]
   5. The national output-fixed capital ratio deteriorated from 0.69
in 1960 to 0.44 in 1972, and that in agriculture, from 0.73 to 0.52.[4]
There was an overall deterioration in the dynamics of labor productivity. A moderate improvement in the rate of decline of capital productivity was registered, possibly due to delayed commissioning of capacities.[5] But the trend of declining marginal productivity of capital remained a major concern.[6]

   In his report about the fulfillment of the 1966-70 FYP, Sava Dulbokov, then chairman of the SPC, attributed the major difficulties in
that period to delays in construction and mastering of capacities and
the wide investment front and large number of unfinished projects.
Other contributing factors were waste of raw materials, accumulation
of above-norm inventories, poor quality of output, shortages of major
consumer goods, shortcomings in trade and services, insufficient expansion of public utilities and housing, and slack efforts in improving
planning methods (particularly norm setting).[7]

The 1971-75 Five-Year Plan

   The 1971-75 FYP (like those of other STEs) was published only
at the end of 1971. Some problems of intra-CMEA coordination were
encountered. The publication data almost coincided with the report
of the execution of the first year's plan. The poor results achieved
in 1971 did not bode well for the execution of the FYP, and one might
speculate whether some scaling down of FYP targets did not already
occur before the plan was finally approved.[8] As usual, the chief
aims of the plan were outlined as continued industrialization, so that
in terms of growth dynamics Bulgaria would be "in a leading position
among the socialist countries."[9] Furthermore, "in terms of pace of
economic development, during the Sixth FYP [Bulgaria would be] . . .
in one of the leading positions in the world."[10] But the Sixth FYP
followed a period of overexpanded investment activity, overheating,

and manifest growth barriers. The targets postulated for the Sixth FYP were slightly below those for the preceding period. In the preceding FYP 12 to 13 billion leva was to be spent on investment, whereas 20 billion was to be spent in 1971-75.[11] But these figures are not easily comparable, since new prices came into effect in 1971. The c.p. expected that the Sixth FYP would be executed by relaying primarily on intensive growth factors: "About 95% of growth of national income must be achieved as a result of higher labour productivity." Annually, only 4 to 8 percent of the increment of national income can be attributed to increased employment. The plan was expected to spur a rise in efficiency and an increase in the rate of utilization of existing capacity. As a result, the planned growth would be "achieved with a relatively smaller increase in the investment volume than in the Fifth FYP."[12]

National Income

Whereas the reported figures for the 1966-70 plan indicated that Bulgaria shared with Czechoslovakia the first place in growth performance, during the Sixth FYP, the annual reported rates of growth of national income were below the planned annual average, below the rates recorded in Rumania and Poland, and also below the annual averages recorded in 1966-70. By 1973 performance had rallied somewhat in terms of original plan and past performance, but that is not a representative year, since the original plan was considerably scaled down in many key activities to relax the accumulated tensions. However, the reported growth rate was still below those of Poland and Rumania. By 1974 a more ambitious plan was postulated, but it, too, was not fulfilled, and Bulgaria trailed behind Poland and Rumania (see Tables B.76-77).

The poor performance of agriculture contributed significantly to the deteriorating economic situation. The FYP called for average annual growth rates of 3.2 to 4 percent (3.5 percent reported in 1966-70). However, the growth rate in 1971 was only 3.1 percent; there was some improvement in 1972 (4.8 percent) and again a deterioration in 1973 (3 percent). But in 1974 agriculture stagnated; this was not supposed to be made up until 1975, with the aid of grain and fodder shipments from the USSR; the average for 1971-75 was to come to about 3.5 percent (see Tables B.76 and B.79).

Industrial Output

The reported figures for the 1966-70 plan show that in terms of average annual growth rates of industrial output, Bulgaria was bested

CONTINUITY AND CHANGE   169

only by Rumania, and it considerably exceeded the other CMEA countries. In the first two years of the 1971-75 plan these rates were below the planned annual average and below that reported for the preceding FYP. By 1971 Bulgaria was only behind Rumania, but by 1972 it fell behind both Rumania and Poland. Despite the increased growth rates in 1973 (slightly above the annual average planned for the period), Bulgaria still remained behind Rumania and Poland. The planned growth rate for 1974 was again ambitious and unfulfilled, trailing both Poland and Rumania. It was expected that by 1975 industrial output would have increased by 55.1 percent over that in 1970—that is, almost at the minimum postulated and considerably below the targeted level and that reported for the Fifth FYP.[13] The estimated average annual rate of growth was 8.7 percent (9.8 percent planned), behind both Rumania and Poland. On the whole, the c.p. was haunted by the deteriorating position of Bulgaria in terms of growth rates of national income and industrial output vis-a-vis the other CMEA countries (see Tables B.76 and B.78).

The 1971-75 FYP stipulated that "decisive priority would be accorded to the sectors upgrading the technical level of production."[14] An attempt was to be made to shift allocations in favor of the power-and-raw-materials base and machinery. The largest beneficiaries of the investment effort were to be fuels and power, machine building, and the chemical industry (continuing along the lines of the previous FYP), followed (with a wide gap) by building materials, food processing, and ferrous metallurgy.[15] Throughout the period the largest rates of increase of output were to be in petroleum (100 percent), machine building (more than 100 percent), chemicals (more than 100 percent), ferrous metallurgy (95 percent), building materials (77 percent), and electric power (59 percent).[16] Significantly, in the consumers' industries the plan called for the increase of output to be accomplished primarily out of existing capacities. Only in the later years would more capital be allocated for reconstruction, modernization, and expansion of these industries.[17]

Within machine building priority was to be accorded to "highly productive machines that would accelerate technical progress and ensure effective exports," with particular stress on computers, automation equipment, electrical-engineering products, metal-cutting machinery, and shipbuilding.[18] By 1975 the share of machinery in total exports was to reach 43 percent.[19] In the comparisons within the CMEA the rates of growth of machine building are an important performance criterion for the various c.p. In the past, starting from a low base, Bulgaria reported relatively high average annual growth rates of machine building (for example, 16.4 percent in 1950-60 and 18.3 percent in 1961-65), but its targeted rate for the Sixth FYP was considerably above that reported for the preceding FYP and those planned by the other CMEA countries (see Table 7.1). The performance in the first two years of the plan was considerably below the average annual target, and Bulgaria was bested by Rumania in 1971 and by the GDR and Rumania in 1972. With the shift of the plan in 1973, growth of machine building picked up momentum but still trailed behind that in Rumania—probably a matter of considerable concern to the c.p.

Already in 1971 the production targets of some key products (such as iron, steel, rolled metals, cement, internal combustion engines) were not met.[20] On the whole, the growth rates of the basic heavy-industry branches were underfulfilled.[21] During 1971-72, growth rates of imports exceeded those of exports (see Table B.72). This indicates not only growing strains in the balance of payments (particularly with the West) but also the increasing tensions and overheating of the economy.

The year of the shift was 1973, the halfway house of the FYP, when the tensions built up thus far due to the overextended investment front were being relaxed and greater attention was being paid to the consumer: "The successful fulfilment of the FYP as a whole depends largely on the successful fulfilment of the 1973 plan and counterplans."[22] Whereas total output of industry was to rise by 9.9 percent, that of light industry was to increase by 10.1 percent. However, high growth rates were also stipulated for some key heavy industry branches (for

TABLE 7.1

Percentage Annual Growth Rates of Machine Building in CMEA Countries, 1966-75

Country	Average 1966-70	Average 1971-75 Plan	1971	1972	1973
Bulgaria	15.3	16.5	15.0	13.9	18.6
Czechoslovakia	5.0	7.7	8.0	8.3	8.3
GDR	5.0	n.a.	6.0	17.0	8.0
Hungary	5.6	5.7-6.0	8.0	5.5	5.6
Poland	6.9	10.55	12.4	13.1	15.0
Rumania	5.8	11.5-12.0	16.0	15.5	20.7
USSR	5.3	11.2	11.0	11.7	12.0

n.a.: data not available.

Sources: I. Ivanov, VT 10 (1972): 2; Statisticheski Yezhegodnik Stran-Chlenov Soveta Ekonomicheskoy Vzaimopomoshchi 1973 (Moscow, 1973), p. 61; M. Golebiowski and B. Zielinska, GP 4 (1974): 227.

example, petroleum, 11.6 percent; chemicals, 22 percent; machine building, 19.2 percent; textiles 6.9 percent; clothing, 19 percent; leather and footwear, 5.6 percent; glass and porcelain, 10.4 percent; food processing, 5.9 percent). Of course, the point was how much of the output of heavy industry was to consist of durables. The 1973 plan anticipated sharp increases in output from increase in the rate of utilization of existing capacities, especially in the raw-materials branches.[23]

Again, the output of some key products was underfulfilled (especially in cellulose-paper, organic chemistry, plastics, cement, construction materials, and processed foods). During its course, the 1973 plan was drastically revised, and targets for the chemical industry were scaled down. The poor showing of the chemical industry was viewed with grave concern in view of the priority accorded to it and the immense resources poured into it.[24] The planned growth rates of machine building were not achieved. In 1971-72 the rates of growth of light industry were much below the overall rate (4.7 and 6.7 percent, respectively); the increased rate planned for 1973 was not met and was still below the overall rate reported (9.1 and 10.6 percent, respectively).[25] Hence, the planned shift in favor of consumer-oriented industries did not fully materialize, and plan execution again bore out the revealed preferences of the c.p. in favor of heavy industry.

By 1974 the c.p. was eager to recoup his losses, and the targets for that year were raised considerably.[26] Light industry relapsed to its previous status. The 11 percent rate of growth of industrial output concealed some significant disproportions. Whereas the chemical industry's output concealed some significant disproportions. Whereas the chemical industry's output was to increase by 27 percent, and that of machine building, by 21 percent, the light and food industries' outputs were to grow by 8.3 and 6.9 percent, respectively. Again, the plan was underfulfilled in almost all areas. National income increased by 7.5 percent (10 percent planned), and industrial output, by 8.5 percent. Output of chemical industry rose by only 14.5 percent; machine building, by 13.6 percent; textiles, by 4 percent, garments, by 7.4 percent; leather and footwear, by 4.5 percent; food processing, by 4.4 percent.[27]

Productivity

The extent of the 1971-75 FYP's reliance on growth of productivity is indicated in the postulation of the highest average annual rate of growth of productivity in comparison to other CMEA countries. The report for 1971 was below plan for that year and for the planned annual average. Bulgaria was bested in this index by Czechoslovakia and the USSR. The 1972 growth rate was even slower, with higher figures reported in Hungary, Poland, and Rumania. In 1973, when the planners' tensions were somewhat relaxed, labor productivity increased and almost reached the average annual planned, but still Rumania and Poland showed higher indexes. The relatively better performance in 1973 probably influenced the more sanguine 1974 plan, which was badly underfulfilled. The estimated average annual increase of labor productivity in 1971-72 (6.5 percent) was way below plan (8.1 percent); it matched Hungary but trailed both Poland and Rumania (see Tables B.76 and B.81). It was expected that in 1975 about 86 percent of the increment of industrial production could be attributed to increased labor productivity, as against the 95 percent originally planned.[28]

Investment

As shown previously, the share of accumulation in national income has fluctuated considerably. In the 1971-75 FYP this share was to moderate somewhat; the reported figures were 24 percent in 1971, 27 percent in 1972, and 28 percent in 1973 (see Table B.18). The Sixth FYP started out with a relatively low increase of investments, but they shot up in 1972. In 1973 the tensions built into the plan by the high investment rate were supposed to be relaxed; instead of the 8 percent planned increase, investment grew by only 6.9 percent. The considerable increase in the 1974 plan reflects the lower starting point and the pressing needs to complete projects in progress. Again, that was underfulfilled by four percentage points. The estimate of the average annual rate of growth of investment was 6.4 percent (5.9 to 7 percent planned). On the whole, the rate of increase of investment was higher in Czechoslovakia, Poland, and Rumania (see Tables B.76 and B.80). As could be expected, the investment program was revised many times, and some of the figures are irreconcilable. They indicate the ferment in this volatile activity, rather than provide a clear-cut quantitative picture.

The 1971-75 FYP relied significantly on increased investment efficiency; this did not materialize (in Rumania the ICOR fell, in comparison with 1965-70). According to provisional estimates, whereas the average annual gross ICOR in Bulgaria for 1966-70 was 4.1 percent, the Sixth FYP postulated 3.9 percent, and the reported data indicated 4.8 percent in 1971, 5.1 percent in 1972, and 4 percent in 1973.

The accent in the plan was on achieving higher returns on investments by means of improving the structural allocation, concentrating resources, diffusing technical progress, reducing gestation periods, improving performance at drafting, execution, and fruition stages, and allocating larger funds to modernization. Stress was laid on labor saving investments, modernization, automation, and replacement. Priority was to be accorded to completing investment in progress and to those investments whose commissioning would provide the economy with the needed goods during the FYP.[29]

In 1971 the plan for investment outlays was exceeded.[30] In 1972 this plan was exceeded by 200 million leva. However, the construction plan was underfulfilled—suggesting considerable overheating and tensions in this crucial area.[31] From the very beginning of the Sixth FYP enormous difficulties were encountered in construction. This was a legacy of the previous FYP, which had left an extended construction front.[32] Serious difficulties were encountered in agriculture and the food industry in building and commissioning new plants in 1971. Aside from tensions on the construction front and shortcomings of building organizations, the investors were blamed for "pushing" their projects into the plan, without even having the initial preparation. In addition, the delivery of equipment was often delayed. The investors were also blamed for not conducting efficiency analysis. The design organizations were not supplying adequate documentation. Finally, both the final design and executed projects were much more expensive than the cost estimate incorporated in the plan.[33]

The enormous delays in commissioning capacities in 1971-72 (4.8 billion leva-worth of fixed assets commissioned, as against 6.8 billion planned) were threatening the entire FYP. In 1972 the plan for commissioning capacities was fulfilled by 72.8 percent. At the end of 1971 unfinished construction was 110.4 percent of completed, and at the end of 1972 the figure was still higher.[34] The plan had called for 35 percent of investment to be allocated to modernization, but in 1972 only 18 percent of investment funds were spent for that purpose.[35] Toward the end of 1972 emergency measures were being taken to complete the most important projects in construction. Other projects were temporarily halted and manpower and materials reallocated to finish the crucial projects. The completion of a number of electric-power and raw-materials-producing plants, whose output was envisaged in the plan, was delayed.[36] For example, these delays pertained to, among others, the chemical combine for mineral fertilizers in Povelyanovo, the chemical combine in Svishtov, the laminated cardboard plant in Nikopol, the fifth and sixth sintering conveyors in the Kremikovtski Metallurgical Combine, the Devataki-Vrats Gas Pipeline, and the Krichin Hydroelectric Power Plant.[37] The reasons for these delays included faulty planning and designing of projects, delays in material supply, and shortages of construction workers.[38]

Early in 1973 a special conference was called to discuss the "serious weaknesses and alarming state of certain key construction projects." Premier Stanko Todorov laid the chief blame on suppliers of machinery and equipment, designers, investors, and construction organizations. He also blamed the planning practice whereby new projects are "forced" into the plan, with minimum resources allocated for the first year, thereby extending the construction front. This deeply rooted practice contributes to cost underestimation, substandard project designs, extensions of gestation periods, lack of synchronization between construction and supply of equipment, and so on.[39] There was general mismanagement in investment planning and construction. Considerable shortages of material and labor resources were encountered, while delivered equipment stood idle at the construction site and was wasted. The Ministry of Heavy Industry, the main investor of key projects, was blamed for failing to coordinate design, construction, and procurement of equipment for its projects. The construction industry was censured for poor organization of construction and inefficient use of the available labor force. Despite the labor shortages in this field, the planned figures were exceeded, and the labor productivity targets were not met. Labor turnover was very high.[40] The construction industry was backward and highly material intensive—in the interest of the designers and builders. It was estimated that the construction industry used 100 percent more steel and cement than was actually needed.[41] The relative consumption compared very unfavorably with that of the USSR.[42]

To mitigate the problems on the investment front, the 1973 plan naturally called for precedence being given to completion of projects under construction, especially for chemical fertilizers, calcinated soda, staple fibers, corrugated steel, ore mining, nonferrous metals, power, and a number of agricultural and transportation projects.[43] The main stress was on concentration of investment resources on fewer projects.[44] Additional funds were appropriated for reconstruction and modernization.[45] But, again, the construction plan was underfulfilled, and some of the key projects that were expected to be commissioned in 1973 were delayed.[46] The construction of projects for the consumers' industries was particularly behind schedule. Material and labor resources were being shifted away from these projects to those of heavy industry. Modernization investments were also underprivileged.[47] The 1974 plan called for considerable increase of investment, again with the admonition that it be centered on completion of projects and on modernization. But, again, execution was far from satisfactory. The plan forecast 4.452 billion leva allocated to investment, but only 3.9 billion was spent. Fewer projects than originally planned were commissioned.[48] In 1971-74 the total completion-investment outlay ratio rose in countries with decelerating investment push, such as Hungary and the USSR.[49]

The 1975 plan renewed concentration of resources on projects about to be commissioned, on reducing the number of new projects, and on putting finished projects into operation. By 1975 unfinished construction was to be reduced to 79 percent of the volume of capital investments, whereas the 1971-75 FYP had called for it to be between 55 and 60 percent.[50] Construction difficulties continued through 1975. There were severe complaints of shortages and "unrhythmical" supply of building materials, underinvestment in construction, antiquated construction technology, and low worker skills.[51] In fact, as a result of delays in commissioning, fulfillment of some FYP indexes was endangered. Output of plants that were to have been commissioned was not forthcoming. There were continuous shortages of specific kinds of rolled metals in 1974 and 1975.[52] At the Kremikovtsi Metallurgical Combine (the largest in Bulgaria) the volume of output was being unfulfilled. For example, less than 50 percent of the planned quantity of tin plate was produced in the first eight months of 1975. A similar situation prevailed in galvanized sheet, black steel, and seamless pipe. One of the reasons was a delay of several months in commissioning the cold rolling line. After commissioning, certain assemblies remained idle, and it was found that a portion of the preceding stages needed reconstruction (which could not be completed until 1978).[53]

Living Standards

The rate of increase of real wages in the Fifth FYP was about 5.4 percent. This rate was not expected to be maintained in 1971-75, which postulated a 4 percent rate. During the first two years, real wages grew very slowly (3 percent in 1971 and 2 percent in 1972). By 1973 the rate of growth was still below the low average planned. In 1973 nominal wages increased by 3.8 percent, but on June 1, 1973, consumers' prices were modified.[54] In 1975 average nominal wages rose by 3 percent, and per capita real incomes rose by 5 percent.[55] By 1975 real per capita income was expected to be 33.2 percent higher than that in 1970—that is, an overfulfillment of the targeted level by 3.2 percentage points and only slightly below the rate of increase reported in the 1966-70 period.[56]

As a result of the poor showing throughout the period, there was a proliferation of accusations of mismanagement, and slack discipline, together with mounting exhortations for mobilizing planning and campaigns to make order in the house. Such campaigns are a built-in feature of the Bulgarian way of life, but they are usually on the increase in a deteriorating economic situation, and their intensity is a good indication of the acuteness of shortfalls.

In the very first year of the plan complaints were registered about slack discipline and various shortcomings in management. The managers of associations and enterprises were blamed for continuing the practice of storming, so that extreme difficulties were encountered in executing the third- and fourth-quarter plans.[57] As the performance deteriorated, the tenor and intensity of complaints and exhortations grew stronger. The Agitprop Department intensified its campaign to combat workers' apathy to "socialist competition" and to check distortions and manipulations.[58] By mid-1974 there were continuous and strong complaints about violation of discipline: Plan fulfillment reports were being falsified with the knowledge and consent of superiors; the plan targets were inequitably distributed; bargaining about targets continued; employment quotas were exceeded; an uneven flow of output persisted.[59] Later, a special plenary session of the CC TU was called to overcome difficulties and mobilize workers. An appeal was made to overfulfill the 1974 targets, so that the unsatisfactory performance in the first three years of the plan could be made up.[60] It appears also that economic mismanagement was the primary cause for the large-scale purge of officials initiated at the July 1974 plenum. The purge affected not only the Politbureau and the highest administrative organs, but spread through all ranks of the bureaucracy and through upper- and middle-rank management. The scapegoats for the leadership's misconceived policy were found in all walks of life. The SPC was not spared. Ivan Iliev, who had headed it since January 1973, was replaced by mid-1975 by Kiril Zarev, former minister of labor and social welfare.

QUALITATIVE INDEXES

One of the similarities of STEs' reforms has been their advocacy of a shift from extensive to intensive growth and from quantitative to qualitative indexes for evaluating performance. But the crucial questions are whether the system has actually adapted itself to such an evaluation in practice and whether under conditions of a sellers' market it can do so. Some of the reasons for the extremely poor Bulgarian qualitative performance are as follows:

1. The high-pressure economy stresses maximization of output volume, pushes qualitative improvement of production to the background, and gives rise to inefficiencies and bottlenecks that pervade the system.

2. Postulation of targets tells us something about the planner's intentions, the acceptable rate aimed at, and the recognition of limitations, but the executant knows that not all targets are equally impor-

tant, and he will discriminate among them if they conflict or if their execution is endangered.

3. The system instills the executant with a short-time horizon; his primary task is to execute the annual (or even quarterly or monthly) targets.

4. The rhetoric and exhortations toward greater efficiency notwithstanding, the various levels of management are still in fact judged in terms of physical plan fulfillment; incentive systems might be tied into qualitative indexes, but they are largely counteracted by the severe penalties for not fulfilling the physical plan.

5. Other reasons include the poor technical and organizational level of production, the proverbial inefficiency of organization, bottlenecks in supply, labor laxity, and so forth.

As a rule, although the quantitative plan indexes are usually fulfilled, the qualitative ones are not. For example, in 1970 there were continuing complaints about underfulfillment of indexes, such as costs, profitability, labor productivity, and use of fixed assets. During the first seven months of 1970, returns on fixed productive capital dropped by 0.42 leva in comparison to 1969. The 1969 profitability level was not reached in important branches, such as chemistry, metallurgy, machine building, light industry, transport, and others.[61] Again, in 1971 the indexes of labor productivity, cost reduction, increased profitability, and product quality were not met.[62] Early in 1970 Zhivkov reportedly complained about the pace of cost reduction being "extremely unsatisfactory."[63] In 1968 such branches as electric power, cellulose-paper, and lumber reported higher costs than in 1967. During the first nine months of 1969 the reported cost reduction was about 70 million leva lower than that during the same period in 1968. In particular, in 1969 costs increased in nonferrous metallurgy, building materials, textiles, leather, and so on. This was attributed particularly to production of semifabricates by final producers, increased wages and premiums, uneven flow of supplies, replacement of unavailable cheaper materials by more expensive ones, increased waste products, lower yields, and higher shipping costs.[64]

Three particularly irksome problems were distinguished: the declining efficiency of investments, waste of materials, and underutilization of capacity. For example, until 1960 growth rates of industrial output outstripped those of fixed assets; since then the trend has been reversed.[65] Investigations indicated that in Bulgaria 8 to 27 percent more material inputs are used to produce given goods than in other countries. Bulgaria ranks very low in effective use of metals and plastics in machine building.[66] Bulgarian-made machinery is generally 5 to 15 percent heavier than that made elsewhere. This applies also to the goods in which Bulgaria specializes within the

CMEA. For example, the EV-210 battery-operated lift truck is 245 kilograms heavier than its West German counterpart, and the Elka-21M electronic calculator is 13 kilograms heavier than its Japanese counterpart.[67] There is considerable overconsumption of fuels at thermal power plants (in the first half of 1975 overconsumption amounted to 73,550 tons of fuel units). About 10 percent of total production of electric power was lost in the first half of 1975 in transmitting and transforming (this is more than the entire output of the 630-megawatt Bobov Dol Plant during that period). These losses have been ascending in the past few years. The losses were attributed primarily to backwardness of the transmission network, negligence in repair, design errors, and unsatisfactory operation.[68] Computations indicated that a 1 percent reduction in the use of materials would amount to annual savings of about 190 million leva.[69]

Empirical studies conducted in 1972 showed that in many branches plant and equipment is utilized by only 55 to 70 percent.[70] For example, there is considerable underutilization of capacity in ferrous metallurgy, mainly due to poor synchronization of capacities in units that follow each other in the technological process, to increasing obsolescence of certain processes, and to lack of spare parts for maintenance and repair.[71] Large automated power plants stand idle due to delays in fuel deliveries. Railway cars and entire trains are idle for days and weeks. Construction equipment is badly underutilized.[72] In the Madera Truck Plant in Shumen 60 to 65 percent capacity was utilized in 1975.[73] It was estimated that if all production capacity had been fully utilized in 1971-72, the additional "financial accumulation" would have amounted to more than 600 million leva.[74]

Labor Productivity

The goal of increasing labor productivity has been at the forefront of the c.p.'s attention for many years and has assumed particular importance since the early 1960s. Recently, in a keynote speech, Zhivkov emphasized this goal as a means of bridging the gap between Bulgaria and developed industrial nations. Although this could not be accomplished even within a FYP, all efforts were to be directed toward this goal.[75]

But recent reports indicate a retarding growth rate of labor productivity and unfulfilled plans (see in particular Tables B.57-58, B.76, and B.81). For example, in 1969 the construction industry fulfilled the labor productivity plan by 96.5 percent and the wage plan by 99.5 percent. The actually worked man-days per worker dropped from 276 in 1968 to 270.8 in 1969. There was an increase in absenteeism. Work stoppages increased from 12.7 man-hours per worker in 1968 to 14.5 man-hours in 1969.[76] In the electric-power, non-ferrous-metallurgy, timber and woodworking, garments, and food industries the entire increment of output in 1970 was attributed to increased labor productivity.[77] It is interesting that together with the underinvested, low-priority light and food industries, one of the lowest increases in labor productivity was reported by the overinvested machine-building industry, on which so much depends to raise the technical standards of the rest of industry.[78] A far more ominous aspect is the declining trend in labor-productivity growth. In comparison with subsequent years, the 1970 results were relatively satisfactory: "Since the end of 1972 ineffective utilization of manpower has become manifest. In some sectors increased output is achieved essentially by increasing the number of workers. For this reason, with the exception of 1971, the growth of industrial output via higher labour productivity declined in 1972 and dropped even further in the first eight months of 1973."[79] The concern with the deteriorating situation was obvious during the National Party Conference in 1974, whose agenda centered on increasing labor productivity as a means of improving Bulgaria's competitive position in foreign markets.[80]

Output Quality

Poor quality of producer and consumer goods has long plagued the Bulgarian economy and has apparently reached intolerable proportions.* While the c.p. is not insensitive to the poor quality of consumer goods and services, he is especially concerned with the poor quality of capital and other producer goods that affect the growth momentum and of goods that affect his terms of trade, particularly with the West. In 1973 a number of specialized goods in electrical engineering and electronics were not exported due to poor quality. Apparently, low-quality output "creates certain difficulties in expanding production cooperation with other CMEA countries."[82] Bulgaria is under increasing pressure from the USSR to upgrade its exports.[83] In addition, the poor quality of output and repairs became a problem of "exceptional importance because the combat readiness of military units depends on it."[84]

Some institutional attempts were made to tackle this notorious problem. In 1970 the General Directorate on Quality, Standardization and Metrology was instituted. A few months later it was transformed

---

For example, by mid-1972 it was reported that wholesale trade alone had a 35 million leva inventory of unsalable goods—and this is to say nothing of retail trade.[81]

into a committee. By 1973 the Institute of Quality Standardization and Metrology was set up to provide the committee with the formulations for quality control; quality control was promulgated as one of the chief tasks. The persisting complaints are that the cooperators in the production process do not meet quality standards and thus prevent the manufacturer of the finished product from meeting standards, even with the best of intentions. All sorts of manufacturing difficulties are continuously being overcome at the expense of quality. The system of quality control is far from adequate. [85]

Dissatisfaction with the attempted solutions for improving quality was evident from the publication of yet another decree in March 1975, which stressed again the setting up of "comprehensive standards" and warned of severe punishments for producers of goods that do not meet quality standards. A very telling passage dealt with meeting CMEA quality standards and upgrading the quality of Bulgarian goods that as yet do not meet requirements for integrating Bulgaria's economy with the USSR. The decree also specified that "better conditions" should be provided for producers of high-quality exportables. [86]

Measurement Problems

The determination, classification, and computation of "economic-intensification" indexes is faced with many problems. Besides reporting industrial output by the plant method, the associations have to report their final product, regardless of the plants that might be classified in different branches or sectors, to avoid double counting. Sufficient information on production capacity, availability, degree of use, and reserves for more effective use are all lacking. In most associations standard costs are still used ineptly, if at all. A serious problem is the determination and reporting of output quality. The range of quality groups is relatively narrow. The classification varies: Often a product is classified in a higher group by the producer, only to be reclassified into a lower one by retail trade or the foreign trade unit. There is little reporting on technical progress, such as the time required for introducing new products or processes, comparisons with world standards, and the extent to which new equipment is being used. Although calculations of investment efficiency and the recoupment periods have been conducted for some time, there is little verification of the efficiency of projects put into use. The market studies for consumer goods are inept. There is a dearth of statistics on the raw-materials base, the structure of above-norm stocks, and comparisons of material-intensity of production. Various aspects of profit generation and disbursement are not reported. Price statistics are inadequate. Price indexes, measuring ratios between imported and domestic goods, between industrial and agricultural products are lacking.[87]

The system continues to promote distrotion of information. For example, whatever the difficulties of determining production capacity, the periphery takes advantage of the situation, for it is not interested in revealing such reserves. Then again, the rationale of indexes expressed in value terms depends on prices and costing methods. A rise (or decline) in profits, rate of profit, value added, or labor productivity (either in terms of gross output or value added per employee) does not necessarily reflect contribution to national income and might be a spurious index of efficiency. It is easier to report a rise in profit by increasing the volume of output, manipulating the product-mix and disguised price increases, rather than by exerting efforts to reduce costs, to experiment with new techniques. Intensified pressures promote further distortion of information, higher "reserves," and unreliable fulfillment reporting.

LABOR PROBLEMS

If output is reduced to a function of quantity of labor employed and its productivity, the higher the rate of growth of productivity, the smaller the quantity of labor required to achieve a percentage point increment in the growth rate. Ipso facto, if a smaller share of the growth of output is to be propelled by gains in productivity, larger employment is required to achieve the same result. It is in this context that the emerging labor barrier adds to the list of pressing problems and restricts maneuverability.*

In Bulgaria the traditional sources of labor supply are slowly drying up, with acute exhaustion predicted after 1975. Already during the 1966-70 FYP, serious changes were felt in the supply of labor. Some branches and regions (mainly large urban centers) began to experience some shortages, particularly of skilled specialists and mature men, while there was still considerable disguised unemployment in industry and the potential female workforce was underutilized.**
In 1974 industry was operating with 2 percent less labor than planned; construction, with 2.9 percent less; and transportation, with 4.4 percent less. According to demographic computations, the increase of working force in the latter 1970s would be minimal, and in the period through 1990 an absolute decline was expected.[90]

---

*This problem has other dimensions that cannot be dealt with here.[88]

**For example, in machine building the share of women in the total workforce rose from 18.5 percent in 1960 to 30.8 percent in 1967.[89]

Since the late 1960s labor problems were increasingly singled out for interfering with production: lack of discipline, high turnover, underutilization of machinery, damages to machinery, poor labor qualifications, and misallocation and underutilization of skilled labor. Widespread alcoholism was implied by the creation of 1,400 so-called sobriety cells in industrial plants since February 1973.[91] Managers were criticized for luring workers from other districts, instead of retaining their own by improving conditions and for enticing labor away from agriculture. Some enterprises underutilized capacity many years after their commissioning, allegedly because of the lack of skilled labor.[92] Quite often, completion of factory housing lags behind commissioning of productive capacities, so that difficulties are encountered in attracting the labor force. The serious delays in construction were also blamed on labor shortages.[93]

Bottlenecks in some activities are accompanied by disguised unemployment in others; and this seems to be mainly due to system-made incentives to "store labor."[94] Bulgarian economists admit the existence of considerable disguised unemployment in industry and administration and some unemployment persists in certain regions and categories.[95]

A continuing complaint has been the loss due to the high rates of labor turnover. The coefficient of labor turnover of blue-collar workers in industry fell somewhat, from 57.8 percent in 1967 to 56.8 percent in 1968.[96] However, in 1970 complaints of extensive labor turnover, slack discipline, and low skills continued.[97] In industry alone about a half million workers change jobs annually, with about 30 days lost in the interval between jobs. Nationally, the loss is estimated at about 15 million man-days, equivalent to a year's work by 54,000 men.[98] Turnover partly manifests dissatisfaction with working conditions: disparities in the plant equipment, wages, development prospects, availability of housing and child-care institutions, and the distance between the place of work and the settlement.[99] More strict regulations have tried to cope with the problem. In January 1970 those who left their jobs were forced to pass through a district bureau that would assign them to areas with the greatest labor shortages. However, the decree is not effective simply because it is not adhered to. Management is willing to hire workers even without asking for their work books.[100] Labor turnover is encouraged by managerial tactics to attract skilled workers. A veritable "black market" has come into being.[101]

In the 1970s there was some decline in labor turnover. But in 1973 it was still the case that 56 percent of production workers in machine building changed jobs.[102] The overall share of workers who left their jobs fell from 52.1 percent in 1970 to 39.3 percent in 1973 and to 34.3 percent in 1974. This was attributed partly to the raising of the minimum wage, the increasing of low wages, and higher wages for work under difficult conditions in 1973-74. These categories of workers were particularly subject to turnover.[103]

Undoubtedly, the repeated complaints about workers' idleness, lack of discipline, unauthorized absenteeism, and general indifference to work aim at mobilizing further efforts and/or shifting responsibility for blunders from the c.p. to the workers. They are indeed a real problem caused by the general apathy, indifference, and alienation provoked by a highly bureaucratized system—but they are also a stigma in most modern, highly mechanized production processes. The regime's much vaunted commitment to education and the resources poured into it—aside from the indoctrination and propaganda purposes —must have had a palpable effect on improving skills. But it is important to remember that the proletariat is of very recent vintage and of a preponderantly peasant mentality.

There is also the problem of the quality of industrial experience that the former peasants are acquiring in the process of social transformation. Learning takes place during activity through the attempt to solve a problem. Learning associated with repetition of the same problem is subject to sharply diminishing returns. Probably the forces in STEs more or less favoring perpetuation of production and managerial processes impoverish the learning process to the extent that stimulus situations are not steadily evolving.[104] The low wages and poor working conditions divert workers' attention from performing on the job to learning "how to play the game" and are likely to have adverse effects on the system's dynamic efficiency. The workers' intensity of exertion is only one dimension of productivity performance. It is difficult to measure the relative slack of STE workers' exertion. The employees seem to manifest a strong preference for leisure rather than for hard work, partly because of defective incentives.

MALFUNCTIONING OF SUPPLY

Supply deteriorates with intensification of industrialization drives and as sequences of technical and organizational barriers are encountered. As construction and production activities are pushed harder, the demand is magnified for domestic and imported producer and consumer goods, not only because more inputs are needed to foster more output and because marginal costs tend to rise under conditions of rush (and the process increases purchasing power), but also because producers tend to hoard more stocks against increasing deterioration of supply. Supply from lower stages of production is likely to lag behind, with a tendency for underfulfilling output and investment plans in the basic materials and extracting industries and for overfulfilling the production quotas at higher stages of processing. Strong pressures are likely to be exerted on the balance of payments and the difficulties in foreign trade may make it impossible to redress the situation by indirect production. The real remedy is to decelerate the tempo of expansion. Supply improves with a slowdown of the growth rush, not only because demand for inputs lessens, but also because new capacities started during the rush enter the production process.[105]

As Bulgaria became more industrialized, its supply problems intensified. The overdeveloped supply system has a power structure of its own and contributes to many apparent shortages. Contradictory directives impeded planning, mainly because of multilevel subordination.[106] There seems to be a pervasive belief that streamlining the cumbersome bureaucratic apparatus by modern organizational and push-button methods will solve the problems; this neglects the fact that the bulk of the supply problems is created by the framework of macrodecisions. The unreliability of supply, shortages, and maldistribution are constantly blamed for plan underfulfillment and for undermining the functioning of the "new system."[107] Poor maintenance of machinery, lack of spare parts, and other breakdowns of the supply system continue to be the culprits of many work stoppages and overtime.* Hence, enterprises persist in hoarding materials, despite the lower profits (and higher costs) and consequently lower premiums.

Despite measures taken by the MSSR and the bank, above-norm stock formation has been accelerating. During the first half of 1975, above-norm inventories rose by 153 million leva over the same period in 1974. The primary culprits were the well-known priority industries. In the first half of 1975 above-norm inventories were 75 million leva in machine building, 55 million in electronics, 55 million in chemicals, 40 million in construction, and 51 million leva in light industry. Investigations in the electronics industry indicated that both inventories of materials and finished goods were due to changes in production programs, slow mastering of new techniques, and unrealistic planning.[109]

Concurrently, the deeply ingrained practice of storming has not been eradicated.[110] For example, many associations and enterprises apportioned their 1971 plan by leaving considerable slack in the first half of the year, which was easily overfulfilled with high premiums paid, while fulfillment of the third- and fourth-quarter plans, as well as the annual plan as a whole, suffered.[111] But storming is aggravated by and in turn contributes to supply shortages.

---

*For example, during the first half of 1970, work stoppages in the Plovdiv district amounted to 497,837 man-hours—392,078 man-hours in the first half of 1969—and 6.4 hours average work loss per worker—5.2 hours in the first half of 1969.[108]

Unwillingness to remove the tensions in the system prompted all sorts of administrative actions. The considerable difficulties in deliveries of subcontracted goods were to be counteracted by higher-echelon compulsion on subcontractors to deliver the required goods. Receipt of premiums became contingent on fulfillment of the plan for subcontracted goods.[112]

After the July 1968 plenum the MSSR was set up as the supreme organ in the supply system. Specialized marketing-procurement agencies for goods, such as timber, steel, and chemicals were instituted. The ministry, its organs, and other economic ministries continued to be accused of departmentalism and favoritism in deliveries.[113] The minister of supply admitted that due to persisting shortages and the monopoly over distribution, the customer was blatantly neglected by the supply apparatus.[114]

In 1972 the new minister of supply complained about particularly grave breakdowns in supply and unrealiable deliveries in terms of quantity, quality, and specifications, both from domestic producers and imports.[115] The campaign to utilize the above-norm unused stocks and by-products failed. Moreover, the needed reserves were not created, nor were the existing reserves of any help, as their composition proved unsuitable to meet specific shortages. In 1973 the focus was again on rhythmical supply of resources in terms of quantity, quality, and specifications, indicating that previous efforts had failed. Supply and marketing organizations were to put "active pressures" on producers, and particularly on foreign trade organizations. A remedy was sought in strengthening control and applying sanctions against violators.[116]

Appropriate stocks at the warehouses of the supply and marketing organizations are needed for smoother operation of the supply system, but warehouses are few, ill-equipped, and poorly organized. The minister of supply pointed to Soviet experience as an example to be imitated.[117] That, as we know, is none too shining in regard to prompt, specific, and timely supply of required materials.[118] The redesign of the supply system parallels the overall recentralization. Although declarations continue to be made about expansion of direct production ties, now "particular attention will be paid to the development of the centralized forms of delivery."[119] This is supposed to be accomplished by computerization. Much is expected from the creation of an Automated Management System (AMS) for producer-goods supply.[120] In 1973 electronic computers were to be used for controlling metal supply and operational information and control subsystems, initiating preparations for introducing the subsystems to elaborate material balances and control of metals. Also in 1973 an AMS was supposed to have been worked out for the supply of bearings, cables, spare parts, and chemicals: "A decisive step [would] be taken towards using computers in compiling material balances for 1974," and an attempt would be made to "improve on the previous method of determining requirements on the basis of standards and norms."[121]

MACHINE BUILDING AND FOREIGN TRADE

The c.p.'s forcing of machine building is a major determinant of structural change, of direct and indirect production, and of qualitative performance of the system. It is also a tension-producing factor. The strategic role of machine building as the hallmark of progress is often emphasized by the c.p. The Sixth FYP postulated that output of machine building would more than double and "in the next two or three FYPs there [would] be possibilities for machine-building to increase several fold."[122] In 1971-75 machine building was supposed to "help reequip the national economy with highly productive equipment, accelerate technical progress, and ensure effective exports."[123] Simultaneously, a structural break was to take place. In fact, by 1974 machine building usurped first place from food processing as Bulgaria's leading industry, and exports of machinery reached 40 percent of total exports.[124] The ambition was to establish Bulgaria as an important machinery exporter in the world market.[125] The Soviet economy's requirements were a determinant of the Bulgarian production profile.[126] The clue for the pushing of exports of machinery so hastily lies, among other things, in competition and rivalry with other CMEA members in terms of this important "index of industrialization," which still finds Bulgaria in one of the last positions (see Table B.73).

This is a multidimensional problem. We are principally concerned with one aspect only: the effects of the strategy on the system's dynamic efficiency. We shall briefly discuss the rationale of the policy itself. Whether Bulgaria benefits from its dependence on the USSR, or even if on the whole it exploits some CMEA partners rather than being exploited by them, as tentative Western estimates and my own preliminary research tend to indicate, the long-term effects of the policy cannot be ignored.*

---

*In the late 1950s Horst Mendershausen offered evidence of Soviet discrimination against Eastern Europe by overcharging for exports and underpaying for imports, in comparison with trade with Western Europe for similar goods. Holzman questioned these findings and suggested that Soviet terms of trade were not exploitative. Revisiting the controversy recently, he stressed that

CONTINUITY AND CHANGE   187

To recall, the Bulgarian economy is heavily dependent on foreign trade, being poorly and lopsidedly endowed with raw materials. It is a primary exporter of agricultural products both to CMEA and to Western countries. It is another question, whether intensified food processing would be more beneficial than the pushing of machine building, in which the country seems to have comparative disadvantages. For the past few years the intention of close economic integration with the USSR has been the goal—of particular significance for the attempts to infuse the Bulgarian economy with technical progress borrowed from trading partners. Concurrently, Bulgaria is to specialize in some specific areas of machine building. However, since in the foreseeable future the bulk of its exports will still consist of agricultural products, Bulgaria is attempting to secure not only the

---

> disproving the basis of Mendershausen's contention of Soviet exploitation does not necessarily disprove Soviet exploitation although there does not appear to be any strong evidence for the contention on price grounds. On the other hand, some of the bloc members may well be viewed as "exploited" in the sense that they gain less from trade than they otherwise would because they are obligated to divert most of their trade away from traditional and most profitable channels in the West to Comecon, and they are in some instances forced to specialize in the production of commodities in which they do not have a comparative advantage in order to meet the needs of the bloc.[127]

Holzman applied Mendershausen's procedure to Bulgarian data and found that Bulgaria discriminated against other CMEA countries much more effectively than the USSR, but he found the data inconclusive.[128] Pryor's calculations suggested that, in relative Soviet terms of trade with Eastern Europe, in 1955-58 the USSR favored the GDR and Bulgaria and consistently discriminated against Hungary, Czechoslovakia, and Rumania.[129] To Wiles the evidence indicated that most CMEA partners favored Bulgaria. The USSR did so by large turnover and small unit advantage and Czechoslovakia and the GDR by smaller turnover and larger unit advantage. Czechoslovakia and the GDR were "exploited" (the opposite of favored) by all the other CMEA partners. Bulgaria, and not the USSR, was the foremost "exploiter" or beneficiary of aid rendered from rich to poor countries.[130] G. S. Garnett and M. H. Crawford estimated the Soviet aid rendered to Bulgaria from 1945 to 1962 to have been more than seven times as large as that to Rumania, somewhat lower than that to the GDR, and about twice that to Hungary and Poland.[131]

best terms of trade in that area, but also to force its CMEA partners to pay for the highly capital-intensive investments in agriculture—just as the USSR is doing in the raw-materials area.

Bulgaria ranks second, after the GDR, in per capita reciprocal trade with the CMEA. In 1971-75 Bulgaria maintained its fourth place among Soviet trading partners.[132] But the honor is achieved at the loss of exposure to the discipline of Western markets. The goal of close integration with the USSR is repeatedly stressed in Bulgaria. As in the past, the future automation, mechanization, and overall technical progress would rely on the USSR's achievements. Equipment for new capacities built in 1971-75 was to be imported mainly from the USSR. Extensive integration was envisaged in the development of computer production and technology.[133] In fact, by 1980 electrical engineering and electronics were supposed to account for about 50 percent of all Bulgarian machine building.[134] The minister of machine building pointed out that the Bulgarian research institutes, associations, and enterprises should be closely linked with their Soviet counterparts.[135]

Machinery holds a key position in Bulgarian-Soviet integration. In 1971-75 machinery was to reach about 41 percent of Bulgarian exports to the USSR and 46 percent of the imports from the USSR.[136] Exports of machinery grew faster than overall Bulgarian exports to the USSR.[137] Statistics on Soviet aid are sparse and contradictory as sketched in Chapter 3. Zhivkov pointed out that plants built in Bulgaria with Soviet credits and technical assistance accounted for 95 percent of the entire ferrous metallurgy output, for 80 percent of the output of petroleum-refining and chemical industries, and for 60 percent of the electric power generated.[138] Bulgarian economists have hinted that because of this the techniques adopted tend to be material- and fuel-intensive, contributing to waste.[139]

We can only allude to the impact of Soviet technology and technocratic style on Bulgaria.[140] Just to note, in the past more than 180 projects were built on the basis of Soviet designs and equipment. More than 7,000 Soviet technocrats were on the spot to influence the character of the development. During 1971-75 alone, another such 150 industrial projects were under construction, including 37 machine-building enterprises. During 1971-75 the import of completed Soviet equipment was to exceed 650 million rubles, "under exceptionally favourable terms for us."[141] Most of the "mineral resources were discovered and exploited with the help of Soviet specialists."[142] In the human capital area 9,000 Bulgarian specialists were sent to the USSR to study its technology, and about 5,000 students "have become a living bridge for the transfer of the rich Soviet experience to key areas of our economy."[143] Most scientists and technicians have specialized at Soviet institutions.

Briefly, in the past intra-CMEA prices remained virtually unchanged during a FYP—these were so-called stop or fixed prices. They were set on averaged-out world market prices of the previous period, from which monopolistic and other distortions were supposed to be eliminated and to which transportation costs were added. Such prices were usually behind prevailing world market prices. The East European countries paid high raw-materials prices to the USSR in the early 1960s (based on world market prices inflated by the rising demand during the early to mid-1950s), while the prevailing world market prices were lower. When world market prices of raw materials escalated in the late 1960s and early 1970s, the USSR, much to its dissatisfaction, was being paid the lower CMEA prices, reflecting the relationships prevailing on the world market in the early 1960s.[144]

Within the CMEA Bulgaria is the largest per capita exporter of agricultural produce and foodstuffs.* The Bulgarians have become increasingly more vocal publicly about their discontent with CMEA pricing of agricultural produce, which moves against their interests. Increasingly, Bulgaria has been delivering goods to the USSR on credit.[146] It appears that the terms of trade are deteriorating. However, Bulgaria supports the Soviet stand about rising capital intensity and relatively very low returns on capital invested in the raw-materials industries, by supplying the same arguments for agricultural products.[147] The Bulgarians argue that this issue cannot be solved merely by changes in foreign trade prices that would improve the terms for the exporters. The shortages of domestic investments in the exporting countries should be made up by specific assistance and credits from trading partners, contributed proportionately to their imports of agricultural products.[148] In the meantime, Bulgaria is investing heavily in development of raw materials in the USSR. In 1975 such investment amounted to more than 90 million foreign-currency leva, and was slated to go up to 277 million in 1976.[149]

The pre-1975 intra-CMEA price structure was criticized on the grounds that it hindered specialization. It favored machinery and discriminated against raw materials and agricultural produce.[150] With increasing raw materials and food prices on world markets, the Bulgarians obviously wanted to improve their position within the CMEA. They felt that the price hikes did not go far enough to compensate them for the gains they lost by not selling their produce on Western markets (as some of their partners did).

Early in 1975 it was announced that new prices of raw materials were coming into effect immediately in the CMEA, instead of in 1976 as expected. In the bilateral flows of some countries oil prices rose

---

*In the late 1960s exports of raw agricultural products remained roughly at the same level, but exports of processed foodstuffs increased by 25 percent from 1967 to 1970.[145]

by 130 percent; industrial materials, by 53 percent; machines, by 11 percent; agricultural products, by 28 percent (beef, by 43 percent); and light industry products, by 19 percent.[151] Of course, the actual price increases of oil for various purchaser countries differed. No data were available for Bulgaria, but Hungarian sources claimed that a ton of Soviet crude oil rose from 16 rubles in 1974 to 37 rubles in 1975 and to almost 40 rubles in 1976.[152] Moreover, the CMEA departed from the fixed-price system to an annually sliding one. The 1975 prices were computed on the basis of a three-year world market price average (1972-74), whereas the consecutive annual price changes are to be based on the average for the preceding five years.[153] All this has played into the hands of the USSR. The changing terms of trade in its favor, together with the increasing participation of East European countries in developing the Soviet raw materials base, might presage a relative deterioration in performance of Eastern Europe in comparison to the USSR.

Naturally, Bulgaria has to pay higher prices for imports from the West and is affected by world market price changes that influence imports from CMEA partners. It is not always in Bulgaria's interest to support the Soviet demands for relative increases of raw-materials prices so as to reflect changes in world trade dynamics and to promote monopolistic practices to raise fuel prices. For these are likely to affect the prices Bulgaria pays for imports from the USSR, unless the monopolistic element will be appropriately rectified—and this does not seem likely. Of course, the matter is complicated by the usual problems of tie-in purchases in bilateral trade agreements.

Officials acknowledge that the benefits of trade are lower, and costs, higher, than usually claimed. In many cases import and export costs tend to be understated; export returns do not cover cumulative import outlays; investments are ineffectively allocated on the basis of overstated investment efficiency claims. More reasonable foreign trade and investment efficiency criteria are sorely needed.[154] Frequently, inputs are imported from the West, processed, and exported to CMEA partners, with consequent difficulties in the hard-currency balance. The implication is that often in intra-CMEA trade the results do not justify the costs.

The impact of foreign trade on reform is blunted by several factors. By securing a relatively stable outlet for about four-fifths of exports, Bulgaria can postpone the internal changes required to modernize and upgrade output, which can be limited only to the strength of the increasing demands of CMEA partners. But even here the bureaucratic process plays the role of a buffer. A precondition for successful reform is that the high quality of imports should stimulate the upgrading of domestic production—that is, that competition be introduced through foreign trade. If the substantial share of imports originates from the CMEA, the beneficial competition effect is weakened.

However, the share of trade with the West has a much greater impact than the mere statistics suggest or than the continuous official stress on trade with the USSR indicates. This is not only for the smokescreen effect. The record seems to show that, granted the powerful constraints, by and large the Bulgarians have displayed fair flexibility in trade with the West. Throughout the Sixth FYP the trade deficit with the West has continued and worsened. The contrast between Bulgarian and Rumanian international economic policies is striking. Yet Bulgaria is not immune from the Rumanian endeavor to expand economic relations with the West and to borrow advanced technology, and it must watch closely Rumanian (and Yugoslav) difficulties in transforming foreign trade. Unlike Rumania, Bulgaria is not endowed with the natural resources that it could export to pay for such imports. While tighter integration with the USSR is promoted as an alternative within Bulgaria's scope of constraints, there must be a desire to exploit trade with the West as an avenue of advancing the technical level, which intensifies as satisfactory progress is not being achieved with Soviet help. While the Bulgarians desire Western advanced technology and know-how, their overt preference for trade with the CMEA is spurred by the political situation and the difficulties in marketing Bulgarian industrial goods in the West. And, again, the bureaucratic insecurity and ineptitude better qualify Bulgaria to deal with the CMEA than with the West. Indirectly, the Bulgarians might expect to benefit from Soviet attempts to borrow Western advanced technology. Such an attitude might overrate the ability and capacity of Soviet industry to adopt, adapt, and diffuse Western technology into its own products. Furthermore, there would be a considerable lag in the process of transmittal of this technology to Bulgaria.

Understandably, the c.p. must have mixed feelings about major expansion of trade with the West, even irrespective of political constraints. Such trade makes plan fulfillment more difficult. The demanding buyer is a problem, and the complaints and pressures he generates might not be disposed of so easily as those from CMEA partners. Adaptation to such trade disturbs the structure of output, deranges the way of doing things, and necessitates more radical reforms. This also intensifies tensions in intra-CMEA trade, because, among other reasons, each country tends to channel its "best output" to Western markets. One of the advantages of intra-CMEA trade is that it makes the task of realization less difficult, and planners can dispose of shoddy output, for in bilateral agreements partners have to accept goods they do not want. Whatever the other reasons, Bulgaria's apparent option for intra-CMEA trade might be dictated by the relatively preferential treatment it receives and by the recognition that it can extract relatively better terms from CMEA partners.

There is no good theoretical reason for every country to specialize in and push exports of machinery. At least this is not the prescription of the theory of comparative advantages, which admittedly ignores political realities. Unquestionably, there are advantages in uplifting the skills of the labor force in the process of learning by doing. It is another matter, whether the focus should be on creative adaptation of foreign know-how, rather than on pushing domestic production of machinery on a wide front. The prestige factor cannot be ignored. But a machinery seller, competing on demanding foreign markets, has to be flexible and adaptable—and inflexibility is one of Bulgaria's primary drawbacks. Retardation in performance gives impetus for a reexamination of policy. Apparently, there is a pressure group in Bulgaria advocating a slowdown in the rate of expansion of machine building—it is difficult to discern whether this would be a temporary measure, to relieve tensions, or a lasting policy. This position was openly attacked, the opponents countering that, despite high growth rates, Bulgarian machine building still lags behind that of developed countries. If the rates should be scaled down, the gap between Bulgaria and industrialized countries would widen.[155]

Though it is a planner's performance criterion, expansion of machinery production for export might not be advantageous, for it might be necessary to sell the machinery so cheaply that the returns might not cover the costs of necessary imports and of domestic factors employed. The prescription might be rather a cut of ineffective exports and redirection of resources. According to Kalecki:

> Generally speaking, the so-called problem of the "directions of development" makes sense only when it is treated as the choice of the pattern of foreign trade (as well as the production technology), because only then does it not lead to plan imbalance. Once the problem is formulated in this way, it can be solved by means of the calculus of investment efficiency, provided allowance is made for the absorption capacity of foreign markets for different export goods, and for the technical-organizational barriers to the development of the production of particular branches.[156]

TECHNICAL PROGRESS

With the exception of the United States, diffusion of foreign-generated innovations appear to be at present the most important agent of technical change and a key factor in productivity growth. The growth rate of diffusion varies over time and is affected by the embodied type (import of investment goods, incorporating new foreign techniques) and disembodied (licenses, patents, industrial espionage, international exchanges of research and personnel).[157] In STEs the fundamental problem is the rate at which new techniques are adopted and incorporated into production and spread throughout the system. There seems to be a major difference between the technical change in high-priority activities—which benefit not only from discriminatory allocation of best resources but also from removal of some of the obstacles to new technology—and the rest of the economy.

The first rumblings of reform echoed the calls for technical progress, which have been growing to crescendo proportions, in contrast to other reform desiderata, which slowly waned off. Attention was paid to planning and decision making, with increasingly delineated competence and delegation of decision making to the medium levels of management and the R&D organizations. Reorganizations followed each other with little success.

One of the problems encountered is the disparity between particular research undertakings and requirements of practice. The theoreticians shy away from practical matters. New techniques are not easily translated from the drawing boards to the production process. There is no incentive for generating technical progress at the lower levels, and the c.p. finds it difficult to diffuse the progress originating at the center. The R&D offices tend to spread their work on many projects, without much interest in final practicability. How the product is to be introduced in production, whether its production has been mastered, whether the designs and prototypes have been approved, whether the technical and economic parameters of the new product have been approved—all these questions, and others like them, are often altogether absent from the planning work.[158]

The keynote of the September 1969 plenum was technical progress. Industry was urged to modernize according to the highest world standards, mainly by means of concentration, automation, and specialization. Much was said, and a number of legal documents were adopted. The vast literature is mainly concerned with desiderata rather than with concrete steps to realize them, but its tenor should be conveyed so as to indicate the shifting content of the reform. Though such pronouncements do not necessarily mean that conditions for their implementation are created, they indicate the state of the economy, the activities that require attention, and the success of measures taken so far. Though vague pronouncements give the impression of future solutions and detract attention from other sensitive areas, they also indicate an awareness of the possibilities of borrowing the fruits of research from abroad, of the need to allocate resources to the "development and implementation" activities, and of the exigencies of relying on incentives. But neither awareness nor new laws are enough. The real problem is whether the c.p. is ready and willing to sacrifice other conflicting exigencies and desiderata.

The latest attempt at revitalizing the lagging technical progress concentrated on removing some of the built-in obstacles. The 1973 decree of the CM and CC BCP stressed the following points:[159]

1. There would be "[extensive] utilization of progressive experience and of achievements of advanced countries," without the usual injunction to utilize the experience of the USSR. The weakness of a technical progress policy that focuses on copying the Soviet pattern is obvious. With the exception of important achievements in military and space technology, which, however have not spilled over to industry, the Soviet experience seems unimpressive, at least by Western standards. The purchase and use of prototypes and licenses was particularly stressed, and purchases for goods whose quality indicators were below Bulgarian or world standards were expressly forbidden—an indication that such purchases had been made in the past. Serial production was to start at the latest two years from the date of purchase, and licenses were to be bought only with assurance of complementary inputs. The accent was also on cooperative production. Nevertheless, by 1975 it was still claimed that only on the basis of intensive cooperation in machine building with the USSR could Bulgaria "resolve the problem of modernizing capital assets."[160]

2. Technical-progress undertakings would be part and parcel of the annual plan and FYP, binding on all concerned in their execution, and supported by sanctions for failure to do so. The basic idea was to draw more closely the needs of production to the research apparatus. Although such a step seems sound, there is the danger of falling into narrow practicality and of avoiding long-term projects.

3. The plan was to provide reserves to support the technical-progress undertakings. The executants' s.p.f. earmarked for technical progress were to be increased to provide adequate financing. Both the NPF and DF of associations were to be spent mainly on technical progress and modernization. Production of nonstandard instruments and equipment was to be encouraged. The supply system was enjoined to supply the units engaged in technical progress activity on a priority basis. A special foreign-currency reserve fund was to be set up for additional imports needed in the course of research and implementation. Stress was also laid on on-the-job training. Promising researchers were to be allowed to specialize at the best international institutions.

4. The decree attempted to remove one of the most obdurate obstacles to technical progress in the past: the conflict between fulfillment of a taut plan and implementing technical progress. Thus, it specified that enterprises implementing technical progress <u>should</u> have their output targets temporarily reduced; higher costs should be tolerated; stimulating prices should be provided; more advantageous credits should be granted; budget payments should be reduced; and

so forth. In cases of proven efficiency, the associations could be granted credits in excess of the state limit to finance investments involved in production cooperation with other countries. In addition, the classification (for wage fund purposes) was not to be lowered if, due to technical progress, the number of workers had been reduced. Contrariwise, those that overfulfilled the output targets of new goods and technical progress were to be reclassified into higher categories.

The 1971 revised factory prices did not cope with the problem of reflecting quality and novelty. The decree underlined that output of new goods was to be stimulated by prices reflecting the "economic effect of such goods" in production and use. At the initial stage the producer was to profit, but during serial production both the producer and user were to benefit. As soon as new equipment was introduced, prices of obsolete equipment would be cut to discourage production.

Personal incentives would involve periodic reclassification of personnel into higher wage brackets to reflect performance. Innovators were to be rewarded with state prizes, scientific titles, and all the privileges accorded to scientific workers.

5. Each project was to be coordinated by a temporary "complex collective" chosen from the various organs to see the project through from the research to the production cycles. The State Committee for Science and Technical Progress (SCSTP), and in some cases the CM, was to be "the supreme planning, regulating, coordinating, and controlling organ."

As a background to the Seventh FYP (1976-80) and the general development lines to 1990, the leadership reaffirmed its insistence on technical progress as the prime mover and deplored the past lack of success. The tenor of the 1973 decree was even more forcefully reiterated, especially as regards qualitative improvements of scientific cadres and their output.[161]

These are only the latest of a number of attempts at injecting technical progress on a large scale (their predecessors had failed dismally); success for these is by no means assured. Aside from failing to provide the necessary incentives to elicit technical progress initiative from below, there are built-in weaknesses and imponderables, such as the interpretation of certain nebulous passages and the force of conflicting desiderata. Although the decree no longer specifies that Soviet experience is to be imitated (it speaks rather of world standards), there is a very real threat that in practice it will be interpreted to mean Soviet experience, if for no other reason that that of perpetuation. Moreover, with a taut plan, when impending shortages of materials or foreign currency threaten current plan fulfillment, will not the reserves earmarked for technical progress be reshuffled for current production "to save the plan?" Similarly, when an entire association is bound to fulfill certain taut targets, will not the various reduced targets and privileges of units introducing technical progress be reversed? In the same vein, when the costs of the limited investments are considerably exceeded on the national scale, as they usually are, will the c.p. still allow for technical progress undertakings above the limit? If the relative premiums granted for current undertakings are stronger than those for technical progress, should a conflict arise, it is likely that the former will predominate.

CENTRAL PLANNING

Computerization

The stress on the computerization of the economy and on AMS began in the early 1970s. Having abandoned market-type reforms, the leadership still had to grapple with the exigencies that made them turn to these reforms in the first place. It had to find not only short-term solutions, which it expected from increasing the number of direct orders and submerging enterprises in associations, but also long-term solutions in which the managerial apparatus and the population as a whole could believe. Such a long-term solution could be sought in computerization. In a sense the computerization drive should be viewed not merely as a technocratic approach but also a "need to supply a religion," to divert attention from current problems and the "market heresy," and to promise a bright future. Computerization seems to be favored because it permits a high degree of centralization and control. The memory capacity, the operational speed of the computer, the enhanced expediency of central decision making, and the possibility of elaborating several plan variants are all stressed in the official proclamations about AMS. In the USSR considerable philosophical differences divide the various approaches to the use of mathematical techniques in planning. Roughly, such application could entail a greater decentralization and subordination to the central plan via parameters (indirect centralism) or lead to centralization of decision making with highly automated control systems (embracing current operative decisions).[162] While avoiding the extremes of ultracentralization, the official approach in Bulgaria gravitates to the second philosophy.

It might not be too farfetched to suggest that the USSR is using Bulgaria as a willing guinea pig to test the feasibility and advisability of a nationwide AMS. Although such an experiment could be conducted in one of the Soviet republics, it would not be so "pure" as in Bulgaria, for the former are closely interlinked by a variety of ties, and the distortions introduced by the nonexperimenting republics could be sufficiently powerful to undermine the validity of the experiment. Bulgaria, on the other hand, is a small country with a less complex economy that, although fairly closely dependent on the USSR, is not directly subordinated to Gosplan, and such a system could be tried there with relative ease. By Bulgarian accounts, the process of establishing the computer centers is considerably assisted by Soviet hardware, software, know-how, and training of specialists.

The process of computerization in Bulgaria was visualized as concentrating on large computer centers, staffed with teams of mathematicians, designers, programmers, and operators. It was foreseen that within the 1971-75 plan the network of computer centers should be established.[163] Apparently, such centers were already built in the large district capitals, to be integrated with the center in 1976-80.[164] The information flow from bottom to top is to provide detailed data for elaborating state plans. Statistical information on plan fulfillment will be collected through this channel. The two-way channel should provide "daily" information on plan fulfillment and changes in various indexes, both of which could easily be spread around the various agencies. The unified system is being built in the form of a pyramid, following the structure of the administrative apparatus. The base of the structure will be formed by enterprises and other units, which will feed primary information to the regional centers. A part of the data will be provided to the local authorities to enable them to direct and control the units. Another part—the most important indicators—will travel all the way to the peak of the pyramid: the State Information Administration.

There is marked pressure from above and considerable resources allocated for computerization of the managerial and production process. During 1971-75 the number of electronic computers in Bulgaria almost doubled. However, they were relatively underutilized: The data-processing time in relation to total service time reached only about 50 percent. Managers were not particularly interested and had little training in computer use.[165] Some of the most obvious problems included lack of trained personnel and of unity in design and implementation of the program.[166] Of course, one of the more serious problems is lack of modern hardware and software.[167] AMS have been introduced in a number of heavy-industry enterprises. Their development is at a more advanced stage at the Burgas Petrochemical Plant and the Chemical Combines at Vratsa and Dimitrovgrad, where so far the following shortcomings were noted:

1. The potential of computers is either underestimated or overestimated, so that either "magical" solutions are expected or equipment lies idle.
2. Untrained personnel and "old-style" work methods predominate.

3. Installations and programs are ill-synchronized, endangering the unified systems within branches and on an economywide scale.
4. The redesign of the information system is lagging.
5. The equipment's idleness is often due to a lack of preparation and execution of installation work; unimportant questions are being solved due to a lack of methods and programs; enterprises try to secure computers without having sufficient use for them.
6. Many organizations consider computerization as a burden imposed from above; they execute the installation unwillingly. In other cases they undertake installation energetically without paying much heed to efficient use.[168]

On the whole, considerable difficulties were encountered in introducing AMS, and in many cases the results were disappointing. It was admitted that the implementation of the idea proved to be a much more complicated and difficult undertaking than many enthusiasts had initially believed.[169] Many technocrats tend to treat automated systems "as a goal in itself, a kind of fashionable undertaking affording an opportunity to show off technical progress."[170]

Management is usually inexperienced and unprepared in computer technology. It can participate neither in designing the system nor in presenting coherently the problems that the computer is supposed to solve. This lack of involvement, often caused by the knowledge barrier, frequently results in systems that are particularly ill-adapted to the requirements of a given industry or enterprise. Even the specialists in charge of introducing the system often feel unqualified and are learning by doing. Furthermore, their cooperation with management is tenuous, since they feel that the managers lack the basic knowledge to understand the system. Under such conditions the managers who have no grasp of the computer system installed fall into the unenviable position of relying entirely on their subordinates and have no means for controlling them. Therein lies the danger of isolation of the erstwhile manager from the management process shaped by the computerized system: The potential loss of power is a real threat to the manager.[171] Hence, although he will execute the order received, he will do it neither willingly nor diligently, and he will certainly not take the initiative. This is a serious stumbling block with which the designers of the AMS have to cope.

Another and far more serious obstacle to computerization, as it pervades the entire system, is the entire complex of shortcomings of the information system. No matter how modern and technically effective the hardware installed, how well thought-out and adapted the software, how well-trained the technicians, and how well integrated the entire system, the computerization can only be as effective and the decisions as accurate and to the point as the information

CONTINUITY AND CHANGE

fed into the system. In view of the all-pervasive disinformation, there is serious doubt that the computerization, at least within the existing system, will palpably improve the efficiency of the Bulgarian economy.

Plan Construction

Concurrently with the attempts to lift the Bulgarian economy by means of infusion of technical progress and streamlining data processing and decision making, continuing changes are being made in the planning system to make it more "scientific" (to use a favorite term of East European planners). There is increasing and repetitive stress on the advantages of "more not less" centralized planning, with one of its main functions the promotion of technical progress. But the stress is on organizational and technical aspects of planning, while the multidimensional aspects of interactions between levels and self-interests of economic actors are avoided.

In preparation for the 1971-75 FYP, joint sessions of the CC BCP, its secretariat, and the CM, under the chairmanship of Zhivkov, supposedly reviewed economic models and sectoral concepts: "The formulations and instructions ensuing from these discussions were the basic point of departure in the drafting of the unified plan."[172] Apparently, "attempts were made to use modern mathematical techniques and computers in formulating the economic models and some of the concepts. This enabled us to provide a better substantiation for the models."[173]* Soviet specialists from the Main Computer Center of Gosplan apparently helped in elaborating a macroeconomic model, a dynamic intersectoral balance, and some branch models for electric power and building materials for 1971-75.[175] It is claimed that "all this enabled us to perceive more fully the various links and to establish more accurately the proportions within the national economy." An effort was apparently made to "reduce subjectivism in planning" and to take cognizance of realistic objective conditions, particularly

---

*One of the novelties involved in drafting the 1971-75 FYP was the introduction of certain guiding indexes, such as "concepts of development," elaborated by various branches and forecasts of development of various branches and subbranches prepared by research centers and departments. In fact, it was specified that the anticipated plan fulfillment for the current year, 1970, was no longer the starting point for drafting the plan. Rather, the Sixth FYP was to be based on the new trends unearthed by forecasts and various concepts of development.[174]

in norm setting: "Thus the formulated Sixth FYP is a realistic plan."[176]

But in practice the essence of the system is still a bargaining game played with incomplete and distorted information. There are successive revisions of aspirations as new developments and specific constraints are recognized. The participants are trying to outguess each other and their actions are guided, at least in part, by the aim of influencing the actions of other actors. Thus, successive correctives are issued. Planning as practiced increases rather than reduces the uncertainty components. This is one of the adverse consequences of a taut plan. Under the circumstances the stress on prediction of long-term developments is likely to have limited impact. A taut plan is invariably subject to shift in policy. The c.p. would rather not commit himself in any concrete sense to a development path (clearly, political factors enter here, too). His real time-horizon is limited, without any meaningful perspective plan. This is not to say that the better techniques and forecasting methods are sterile exercises but merely to suggest that the scope of improvements is limited.

The Ministries

The strengthening of associations has created a considerable vacuum in the functioning of the ministries and left them rather powerless in operative planning and administration—at least on paper, that is. It is not entirely clear to what extent the present ministries tend to interfere in practice in the operative functions of their associations. The official statements vest the ministries with long-range development planning and decision making for their branches. In the past this usually has been an unimportant function, and despite the rhetoric, it seems to have remained that.

In the early 1970s one might have presumed that this could be a first step toward the withering away of ministries, at least as we know them from the traditional system. By 1969 there were four industrial ministries: Machine Building, Chemical and Metallurgy (also including the hitherto independent Central Geological Bureau), Fuels and Energy, and Light Industry. Food industry was incorporated in the Ministry of Agriculture. However, by mid-1973 the ministries showed a tendency to multiply—a sure sign that the economy was doing badly. The Ministry of Construction was established. The former Ministry of Heavy Industry was split into the Ministry of Chemistry and the Ministry of Metallurgy, and the former Machine-Building Ministry was divided into the Ministry of Machine-Building and Metal Processing and the Ministry of Electronics and Electrotechnology. The establishment of three new industrial ministries might be an indication that the enlarged and more powerful associations are not doing as well as was expected. These new ministries are also dealing with particular trouble spots in recent development and with fields in which Bulgaria hopes to specialize within the CMEA, so in Bulgaria, thus far, there seems to be no grounds for expecting a general fading away of ministries and their supervising and interfering functions.

Complexes

All these "improvements" in central planning failed to meet expectations: "The existing forms and methods of central planning are still inconsistent with the new requirements. The difficulties which arose in the implementation of the Sixth FYP prove this."[177] Despite some experience in formulating economic-mathematical models and forecasting, "the basic shortcoming in planning is the fact that in the solutions of all problems we are applying almost exclusively the sectoral approach to determine the growth rates in individual sectors. The program-target, the complex approach has still not been used. This is having adverse effects on all indicators of economic development."[178] The accent is on viewing the economy as an interdependent and dynamic system consisting of a number of subsystems or complexes, and on integral planning, and rightly berating the narrow monosectoral, or branch, approach as hindering restructuring of development and leading to plan dislocations and imbalances. The system should be improved by concentrating on the "complex" as the basic structural unit in the organization of economic processes, while exalting the center's commanding post and control over the economic process.*

Throughout Bulgarian literature Zhivkov is echoed in his pronouncement that "the main approach to planning must be the target-programme approach," meaning in practice that the country's development plan should not be based on the plans of individual branches, but on the programs for development of the individual complexes or subsystems with Western-type "management by objective."**

---

*The December 1972 plenum was reverently referred to for developing the "original idea of establishing national complexes," which entail "profound changes in organization, structures, and management techniques. It would be hardly an exaggeration to say that we are facing a new revolution in theory and practice of socialist management, whose consequences would be difficult to predict at this point."[179]

**The complex programs "are qualitatively different from existing plans. This distinction must be sought mainly in the relatively better diagnosis of possibilities and richer information, compared

The real problem does not seem to lie in an abundance of loose thinking. Nor is it confined to giving practical expression to the fairly vague notions. One wonders whether the stress on complexes is mainly dictated by the "logic of planning." In this instance control may play a crucial role, together with purges to strengthen the center's grip. The c.p. finds it more expedient to oversee the activity and force the compliance of a smaller number of conglomerates. Moreover, it is also easier to overrule the periphery's preferences by the plausible and expedient argument that its proposals do not fit the complex program.

Predictably the design and introduction of complexes has proved to be a far more "complex task" than many of the enthusiasts had originally envisaged. The basic organizational and management structure remains unresolved. Apparently, the "eternal conflict between functionalists and sectoralists" has come to the forefront.[181] Understandably, "a management system whose simplicity is manifested above all in the elimination of all unnecessary elements, horizontal and vertical, and the development only of components needed for the normal functioning of the system" runs counter to the vested interests of the existing bureaucracy.[182] Probably, the new organizational form that will finally emerge will incorporate a good part of the mammoth bureaucratic apparatus, and the old ways of doing things will reassert themselves. Integral economic planning is probably incomprehensible to the traditional planners.

The reshuffling of the existing organizational structure is by no means the only problem. There are also the questions of the center's preservation of essential controls over the system and of checking the monopolistic behavior of large components. There is also likely to be interference with the distribution of targets and norms by the management of the complex. A system of checks has to be designed to curb the potential power of the top management of complexes. Understandably, Zhivkov spoke of the need to strengthen central planning over the complexes, enforcing limits and norms for the basic resources at their disposal (singling out investments), and directing their activity by imposing "more important physical and value directives for the production of final products." Within that scope the management of the complex is to be empowered to "use resources so that they may be effectively utilized."[183]

---

with the long-term plans; the complex integrated nature of activities and measures within the stipulated period; the organic link among all economic-technical, production, organizational, scientific technical, and social measures involved in the solution of a specific problem; their more concrete nature; and so on."[180]

CONTINUITY AND CHANGE   203

Whatever the political underpinnings, the interesting development is the focus on rationalization of the existing system and on overcoming some of its most intolerable inefficiencies by experimenting largely with changes of an organizational nature and nonmarket-type reforms, while ensuring the center's retention of tight controls over the system. Naturally, there is an obvious need to overcome or reduce the familiar problem of "departmentalization and fragmentation"—the almost impenetrable division of bureaucrats in charge of economic processes who pursue empire-building aims. This requires, among other things, a complete overhaul of the vertical planning system, change of personnel, and elimination of the socialist patronage system. To overcome the narrow approach to planning one needs more than organizational changes (while stressing one aspect of this, they do so at the expense of others).

Postulates for Future Development

This is not the place to discuss the detailed targets of the 1976-80 FYP. Since the c.p. tends to plan more realistically after a period of poor performance, one might expect the Seventh FYP to be less strenuous. As a background to the Seventh FYP a set of Theses on the development of the economy up to 1990 was published. Despite their general vagueness, the Theses upheld many of the aims announced during the early 1970s. They reiterated the stress on raising living standards (see Chapter 8). The chief desiderata were the utilization of all resources more effectively, the promoting of intensive growth (increasing labor productivity by 200 to 250 percent from 1975 to 1990), the reduction of raw material and power inputs per unit of output, the promoting of more "harmonious" economic development, the implementation of technical-progress achievements, the giving of priority to reconstruction and modernization in investment undertakings (with at least 55 percent of outlays earmarked for these purposes), the improvement of skills and the reduction of employment in administration and agriculture, the furthering of integration with the USSR and the CMEA, and so on. The Theses also subscribed to the further reduction of the gap between growth rates of producer and consumer goods. The subbranches of the machine-building and chemical industry, in which Bulgaria specializes within the CMEA, were emphasized. The Theses also pointed to the lagging of primary industries behind the higher stages of processing by stressing that production of the electric-power branch should exceed that of the other industrial branches and that output of rolled ferrous metals and pipes should be ahead of metal-using branches and construction.[184]

In his report to the Eleventh BCP Congress, Zhivkov pointed out that the economy's efficiency could be raised by concentration, specialization, and modernization of production; efficient use of inputs; improvements in output quality; and raising skills. In general, the tenor was that the growing technological sophistication would require greater centralization.[185] Yet another in the series of administrative reorganizations seems to be in the offing. It was pompously announced as a "new economic mechanism" to increase rights and responsibilities of economic complexes (the basic production units) and ministries.[186] The role of associations was considerably underplayed in the new regulations on economic organizations, which exalted the complexes (to be managed by ministries) but did not give them clear-cut shapes. An interesting innovation was the definition of combines, composed of production units tied in with each other on the basis of complex processing of raw materials in consecutive stages.[187]

The establishment of such combines on an industrywide basis was to take place in 1975 in the chemical industry. The industry was to be organized into 11 autonomous combines (and one enterprise in charge of oil and gas extraction and exploitation of USSR-Bulgarian gas pipeline) under the direct aegis of the ministry. In this scheme the associations were bypassed entirely, some of them to be dissolved, and three to be "temporarily preserved." The reorganization was apparently prompted by the need for concentration. The new combines were to be formed around a basic plant, with an eye to localization. Most current production decisions were to be vested with the combine, while investment, specialization, and foreign trade decisions would rest with the ministry.[188] All this seems to presage an end of the association's rule, but it is only one more step in the endless succession of reorganizations that have been "particularly painful" in the past.[189] If pushed to the logical conclusion of integration along the lines of vertical concerns, this might offer a more viable model of organization for devising a more realistic mechanism of functioning within existing constraints of an economy without slack.

NOTES

1. D. Fratev, Stat. 4 (1972): 72.
2. BTA, September 28, 1966.
3. Statistical Pocketbook 1971 (Sofia, 1971), p. 140.
4. IZ, August 14, 1974, p. 2.
5. Compare United Nations, Economic Survey of Europe in 1970, Part 2 (New York, 1971), p. 117.
6. IZ, August 14, 1974, p. 2.
7. S. Dulbokov, PS 1 (1972): 7.

8. Compare I. Iliev, NA 24 (1971): 23.
9. Ibid., 4.
10. Dulbokov, PS, p. 6.
11. I. Iliev, NA, p. 3; RD, December 17, 1971, p. 2; RD, November 26, 1966.
12. I. Iliev, NA, p. 3. Compare Dulbokov, PS, p. 12.
13. I. Iliev, RD, October 30, 1974, p. 2.
14. Dulbokov, PS, p. 9.
15. RD, December 17, 1971, p. 2.
16. Dulbokov, PS, p. 9.
17. M. Dakov, PS 9 (1972): 107.
18. RD, December 17, 1971, p. 2.
19. I. Ivanov, VT 10 (1972): 2-6.
20. I. Iliev, NA, p. 23.
21. RD, February 2, 1972, p. 2.
22. N. Zhishev, IZ, February 14, 1973, p. 1. Compare S. Todorov, Stroitelstvo 3 (1973): 1.
23. S. Dulbokov, RD, December 19, 1972, p. 2; I. Takchiev, PS 1 (1973): 13.
24. Compare D. Popov, FK 1 (1973): 8.
25. M. Golebiowski and B. Zielinska, GP 4 (1974): 227.
26. RD, February 12, 1974.
27. SI 12 (1974): v-vii.
28. I. Iliev, RD, p. 2.
29. Dulbokov, PS, p. 13; RD, December 17, 1971, p. 2.
30. RD, February 2, 1972, p. 2.
31. RD, January 30, 1973.
32. Compare TD, February 16, 1971, p. 1; IZ, February 13, 1969, p. 1; E. Sibinov, IZ, December 26, 1968, p. 1.
33. D. Spasova and N. Zagorsky, KS, March 14, 1972, pp. 1-2.
34. S. Stamenov, Stroitel, March 28, 1973, p. 1.
35. RD, March 28, 1973.
36. S. Kalarov, ZZ, October 12, 1972, pp. 1-2.
37. Stamenov, Stroitel, p. 2.
38. IZ, October 11, 1972.
39. Stroitelstvo 3 (1973): 1.
40. Stamenov, Stroitel, pp. 1-2; Ts. Ivanova, IZ, September 26, 1973, p. 10.
41. Stamenov, Stroitel, p. 2.
42. IM 9 (1974): 4.
43. Todorov, Stroitelstvo, p. 1.
44. Dulbokov, RD, p. 2.
45. D. Popov, RD, December 19, 1972, p. 3.
46. RD, January 31, 1974.
47. KS, July 27, 1973, p. 2; RD, January 31, 1974.

48. RD, February 1, 1975.
49. United Nations, Economic Survey of Europe in 1975, Part 1, Chapter 2, prepublication text, ECE (XXXI)/Add. I, p. 150.
50. I. Iliev, RD, p. 2; RD, December 17, 1971, p. 1.
51. Stroitel, July 23, 1975; D. Tsvetkov, IZ, May 7, 1975, p. 11; S. Veleva, Stroitel, August 27, 1975.
52. D. Asenov, IZ, July 23, 1975, p. 4; T. Tsolov, OF, January 23, 1975.
53. M. Ivanova, Trud, October 2, 1972.
54. Statistical Pocketbook 1970, p. 140; A. Szabo, Koz. Sz. 10 (1974): 1184.
55. RD, February 1, 1975, p. 2.
56. I. Iliev, RD, p. 2.
57. I. Iliev, NA, p. 24.
58. D. Dimitrov, RD, August 14, 1973.
59. RD, May 29, 1974.
60. Trud, August 17, 1974.
61. I. Vasilev, NA 18 (1970): 17–18.
62. I. Iliev, NA, p. 17.
63. K. Kostov, NA 6 (1970): 14–15.
64. Ibid., p. 16.
65. G. Atanasova, NA 16 (1968): 22.
66. Kostov, NA, p. 18.
67. IM 9 (1974): 48.
68. V. Zanchev, IZ, August 27, 1975, p. 1.
69. B. Belchev, IZ, June 20, 1973, p. 1.
70. R. Yanakiev, ZZ 10 (1973): 3. Compare I. Dimitrov, PZ 17 (1970): 54.
71. S. Petrov, IZ, June 4, 1975, p. 3.
72. A. Buchvarov, IZ, April 30, 1975, p. 12.
73. G. Grigorov, Trud, May 29, 1975.
74. Belchev, IZ, p. 1.
75. T. Zhivkov, RD, December 24, 1971.
76. E. Chervenyakov, Stroitel, February 11, 1970, p. 11.
77. RD, January 29, 1971, p. 2.
78. For statistical illustration of performance in 1970, see ibid., p. 2. The reported rates of increase of labor productivity vary sharply in different branches.
79. IZ, October 3, 1973, p. 1.
80. T. Zhivkov, RD, March 21, 1974.
81. I. Karabozhikova, IZ, May 24, 1972.
82. V. Marinov, NV 8 (1975): 28.
83. Compare T. Zhivkov, RD, March 21, 1974.
84. NArm, May 1, 1974.
85. Interview with A. Dimitrov in Pogled, January 10, 1972, pp. 1–2.

86. RD, March 3, 1975.
87. Stat. 6 (1971): 15-21.
88. See G. R. Feiwel, SSt, July 1974, pp. 344-62.
89. For statistical data on the labor-participation rate of women, see T. Danev, PS 7 (1969): 24.
90. Y. Bozhilov, IZ, April 16, 1975, p. 1.
91. K. Koparanov, Trud, March 24, 1975.
92. RD, December 2, 1972.
93. Compare T. Zhivkov, RD, March 21, 1974.
94. For statistics on above-plan employment, see I. Dimitrov, PZ 17 (1970): 55.
95. A. Dobrev, PT 1 (1970): p. 78.
96. G. Iliev, BP 1 (1970): 5.
97. Vasilev, NA 18, p. 19.
98. G. Iliev, BP, pp. 5-6.
99. Ts. Ivanova, IZ, p. 10.
100. I. Popov, Trud, October 17, 1971.
101. G. Iliev, BP, p. 6.
102. D. Kosev, BP 2 (1975): 11.
103. S. Filev and S. Radev, IZ, June 4, 1975, p. 11.
104. Compare K. J. Arrow, RES, June 1962, pp. 155-73.
105. Compare M. Kalecki, Z zagadnien gospodarczo-spolecznych Polski Ludowej (Warsaw, 1964).
106. K. Dimov, PZ 6 (1975): 29.
107. Zh. Zhivkov, IM 2 (1970): 4.
108. I. Dimitrov, PZ, p. 54.
109. Z. Zakhariv, RD, November 27, 1975.
110. Interview with I. Iliev in Rudnichar, February 26, 1970, p. 1.
111. I. Iliev, NA, p. 13.
112. I. Donkev and Ts. Petrov, PS 7 (1968): 10. Compare DV 91 (November 22, 1968): 15-16.
113. Interview with A. Pashkov in IZ, May 27, 1970, pp. 1-3.
114. Ibid., p. 1.
115. N. Zhishev, IZ, February 14, 1973.
116. Ibid., p. 1.
117. Zhishev, IZ, February 2, 1972.
118. Compare G. R. Feiwel, The Soviet Quest for Economic Efficiency (New York, 1972), pp. 403-11.
119. Zhishev, IZ, February 14, 1973, p. 2.
120. Ibid.
121. Ibid.
122. Dakov, PS, p. 8.
123. RD, December 17, 1971, p. 2.
124. M. Golebiowski and B. Zielinska, GP 4 (1975): 270, 276.
125. Dakov, PS, p. 9.

126. Ivanov, <u>VT</u> 10, pp. 2-6.

127. F. Holzman, <u>Foreign Trade under Central Planning</u> (Cambridge, Mass., 1974), p. 21. The controversy was discussed in M. Kaser, <u>Comecon</u> (London, 1967), pp. 182-83; F. Pryor, <u>The Communist Foreign Trade System</u> (Cambridge, Mass., 1963), pp. 140-43.

128. Holzman, <u>Foreign Trade</u>, pp. 283-84.

129. Pryor, <u>Communist Foreign Trade System</u>, p. 146.

130. P. J. D. Wiles, <u>Communist International Economics</u> (New York, 1969), p. 399.

131. U.S., Congress, Joint Economic Committee, <u>Dimensions of Soviet Power</u> (Washington, D.C., 1962), p. 474. Compare Kaser, <u>Comecon</u>, pp. 105-06; Wiles, <u>Communist International Economics</u>, p. 399. Some of the literature on special price concessions granted to Bulgaria is cited in P. Marer, <u>Postwar Pricing and Price Patterns in Socialist Foreign Trade</u> (1946-71) (Bloomington, Ind., 1972), p. 14. For a discussion and data on foreign trade price patterns and estimates of terms of trade for individual East European countries, see ibid. For a brief summary of the extensive literature on foreign trade pricing, see Holzman, <u>Foreign Trade</u>, pp. 11-16.

132. S. Dulbokov, <u>Pravda</u> (Moscow), November 2, 1971.

133. Dulbrokov, <u>PS</u>, pp. 7, 15.

134. Marinov, <u>NV</u>, p. 32.

135. I. Popov, <u>Trud</u>.

136. Ivanov, <u>VT</u> 10, p. 3.

137. A. Sokolov, <u>NArm.</u>, January 17, 1974, p. 3.

138. <u>RD</u>, November 24, 1972.

139. V. Kalchev, <u>PS</u> 5 (1973): 13.

140. Compare M. Petrov, <u>RD</u>, January 10, 1974, pp. 4-5; Sokolov, <u>NArm.</u>, p. 3; K. Kolev, <u>NV</u> 12 (1973): 23-24.

141. Ivanov, <u>VT</u> 10, p. 3.

142. Kalchev, <u>PS</u>, p. 13.

143. M. Petrov, <u>RD</u>, January 10, 1974, p. 4.

144. Compare J. Szeliga, <u>Polityka</u> (Warsaw), February 22, 1975.

145. I. Donkov, <u>PS</u> 1 (1972): 23-27.

146. Compare <u>RD</u>, May 2, 1974.

147. Kalchev, <u>PS</u>, p. 14.

148. Ibid. Compare A. Zubkov, <u>VE</u> 9 (1972): 76.

149. S. Todorov, <u>RD</u>, December 5, 1975.

150. I. Ivanov, <u>VT</u> 7 (1973): 5.

151. United Nations, <u>Economic Survey . . . 1975</u>, p. 77.

152. I. Foldes, <u>Nepszabadsag</u> (Budapest), February 23, 1975; <u>Nepszabadsag</u> (Budapest), January 14, 1976.

153. Szeliga, <u>Polityka</u>; Foldes, <u>Nepszabadsag</u>.

154. I. Ivanov, <u>VT</u> 6 (1973): 2-10.

155. <u>IZ</u>, October 10, 1973, p. 3.

156. Problemy teorii gospodarki socjalistycznej (Warsaw, 1970), p. 116.
157. For a discussion and quantitative assessment of the components of technical change in various countries and an interesting analysis of Soviet growth determinants, see S. Gomulka, Inventive Activity, Diffusion and the Stages of Economic Growth (Arhus, 1971).
158. RD, December 14, 1972, p. 5.
159. Trud, August 1, 1973, p. 1.
160. S. Sharenkov, MO 2 (1975): 27.
161. RD, January 22, 1976.
162. Compare Donkov, PS 1, pp. 197-216.
163. G. Sotvov, TG, February 12, 1972, p. 1.
164. T. Petev, OF, October 23, 1970, p. 1.
165. Buchvarov, IZ, pp. 12-13.
166. Y. Toshkov, IM 2 (1972).
167. For a description of the second-generation computer hardware in use at the Electronic Computer Center of the Ministry of Transportation, see K. Khristov, TD, June 27, 1970, p. 2.
168. P. Kiratsov, NV 10 (1972): 22-26.
169. N. Papazov, RD, February 12, 1974, p. 4. Compare V. Spiridinov, NA 17 (1972): 12-19.
170. Spiridinov, NA, p. 19.
171. I. Mandadziev and E. Atanasova, SK 4 (1968): 40-44.
172. Dulbokov, PS, p. 5.
173. I. Iliev, NA, p. 3.
174. S. Kalinov, PS 5 (1970): 23.
175. Dulbokov, PS, p. 5.
176. I. Iliev, NA, p. 3.
177. T. Zhivkov, RD, December 14, 1972, p. 8.
178. Ibid.
179. N. Stefanov, PZ 17 (1973): 14.
180. Ibid., p. 16.
181. Ibid., p. 19.
182. Compare ibid., p. 21.
183. T. Zhivkov, RD, December 14, 1972, p. 8.
184. RD, January 8, 1976.
185. T. Zhivkov, RD, March 29, 1976.
186. Todorov, RD.
187. DV 100 (December 30, 1975); 101 (December 31, 1975).
188. G. Pankov, IZ, August 13, 1975.
189. D. Fidanov, Anteni, February 21, 1975.

CHAPTER

8

ACCENT ON LIVING STANDARDS?

The economic merit of a system depends on the valuation scale adopted. Here the larger questions of the rationality of the STEs and the nature of ends they pursue will not be raised. The issue in point is how the consumer has fared. The population's disillusionment with the system's sluggish advance of living standards matters, if for no other reason than because it affects the performance of the primary factor of production. In a broader sense, improving living standards is a condition for really successful economic planning. For a plan to be effectively implemented and for economic actors to be stimulated to do their very best, there must be broad popular support for the policies pursued. The consumption-accumulation dilemma has a long history. According to Malthus:

> No considerable and continued increase in wealth could possibly take place without that degree of frugality which occasions capital formation . . . and creates a balance of produce over consumption; but it is quite obvious . . . that the principle of saving, pushed to excess, would destroy the motive to production. . . . If consumption exceeds production, the capital of the country must be diminished, and its wealth must be gradually destroyed from its want of power to produce; if production be in great excess above consumption, the motive to accumulate and produce must cease from the want of will to consume. The two extremes are obvious: and it follows that there must be some intermediate point, though the resources of political economy may not be able to ascertain it, where taking into consideration both the power to produce and the will to consume, the encouragement to the increase of wealth is greatest.[1]

The problems of increasing living standards and the costs of doing so—so conveniently neglected for many years in Bulgaria, as in other STEs—have been surfacing with growing intensity since the early 1960s, and they asserted themselves particularly in the 1970s. These problems were propelled to the forefront for a number of reasons. The population became increasingly restless as it was eagerly awaiting the fruits of industrialization. The greater flow of information and cultural exchange between East and West focused on the low domestic living standards and whetted the population's appetite for more and better consumer goods and services. The regime relied increasingly on the technocrats and had to recompense them with the desired amenities. The growing exhaustion of the sources of "extensive" growth prompted the c.p. to seek out improvements in productivity—partly through the motivation to work, which could be stimulated by providing the worker not only with higher wages but with more and better consumer goods. The need to improve living standards was one of the reasons for contemplating reform of the system of functioning so that the increased efficiency could be used for enhancing consumption and the greater flexibility of production would favor adaptation of the available supply to the structure of demand. But this is not enough. Just as consistent reform requires a relaxation of planner's tensions, improvement of living standards cannot be accomplished without revision of the main growth policy decisions. To recall, Bulgaria was characterized by one of the highest shares of accumulation in national income (see pp. 29-32). It was recently admitted that in the past "it was theoretically admissible and practically possible and necessary" for wages to lag behind the minimum living standards.[2]

Even in Bulgaria, whose population is relatively conformist and inarticulate, due partly to the stronger police-state tactics, there have been rumblings of discontent with the low living standards. The expression of discontent was not confined to the usual dissatisfaction, absenteeism, slack performance on the job, refusal to buy, and so on; it also assumed politically explosive forms. For example, early in 1971 workers of the Nayden Kirov factory in Ruse staged a protest, and the speeches at the district party conference indicated that this was not an isolated case.

The response of the leadership to the demands for improving living standards has vacillated considerably in recent times. For example, the pre-Congress (April 1971) slogan of "concern for man" was dropped in favor of "increased labor productivity." But shortly thereafter, at the December 1972 plenum (hereafter referred to as the Plenum) Zhivkov stressed that economic plans should "satisfy the growing material and spiritual needs of the people." The vicissitudes in living-standards policy roughly parallel the fluctuations in growth-rate performance. In general, when an overambitious and unrealistic FYP is set, its beginning years are characterized by a high rate of investment, imbalances, and neglect of the consumer. These are often mitigated by a shift in policy, reducing the rate of investment and giving greater weight to consumption. This is again reversed by a resumption of the investment momentum. The length of the periods and intensity of shifts depend, among other things, on the abruptness of the increase of the investment rate. The most recent shift in favor of living standards (1973) is partly an expression of the intolerable tensions built into the Sixth FYP, which had to be relaxed midstream. Zhivkov was prompt to point out that the program for improvement of living standards, enunciated with great fanfare at the Plenum, should neither be interpreted as a revision of the growth strategy, nor should it be confused with "consumerism," "consumer socialism," or other concepts of the "mass consumption society" or "affluent society."[3]

Fundamental forces are at work to ensure that the system will relapse to a lopsided industrialization rush at the expense of the consumer after the intolerable situation has been partially redressed. These fluctuations seem to proceed along a rising long-term path. The industrialization rush accompanied by lowering of real wages belongs to the past; this is one of the major changes. Barring dramatic shifts, it is not only difficult to compress living standards for any period, but stagnation (or slow rise) would activate forces to overrule the c.p.

CONSUMPTION STRUCTURE

In recent years there has been a marked improvement in living standards in Bulgaria, particularly if measured by "extensive"-type indicators. It is customary to measure the dynamics of the standard of living by the growth of per capita consumption. From 1952 to 1970 this indicator rose by 225 percent, by 80 percent from 1960 to 1970, and by 36 percent from 1965 to 1970.[4] Even if the official claims are exaggerated, the Bulgarians are indeed better off than they were some years ago, and, as shown in Tables 8.1, B.62, B.64, and B.65, they are better off than the populations of some other CMEA countries.[5] Comparative data on per capita consumption are not very revealing. Consumption data in the CMEA tend to be inflated due to the incidence of turnover tax, price-setting vagaries, and the possible inclusion of collective consumption, which does not always represent welfare-oriented activities. In addition, the index would have to be adjusted for the loss of welfare due to poor quality of (domestic and imported) goods, restricted choice, and general shopping frustrations. The system displays some advantage in restricting the modern West-

TABLE 8.1

Per Capita Personal Consumption in CMEA Countries in 1970
(in West German marks)

Country	Consumption
GDR	3,954
Czechoslovakia	3,017
Hungary	2,678
Bulgaria	2,479
USSR	2,030
Poland	1,924
Rumania	1,732

Source: L. Gyorov and V. Tsvetkov, FK 9 (1973): 5.

ern "bamboozling" of the consumer through spurious and unnecessary product differentiation and "brainwashing" through advertising. Yet the hardships inflicted on the CMEA consumer due to producers' unresponsiveness to consumers' demand are an indictment against the system.

The rates of improvement of living standards are lower than those that could have been achieved had development policy consistently respected improvement of consumption, had planner's tensions been removed and a buyer's market instituted, had the distribution system been streamlined, and generally had a more efficient functioning system, responsive to consumers' wants, been introduced. Moreover, the Bulgarian statistics tend to stress the fairly rapid dynamics of consumption; only occasionally are the structural disproportions within the aggregate pointed out. It is not only the key ex ante macroeconomic allocation and mode of planner's behavior during plan implementation that matter. The system "bamboozled" the consumer in the microcomposition of the total, and this is not dictated by the growth strategy but is a result of the traditional functioning system, and in final analysis it encroaches on the growth momentum.

Improvement in living standards can be evaluated in comparison with the past, with other countries, and with shifts within the structure of consumption. There has been increasing justified dissatisfaction with the first two criteria: "So far we rated our successes in enhancing living standards by drawing comparisons with the past, when our population lived—in the full sense of the term—at poverty level, or by comparing the situation with that of neighbouring backward countries."[6] To measure relative standards of living it is insufficient to compare flow of current consumers' outlays. Stocks of consumers durables are an important yardstick. One could argue that housing and automobiles are probably the main determinants of welfare improvement, as evaluated by the consumer. The changes in the structure of consumption, illustrated in Tables 8.2 and 8.3, indicate little room for complacency, for there was no pronounced movement away from the consumption of foodstuffs to that of durables and services. In the late 1960s and early 1970s the share of foodstuffs in total retail sales declined in most CMEA countries. From 1965 to 1973 this share fell from 47.3 to 44 percent in Bulgaria, from 56.7 to 48.9 percent in Czechoslovakia, from 51.6 to 48.2 percent in the GDR, from 49.8 to 47 percent in Hungary, and from 57.7 to 54.4 percent in the USSR. (In Rumania it rose from 48.1 to 50.1 percent.) However, expenditures on food remain high in the CMEA, so that consumers spend 40 to 50 percent of their income on food; 13 to 19 percent on clothing; 5 to 9 percent on rent, heat and electricity; 5 to 10 percent on household furnishings; 5 to 9 percent on transportation; and 10 to 12 percent on culture and recreation.[7] Moreover, in Bulgaria, within the structure of consumption of foodstuffs from 1956 to 1971, there was only a 7.1 percent drop in consumption of flour; meat consumption increased by 55.6 percent, considerably outdistanced by the increases in consumption of sugar, eggs, and fruit.[8] By 1973, within the CMEA, Bulgaria's was among the lowest consumptions of meat, only above that of the USSR and probably that of Rumania (see Table B.68).

Although the data in Table 8.3 are not strictly comparable, they indicate that in stock of durables Bulgaria was somewhat behind other

TABLE 8.2

Changes in the Structure of per Capita Personal Consumption, 1960–70
(in current prices)

Consumption Structure	1960	1965	1970
Foodstuffs	49.7	47.3	43.1
Beverages and tobacco products	11.1	12.4	11.6
Clothing and footwear	16.5	15.8	16.4
Durables and services	13.0	13.4	15.8
Other	9.7	11.1	13.1
Total	100.0	100.0	100.0

Source: T. Kadiyan, Stat. 6 (1972): 68.

TABLE 8.3

Estimated Stocks of Selected Consumer Durables in Some CMEA Countries, 1960-70
(units per 100 households)

Country	Years	Radios	Television Sets	Washing Machines	Refrigerators	Passenger Automobiles
Bulgaria	1964	72	7	13[a]	3	n.a.
	1968	88	35	14[b]	17	6[d]
Czechoslovakia	1960	80	20	50	10	n.a.
	1965	103	50	60	30	n.a.
	1970	150	80	80	50	17
GDR	1960	90[c]	17	6	6	n.a.
	1965	87	49	28	26	8
	1970	92	69	53	56	16
Hungary	1960	72[c]	3	15	1	n.a.
	1965	80[c]	27	37	8	10[f]
	1970	75[c]	53	50	32	21[f]
Poland	1966	82	52	72	16	2[e]
	1970	84	72	79	37	5

n.a.: data not available.
[a] Sales in 1963.
[b] Sales in 1966.
[c] Permits issued.
[d] 1970.
[e] 1965.
[f] Units per thousand population.

Sources: NA 24 (1972): 3-5; Rocznik Statystyczny 1973 (Warsaw, 1973), p. 708; Rozwoj gospodarczy krajow RWPG 1950-1968 (Warsaw, 1969), p. 127; H. Kocianova, PE 3 (1973): 241-52.

CMEA countries. It was estimated that by 1975 there would be 66 washing machines per 100 households in Bulgaria, 84.5 in Czechoslovakia, 65 to 70 in the GDR, and 72 in the USSR. The corresponding figures for refrigerators would be 58 in Bulgaria, 77 in Czechoslovakia, 75 to 80 in the GDR, 62 in Hungary, and 64 in the USSR.[9] The retail sales of durables in Bulgaria in the 1960s were lower than in other CMEA countries, with the exception of Rumania and the USSR in some items (see Table B.69). In Bulgaria the number of privately owned automobiles rose from a few thousand before 1960 to about 200,000 in 1969 and 340,000 in 1974. The sales gravitated about 50,000 annually in 1973-75. About 70,000 cars were expected to be imported in 1976, with the annual figure going up to 90,000 to 100,000 by 1980. The number of people in waiting line for cars has been growing steadily (from 134,000 by September 30, 1973, to 189,000 by March 31, 1975.[10]

What really matters is that rightly or wrongly the consumers tend to underrate the achievements in raising consumption standards and are not overly impressed with the high dynamics of aggregate per capita consumption. In a high-pressure economy even the real attempts of the government to raise living standards have a reduced impact on the consumer, partly because he is forced to substitute goods he can obtain for the unavailable goods he wants—thus the frustration of unfulfilled shopping aspirations. Moreover, distribution is discriminatory. The best or better goods are appropriated by the privileged groups who also get preferential access to services.

SUPPLY OF CONSUMER GOODS

Improvement of consumer-goods supply depends not only on restructuring allocation of investments in favor of the consumer-goods sector and increasing the output of that sector but also on restructuring foreign trade in favor of larger imports of consumer goods and raw materials for the production of such goods; and especially on reforming the functioning system so as to improve the quality, variety, and attractiveness of the available consumer goods. The Plenum offered the first official negation of the absolute priority of producer- over consumer-goods production. In fact, in some years, in order to readjust the production of consumer goods, output of these goods can grow at a faster rate than that of producer goods "as long as this does not become a permanent trend." In more immediate terms this meant a review of the possibilities of increasing output of consumer goods during the Sixth FYP, which did not fully materialize.[11] The expected increased flow was still to be achieved through noninvestment sources of growth, rather than through the creation of more permanent conditions by expanding productive capacity. Zhivkov was explicit on this score. He relied primarily on the discovery of new reserves (counterplans) "for increasing the output of consumer goods through the most effective utilization of the existing capacities, smooth supply of raw and other materials, and the continuous improvement of labour organization and production processes."[12]

Concurrently, something of a revision of policy as regards import of consumer goods was announced. Thus far, import of consumer goods was notoriously undersized, as shown in Table B.73. By 1973 import of consumer goods would constitute 10 percent of total imports, and increase to 15 percent by 1975 and to 20 to 25 percent in 1976-80.[13] But the restrictions on additional imports of raw materials for the production of consumer goods were not relaxed. Such additional imports were only allowed to the extent that they could be paid for by additional exports of goods produced from by-products or waste materials.[14] Developing countries were pointed to as a good potential source of imports of raw materials for light industry and of finished consumer goods.

Export policy was to be reviewed, with particular stress on increasing exports of producer goods and limiting those of consumer goods whenever particular shortages should arise in the domestic market. In the past there had been complaints of shortages of specific foodstuffs, while these same foodstuffs were being exported to other CMEA countries.[15] Zhivkov called for a stop to the practice of preempting the domestic market for exports: "because of the inadequate development of capital goods for export, we deliberately followed a line of greater exports of agricultural and light industry products."[16]

During the latter part of the 1960s the well-known phenomenon of shortages of many consumer goods began to be accompanied by difficulties in selling others.[17] The periodic shortages of some goods and overstocking of others are caused by production and distribution problems and the inability of the producer to keep in touch and respond to the consumer's needs[18]: "The long lines visible in the morning in front of bread stores and dairies [in Sofia] prove convincingly" that trade organizations are unable to cope with the most elementary problems, even when the goods are available.[19] The nonfulfillment of effective wants aggravates the market imbalances. Consumers demand more than they would under nonshortage conditions, because they do not trust the reliability of supply. They buy not only to satisfy current needs but also to build stocks and overcome the uncertainty of supply. Shortages intensify the "excess demand gap" and a cumulative process of frustrations develops.

Production of consumer goods is constrained by all sorts of shortages of high-quality material inputs, skilled labor, and more sophisticated machinery. In addition, in his aim at producing the most convenient "plan-fulfilling" assortments, the manufacturer of consumer goods is neither constrained by an overly detailed (as regards sizes, colors, and models) central specification nor inhibited by demanding trade organizations.[20] Despite the increasing overstocking of some "unsalable" goods, a buyer's market has not yet been created.[21] The supply of durables is particularly deficient, and those that are available are shockingly outdated.[22]

The Plenum strongly castigated all the shortcomings in consumer-goods availability and production, but it was less forceful in proposing methods for overcoming them. Only a glimmer of solutions was provided, without recognizing or specifying the necessary conditions to elicit the desired behavior. For example, stress was laid on intensifying direct relations between associations and the large stores, more consistent studies of consumer demand, and in all branches (including heavy industry) output-plan drafting with an eye to the domestic market.[23] Still, conditions were not created to compel the seller to react and adapt output to consumer demand and to interest him to introduce really new and better products at terms advantageous to the consumer. This, of course, depends mainly on the state of the market, the relative power of the producer-seller and the buyer, on the information system, and on the motivation to learn and act.

Another sore spot is the supply of consumer services, which is both extremely undersized and of very poor quality.[24] According to Zhivkov: "Services are still considered as a secondary activity. . . . In the case of most types of services the necessary modern facilities have not even been created. The personnel is inadequate and its skills low. . . . The problems of material incentives, price-setting, organization, etc. have not been resolved."[25] In Sofia about 84 percent of services are provided by state-owned and collective establishments. The defects in their functioning are many and varied: unnecessary delays, bill padding, poor quality of workmanship, and so on. Many establishments have no price lists, do not record the dates of transactions, substitute materials, and encourage tipping.[26] Despite the attempts to keep the private sector to a minimum, in services "it is actually expanding at the expense of the socialized sector."[27] Many cases were reported of private artisans undertaking large-scale orders (probably in response to some acute shortages) instead of remaining in the sphere of activity for which they were licensed.[28] "Very often people are forced to seek the services of the private artisans because they know from experience that they will complete the repairs faster and better and the customer will be more politely and courteously served."[29]

It appears that the post-Plenum supply of consumer goods and services did not improve noticeably. The well-known complaints of poor quality, unavailability, lack of supplies and consequent idleness of consumer goods plants abounded.[30] For example, in 1974 the plan of retail sales was overfulfilled by 1.1 percent (an increase of 9.7 percent over 1973). Consumer services rose by 12 percent over 1973. But the plan for a large number of "scarce" consumer goods (mainly textiles, glass, fine ceramics, clothing, appliances, bakery goods, beer, and services) was not fulfilled. Maldistribution of goods among regions persisted. The main problems in services were delays in execution and poor workmanship, mainly due to lack of spare parts for appliances, which has been plaguing the service industry for years.[31] One of the shortcomings in consumer-goods planning was the drawing up of contracts on the basis of old assortment lists, without taking into account consumer demand. The product-mix was not widened. In fact, the variety of textiles was narrowed down in 1974. The managers were blamed for trying to recoup investments on modernization faster by resorting to large-scale production of fewer product groups, thus lowering costs and increasing profitability.[32]

Production of new goods was suffering from the usual problems. These were described in a lighter vein by a Bulgarian journalist: Every year the Plovdiv Fair boasts "superb light industry exhibits which will almost never show up in the stores." Many of these "unique samples" are made of imported materials that are to be replaced by similar Bulgarian materials in serial production. These are usually not available. The production cycle has to be reorganized and "we could hardly boast such flexibility." Introduction of new goods is indefinitely postponed so it should not interfere with the plan. By next year "we will increase our efforts, we have possibilities, we will find materials. . . . I know cases of machines exhibited four to five years ago that are still not being produced."[33]

By the end of 1974 the CM issued Decree No. 113 aiming at improving supply of consumer goods. It began by deploring the unsatisfactory quantity, quality, and variety of goods, such as processed foods, furniture, children's goods, metal products, and fabrics; the lack of adaptability of producers to consumers' demand (tenuous links between industry and trade), and the narrow raw-materials and production base for consumer goods. The decree enjoined that the 1976-80 FYP be prepared with an eye to "conditions for accelerated development of sectors producing consumer goods." This was to be accomplished by means of modernization, increased shift work, switching local and cooperative industry to produce mainly consumer goods and encouraging "cottage industry." Starting in 1975 the supply organs and foreign trade organizations were to supply those producing consumer goods in short supply on a priority basis. Consumer-goods producers were to use 50 percent of above-plan profits for investments, even if they should exceed the approved investment ceilings. Additional incentives were to be provided for timely completion of the more important projects for consumer-goods production. In addition, all producers were to set up auxiliary consumer-goods production in so-called consumer-goods shops, to be set up by June 1975.

The planning procedure for consumer goods was somewhat altered. The SPC was to promulgate volume of output and input ceilings for the ministries producing consumer goods by February of the preceding year. Orders would be placed by trade by March 15 (orders for luxury and fashion goods would be placed from three to six months ahead of the start of the plan period). The economic units that failed to supply trade with the variety and quality of products ordered would have to provide foreign exchange to buy those products abroad. In case of changes in demand during the year, the superiors were allowed to change up to 2 percent of the volume of output and to make such alterations as would be indicated in other aspects of the plan. The measures that were aimed at upgrading output quality fell into well-worn categories of governmental regulations and exhortations; an exception was a stipulation that up to 80 percent of very high quality and 50 percent of other new products mark-ups should be used for incentive purposes. Particular stress was laid on import of raw materials and equipment for consumer-goods production and purchase of licenses. As of January 1975 foreign trade personnel's bonuses were made dependent on the plan of import of consumer goods. Domestic supply of consumer goods was to have priority over exports. Over the 1976-80 FYP the share of imported consumer goods for the domestic market was to increase by 20 to 25 percent.[34] Following in the footsteps of this decree the minister of machine building and metallurgy announced that in 1975 the ministry's total output would increase by 13.6 percent, but the output of consumer goods it produced would go up by 150 percent.[35]

HOUSING

The intensified postwar urbanization gave rise to serious shortages of housing in cities, with concurrent underutilization of housing facilities in rural areas. Aside from its qualitative and spatial shortcomings, even quantitatively housing construction in the CMEA falls considerably behind that in Western countries. Within the CMEA in the 1960s and early 1970s, Bulgaria was trailing most other countries (except for the GDR) in this important index of consumer welfare (see Table B.70). As in other STEs, the housing problems in Bulgaria include, among others: (1) insufficient allocation of resources to housing construction at the planning stage, with further inroads during plan execution; poor design of projects, (2) poor quality of execution, and (3) commissioning of semi-finished dwellings. Among the most irksome problems is the delay in construction even after the tenants move in: "A person spends years hoping to obtain housing, and then spends as many years waiting for the complex to be completed."[36] Modest attempts were made to increase floor space. For example, in Sofia during 1966-70, 2.83 percent fewer dwellings were built than in 1960-65, but the floor area built was larger by 4.2 percent. The floor space per dwelling increased from 56.3 square meters in 1960-65 to 60.3 in 1966-70. Whereas in each of the previous quinquenniums fewer than 8,000 dwellings were built in Sofia, it was estimated that in the next 10 to 15 years about 16,000 dwellings would have to be built annually to meet the needs of the population (at a rate of 9 square meters per person).[37]

In the spring of 1973 it was disclosed that the Sixth FYP housing construction program was seriously behind schedule. Apparently, more housing units had to be built in 1974 and 1975 (60,000 and 70,000, respectively) than in 1972 and 1973 to fulfill the plan for 1971-75. A very broad front of 93,000 dwellings was reportedly under construction in 1973, and even with the expected increased effort only 53,000 were scheduled for completion at the end of that year. During 1971-72 the housing program fell 13,000 units behind. In 1972 only 31,365 units were finished, instead of the 42,000 planned.[38] In 1974 the housing plan was again underfulfilled (only 44,500 apartments were built), bringing the total for 1971-75 to 194,000 housing units. The planned average annual construction of dwellings in 1971-75 was 62,000 to 66,000, but the expected fulfillment was only about 52,400. As a result, in 1971-74 Bulgaria's ratio of finished housing units per 1,000 population (22.5) was the lowest in the CMEA (Poland, 26.6; the GDR, 27.5; Rumania, 28.7; Hungary, 32.6; and the USSR, 36.1).[39]

COLLECTIVE CONSUMPTION

In the postwar period the collective consumption fund grew considerably. The sheer quantitative expansion and the scope of the undertaking is remarkable, even if we discount the frequent inclusion of military and other "secret" state expenditures in this fund. However, the distribution of the consumption fund suffers from many misallocations. Table 8.4 indicates the relatively much faster growth of expenditures on social security and recreation than on education and health, especially during the 1960s. Concurrently there are appalling shortages of nurseries and daycare centers.[40] Public health facilities and health care remain unsatisfactory. There are serious shortages of hospitals and sanitariums, with considerable delays in their construction.[41] The existing facilities are not equipped with modern conveniences and are understaffed.[42]

TABLE 8.4

Growth of the Collective Consumption Fund, 1952-71
(1952 = 100)

Allocation	1956	1960	1965	1971
Education	144.1	190.2	186.3	583.3
Culture and arts	128.7	173.3	231.3	629.6
Public health	158.2	229.3	320.0	598.3
Social security and recreation	146.1	386.1	605.4	1,349.8
Collective total	146.1	282.2	534.6	916.2

Sources: Razshireno Sotsialistichesko Vozproizvodstvo v NRB (Sofia, 1972), pp. 104-05; R. Gocheva, IM 2 (1973): 4.

The Plenum reiterated that "in the future the increase in the population's real income will depend even more tangibly on public consumption funds."[43] An interesting aspect of the approach is the stress on "eliminating equalization in the use of collective consumption." The differentiation is to be achieved by according "certain preferential treatment to the best blue- and white-collar workers and cooperative farmers, by making the various social security benefits indirectly related to individual labor performance, etc."[44] Presumably, some workers (probably the technocrats) will be included in the privileged class that previously consisted mainly of the top leadership and the bureaucracy.

WAGE, PRICE, AND TAX POLICY

The postwar period was marked by a tendency for relative leveling of wages, as indicated in Tables 8.5, B.66, and B.67. This is especially apparent on the economywide scale due to the very low salaries of white-collar workers in nonmaterial production. The variation is more pronounced in industry in view of the higher wages of managerial personnel, but it is not remarkable. Furthermore, there seems to be no shift in favor of greater divergencies. The wage-leveling tendencies are perpetuated by periodic wholesale revisions of wages, which tend to stress increases in minimum wages, rather than in those above the minimum level. While the Plenum recognized the urgent questions of wage differentials, the solution to the problem

TABLE 8.5

Nominal Wages of Blue- and White-Collar Workers, 1948-67

	1948	1952	1957	1960	1965	1967
Average annual wages, blue-collar workers (in leva)	483	656	815	960	1,117	1,270
Index (1948 = 100)	100	136	169	197	231	263
Average annual wages, white-collar workers (in leva)	487	633	815	902	1,095	1,308
Index (1948 = 100)	100	130	167	185	225	269
Wages, white-collar workers (blue-collar workers = 100)						
Economywide	101	104	100	94	98	103
Industry	n.a.	n.a.	121	116	119	122

n.a.: data not available.

Source: G. Gochev, NV 8 (1969): 26.

was conveniently postponed until after 1975.* Zhivkov noted that the porposed wage increases "will cause a temporary disruption in the wage ratios of workers and employees. For a while it will result in wage equalization." But this was unavoidable in view of the shortage of funds "for simultaneous solution of all problems connected with labour remuneration."[46]

Following the Plenum, minimum wages were raised on June 1, 1973, from 60 to 65 leva per month to 80 leva (and the wages of those earning between 80 and 87 leva to 88 leva) per month. The previous minimum-wage increase took place in 1970 (from 55 leva to 60 to 65 leva).[47] On March 1, 1973, wage increases ranging from 15 to 26 percent went into effect for workers in particularly difficult conditions (miners, night shift) and for teachers, medical personnel, and employees in cultural institutions.[48] According to the 1971-75 FYP, the minimum wage was to be increased to 70 leva in 1975.[49] This is

---

*However, in presenting the plan for 1975, the chairman of the SPC noted that "salaries of leaders directly involved in production were increased."[45]

an indication that some immediate action was felt to be necessary. It is too early to tell whether this is just a one-shot move or whether it presages a more lasting shift in policy to benefit the population. The annual cost of wage increases was estimated at 220 million leva, with 70 percent of the working population benefiting. The average increases were to be relatively small (about 105 leva per annum). In Bulgaria more than half of the total working population receives wages close to the minimum (before 1973, between 65 and 87 leva per month; and after, between 80 and 88 leva).[50] In 1966-70 the average annual increase in real incomes in Bulgaria was 5.4 percent (higher than in all other CMEA countries). The 1971-75 plan called for a 4.6 to 4.9 percent annual increase. In 1971 real incomes increased by 3 percent; in 1972, by 4 percent; in 1973, by 7.9 percent; in 1974, by 5 percent; and in 1975, by 5.5 percent.[51]

The Plenum also criticized the wage scales and system of wage increases: Wage increases should be related closer to labor productivity, and "the average annual rate of increase of wages must be higher than before,"* partly to spur improvement of skills and qualifications called for by modern technology.[52] The piece-rate system, which still governs about 60 percent of all workers in industry, was blamed for many shortcomings. Starting in 1973, it was supposed to have been gradually replaced by time pay and premiums.[54] But both workers and managers have vested interests in the piece-rate system, for it provides much room for manipulations and increase of wages. The effective resistance from below to shifting from this system is obvious from the lack of success of the periodic attempts (for example, in 1968).

The gap between workers' and farmers' incomes began to close after 1956, when an average farmer earned only 60 percent of the wage of an average worker.[55] Since then, the average annual growth rate of farmers' real incomes has been rising, and by 1970 the farmers in state-owned farms earned 87 percent of the average wage, and those in cooperative farms 99 percent (see Table B.67). Specifically, nominal wages of blue- and white-collar workers increased by a factor of 2.81 from 1948 to 1968, whereas the earnings of cooperative farm workers rose by a factor of 4.5 from 1952 to 1968. The particularly steep increase of the latter occurred after 1956, with the increased stress on agriculture, as an incentive for better work and to retain the rural population on the farm. The ratio of urban to rural earnings was 2.5:1.0 in 1952, and dropped to 1.58:1.00 in 1968. Furthermore, when taking into account the reduced working time on the farms and the income from private plots, an almost even ratio is ap-

---

*New regulations on wages were issued at the end of 1972.[53]

parently achieved.[56] The Plenum promised a further bridging of the "gap" by bringing the farmers into the pension, social insurance, and other welfare schemes; by introducing industrial wages into the agroindustrial complexes; by providing nontaxable minimum incomes for farmers; and so on.

Income taxes affect disposable income, income distribution, and the will to produce. The general income tax scheme had been established more than two decades ago, and it was time to revamp it. This is not the place to treat this question in detail, but some of the changes introduced or contemplated should be mentioned.[57] For various reasons income taxation did not play a significant role in STEs, which have relied largely on the more expedient and "secure" devices of low wages and turnover taxes. At this juncture income taxation has gained prominence in Bulgaria, primarily as a tool for regulating and siphoning off income of the private sector and restricting its development. A related problem is the regulation and reduction of incomes and preventing tax evasion among scientists, artists, and writers, who get highly differentiated incomes from many sources. Thus, varied rates of progression are used depending on the source of income.

Apparently, minimum wages are somewhat below the average necessary cost of maintenance of the worker and his family. Thus, following the increase of the minimum wage, the nontaxable minimum was raised accordingly.[58] Two conflicting objectives guide the maximum tax limits: (1) restriction of personal consumption and accumulation of wealth in real or monetary assets and (2) minimization of the disincentive effect on labor allocation, performance, and improvement of skills. The new tax table discriminates in favor of engineering and technical personnel, thus tending to make up for wage leveling.[59] The method of tax withholding at source tends to favor those holding more than one job. The new regulations stipulate that incomes earned from more than one job be added to the basic income and progressive tax be assessed on the whole.[60] The limited nature of the present tax changes is recognized: "There is a well-known maxim that the less frequent and fewer the changes in taxation, the better they are tolerated by tax-payers and the less confusion there is in practice. This is one of the reasons why no radical qualitative change in the forms and organization of income taxation is proposed at the present stage."[61]

Pensions are modest, if not inadequate. This is a source of increasing problems in view of the increasing share of pensioners in the population. Some attempts were made to improve this situation.

Although the Plenum reiterated the policy of reduction of retail prices of consumer goods, Zhivkov pointed out that "substantial reduction of consumer goods prices" should not be allowed: "In the future the living standards must be raised above all through higher wages."[62] The June 1973 price reductions were aimed at stimulating population growth, and the price increases at discouraging purchases of imported consumer goods, despite the Plenum's stress on larger imports of consumer goods.

Price stability or even price reduction is not necessarily a blessing or an indication of economic intelligence of the policy maker. As is well known, price stability may contribute to misallocation of resources, shortages, forced substitution, costly and protracted periods of adjustment by extramarket forces, and so on. The c.p. assumes that he safeguards consumers' welfare at fixed prices, and he disregards the law of increasing marginal dissatisfaction from increasingly unobtainable consumer goods. The illegal transactions and black market are an indication that consumers are willing to pay higher prices rather than stand in line or go home empty-handed. And, as always, the distributional effects of price policy cannot be ignored. Low prices do not necessarily favor the poorer segments of the population, nor should they be a device for adjusting these segments' standard of living; such prices often favor special, privileged groups (those who have shopping connections or the retail clerks who take bribes for under-the-counter sales). Rather, low living standards should be increased through income adjustment. Maintenance of more or less stable prices entails extensive and substantial price subsidies, with distorting effects on economic calculation and resource allocation. The distributional, political, and psychological effects of the alleged policy of retail price stability strongly circumscribes the scope of any price reform and introduction of meaningful costing, as pointed out in Chapter 5. But even the regime of stable prices cannot entirely avoid surreptitious multiform price increases. The price offenders are not only the firms that are naturally interested in raising prices, but this tendency is often condoned by the c.p. who finds it a convenient way of extracting purchasing power that exceeds the available supply of consumer goods.

Early in 1968 the gradual introduction of the five-day work week (42.5 hours) was decreed. This was supposed to be accomplished in three stages: (1) a test sample of enterprises in different districts to be transferred in the second half of 1968, (2) enterprises and whole branches (including transportation) to be shifted in 1969, and (3) the remaining industrial enterprises and organizations to be transferred in 1970.[63] The five-day week was only introduced experimentally in the Kabrovo and Stara Zagora districts in 1968. Probably one of the main reasons for the nonimplementation was that the expected increased efficiency from the reformed new system did not materialize and, de facto, growth continued to be propelled by increased employment, rather than increased productivity. Following the Plenum, the issue of the five-day week was revived. It was supposed to be introduced gradually in 1973 and 1974 by districts. Early in 1973 only 17 percent of the labor force was on the reduced work week. Administrative and managerial personnel was to come under this scheme only in 1975. The scheme was not to apply in agriculture, health, and education. But Zhivkov stressed that the conversion must not entail an increase of employment and "one of the basic problems is the elimination of overtime."[64] The introduction of the five-day work week was again delayed and was to be completed by May 1, 1975, when all enterprises of material production would be switched over, together with some employees in the service sector. The ministries, departments, and peoples' councils were to be converted at a later date.[65]

ALIENATION AND ECONOMIC CRIMES

The Plenum hinted at indifference of the broad masses, at alienation and general lack of participation: "We cannot deny that vestiges inherent in the already eliminated alienation system remain in the individual mentality of blue- and white-collar workers, and in their behaviour, as well as in the administration organs."[66] Zhivkov did not mention workers' comanagement, but surprisingly stressed the need of information from below to afford the top decision makers some understanding of what the people really want, and what their goals and ambitions are. It is only with this type of information that the "state, party, and economic agencies" can make decisions that would reflect the people's best interests. However, the bow to socialist democracy did not specify how the "voice of the masses" was to be heard, and was accompanied by the usual calls for strengthening labor discipline and control; invigorating one-man management, command, and central decision making; and upholding political qualities and loyalty to the party as primary qualifications for high administrative posts, with professional skills of secondary importance.[67]

Aside from alienation and lack of democracy, the system suffers from an erosion of morality at all levels that can be partly attributed to the high-pressure economy and unrealistic planning. The most obvious manifestation is the widespread practice of falsification of plan-fulfillment reports. The cover-up operation by the c.p. is quite subtle. After a disastrous year or FYP, the report mentions that the targets have been "in the main fulfilled." Data irreconcilable with the published targets are reported. The report is in value terms rather than physical units. Rates of increase for certain items are not reported. Percentages of fulfillment correspond to the scaled-down plan that is not published. And many other well-known techniques are used by the c.p. to camouflage poor performance. At the lower levels falsification is less sophisticated. For example, by

June 1973 the director of the "23 December" combine (in Kotel) noted that the plan for the first half of 1973 would remain unfulfilled. He summoned the heads of shops, the finished-goods warehouse, and the control department, and commanded that fictitious receipts of goods produced and received by the warehouse be drawn up. During the same period the factory for nonalcoholic beverages in Sliven was 700,000 bottles short of plan. The heads of production and warehousing drew up falsified receipts, and the enterprise director got the director of the trade unit to forge a receipt for 350,000 bottles.[68]

Criminal negligence is also quite a common phenomenon.[69] In 1974 investigations of economic crimes and negligence revealed that the control organs were relatively lenient and weak. Financial-auditing bodies uncovered such infractions as the damaging of materials, losses of crops, high mortality of livestock, and waste or theft of raw materials, but they did not pinpoint responsibility. Only a minimal number of cases, where financial-audit bodies have uncovered damages, are persecuted.[70]

Following the Plenum and concurrently with the purges mentioned in Chapter 7, increasing concern over economic crimes and "several large encroachments on socialist property" during the past few years have been revealed. Workers and managers in different enterprises and organizations were involved.[71] In discussing this state of affairs, the press listed as the main causes slack control,[72] general apathy, lack of criticism, and condonement of unprincipled behavior.[73]* However, the real causes are much deeper.

Economic crimes flourish particularly during periods of intense industrialization. First, because the basic task is production-plan fulfillment, the problems of costs and quality are of secondary importance; hence, control in these areas is usually neglected. Furthermore, owing to the presence of a seller's market, the buyers are not overly demanding; consequently, any misuse of materials that lowers output quality is not revealed at the time of sale. Second, owing to the seller's market, too, the disposal of stolen materials at high prices is facilitated. Third, as a result of the large share of investment in national income, real wages are depressed in relation to the economic potential.

Whereas the first two factors facilitate economic crimes, the third stimulates them. Society tends to regard these crimes indulgently as a source of supplement to generally low incomes. The situation gives rise to a network of middlemen, dealers in stolen goods, who,

---

*It was reported that four high-ranking provincial party officials were found guilty of corruption, but, aside from party or administrative sanctions, no legal action was taken against them.[74]

together with the growing number of others whom practice has inured to theft and who benefit from society's indulgence, tend to perpetuate the crimes.

At later stages, despite the increasing attention paid to costs and quality and the greater choosiness of trade, economic crimes to not abate. Instead, new forms of crimes arise where the individual dilettante is replaced by a group that is much more skilled and better equipped.* Such groups may involve employees throughout the entire enterprise or perhaps only those in sensitive positions.** Because they involve accounting manipulations, the crimes are committed in an atmosphere that considerably reduces the risks of discovery. On the other hand, development of control does not keep pace with the increase of economic crimes. It is safe to assume, in fact, that in some cases the leaders of these groups are sufficiently influential to oppose successfully the improvement of control.† Furthermore, such opposition is supported by those managers who, although unconnected with economic crimes, aim at reporting higher profits (or larger output) by lowering quality.[80]

---

*For example, Mototekhnika, which is responsible for sales of automobiles to the public, was the locus of a scandal involving its upper management. Apparently, these officials were responsible for having friends and relatives purchase several cars within a short period of time, presumably for resale, bypassing the usual four-year waiting line. In addition, damaged cars were sold below market value, minor damages quickly repaired and resold to third parties at exorbitant above-market prices, profiting mainly the officials in question. Still another case was that of sale of disassembled cars for parts, which were then promptly reassembled and returned into circulation.[75]

**For example, a case of corruption and theft was reported at Lovech, involving the director and chief accountant of the enterprise.[76] Since December 1972, a ring of heating oil thieves in the Fuel Sofia City Trade Enterprise was uncovered involving 1,000 people and 450,000 leva.[77] In the past foreign trade was the locus of a number of scandals. The most recent involved 47,000 leva in bribes from a Western firm to employees of Rodopaimpex who were responsible for a 755,000 leva loss to the state.[78]

† In late 1971 and early 1972 the curious case of one Vassil Stefanov was exposed. He had a meteoric rise since 1963 to top posts in Sofia's administration. He used his powerful position to distribute favors, collect bribes, and in general to accumulate huge personal wealth and to organize an entire ring of accomplices. Curiously, although his corruption was known for many years, he was allowed to continue in these high posts. About eight other high-ranking party officials were mentioned as part of his gang.[79]

The alarming aspect of this situation is the general complacency with which such crimes are regarded. The prevailing attitude is "take as much from the state as you can and give it as little as possible." All sorts of falsifications are condoned "sometimes quite overtly without a feeling of deviation from socialist legality." The perpetrators defend themselves on the grounds that these violations are necessary and inevitable in their specific cases. The abundance of legal acts and their interpretation, together with the overlapping of jurisdiction of many departments not only facilitate such violations, but often make them necessary. [81]

NOTES

1. T. R. Malthus, Principles of Political Economy (London, 1936), pp. 6-7.
2. A. Dobrev, PT 1 (1970): 78.
3. T. Zhivkov, RD, December 14, 1972, p. 2.
4. Statistical Yearbook of Bulgaria 1971 (Sofia, 1971), p. 279.
5. Compare U.S., Congress, Joint Economic Committee, Economic Developments in Countries of Eastern Europe (Washington, D.C., 1970), pp. 41-64.
6. T. Zhivkov, RD, p. 1.
7. SH, April 4, 1975, p. 5.
8. T. Kadiyan, Stat. 6 (1972): 69.
9. SH, p. 5.
10. Anteni, September 26, 1975; RD, November 27, 1969; RD, January 17, 1975.
11. For a comparative record of CMEA countries, see Statisticheski Yezhegodnik Stran-Chlenov Soveta Ekonomicheskoy Vzaimopomoshchi 1973 (Moscow, 1973), pp. 66-68.
12. T. Zhivkov, RD, p. 6.
13. Ibid.
14. Compare interview with S. Zhulev in ZD, April 26, 1973, pp. 1-2.
15. Compare A. Mozolowski, Polityka (Warsaw) 30 (1973): 10.
16. Compare T. Zhivkov, RD, p. 6.
17. A. Tskigov, FK 1 (1970): 26.
18. Compare G. Vakliev, IZ, February 23, 1972, p. 12; I. Dimov, NV 10 (1969): 43.
19. M. Genov, IZ, August 22, 1973, p. 11.
20. For a comprehensive discussion of shortages, anomalies of supply, and examples of violations of assortment plans, see the report of a conference dealing with these problems: H. Levi and K. Georgieva, IZ, October 17, 1973, pp. 6-7. For an overview of the

modernization and production program in light industry, see K. Koev, IZ, September 19, 1973, p. 14.

21. Compare I. Dimov, NV, pp. 42-43.
22. V. Marinov, VN, February 19, 1971, p. 1.
23. V. Kotsev, Trud, January 9, 1973.
24. Compare I. Vasilev, NA 18 (1970): 19.
25. T. Zhivkov, RD, p. 6.
26. Marinov, VN, p. 2.
27. For a review of the legislation and examples of the proliferation of private services, see N. Genova, NS 6 (1973): 36-37.
28. Marinov, VN, p. 2.
29. Genova, NS, p. 37.
30. Compare K. Dimov, PZ 6 (1975): 26-28.
31. T. Minev, ZZ, May 14, 1975.
32. N. Simeonov, IZ 5 (1975).
33. V. Gadzhev, Anteni, February 21, 1975.
34. DV 97 (December 13, 1974): 2-4.
35. Y. Petrov, IZ, January 29, 1975.
36. Y. Barukh and B. Andrekov, OF, August 1, 1975.
37. IM 9 (1974): 19.
38. RD, March 28, 1973. For comparative data in other CMEA countries, see Rocznik Statystyczny 1973 (Warsaw, 1973), pp. 714-15; A. Szabo, Koz. Sz. 10 (1974): 1184.
39. Nowe Drogi (Warsaw) 4 (1974): 58; M. Golebiowski and B. Zielinska, GP 4 (1975): 266.
40. R. Gocheva, IM 2 (1973): 8.
41. T. Genchev, NM, July 25, 1974, p. 2.
42. T. Zhivkov, RD, p. 6.
43. Ibid., p. 5.
44. Ibid.
45. I. Iliev, RD, October 30, 1974, p. 2.
46. T. Zhivkov, RD, p. 3. Compare I. Vasilev, NA 24 (1972): 3.
47. RD, August 31, 1969; September 1, 1969.
48. RD, March 10, 1973.
49. RD, December 17, 1971, p. 1.
50. Interview with K. Gyawiov in Trud, March 12, 1973, p. 1.
51. M. Golebiowski and B. Zielinska, GP 5 (1972): 265; GP 4 (1973): 244; GP 4 (1974): 233; GP 4 (1975): 263; United Nations, Economic Survey of Europe in 1975, Part 1, Chapter 2, prepublication text, ECE (XXXI)/1Add.1, p. 61.
52. T. Zhivkov, RD, p. 4.
53. DV 4 (January 12, 1973).
54. T. Zhivkov, RD, p. 5.
55. For a detailed statistical record, see Kadiyan, Stat., pp. 63-64.

56. G. Gochev, <u>NV</u> 8 (1969): 24-25.
57. For a review of the more general aspects and proposals for further changes in income taxation, see G. Petrov, <u>FK</u> 7 (1973): 30-42. Changes in the income-tax law are discussed in L. Gyorov and V. Tsvetkov, <u>FK</u> 9 (1973): 38-48.
58. G. Petrov, <u>FK</u>, p. 34.
59. Compare Gyorov and Tsvetkov, <u>FK</u>, p. 48.
60. Ibid., p. 44.
61. G. Petrov, <u>FK</u>, p. 33.
62. T. Zhivkov, <u>RD</u>, p. 5. Compare Zh. Zhivkov, <u>IM</u> 1 (1973): 3.
63. <u>Sbornik postanovleniya i razporezhdaniy na Ministerskaya Suvet na Narodna Republika Bulgaria</u> 2 (1968): 9-14.
64. T. Zhivkov, <u>RD</u>, p. 6.
65. G. Evgeniev, <u>IZ</u>, April 30, 1975, p. 3.
66. T. Zhivkov, <u>RD</u>, p. 7.
67. Ibid., p. 8.
68. B. Panev, <u>BP</u> 2 (1975): 21.
69. P. Tsankov, <u>SK</u> 7 (1975): 23-30.
70. G. Chernev, <u>SK</u> 10 (1974): 16-21.
71. I. Vachkov, <u>RD</u>, June 29, 1973.
72. On widespread illegal channeling of funds and the urgency of strengthening financial control, see, among others, I. Panov, <u>IZ</u>, September 26, 1973, p. 11.
73. <u>RD</u>, April 2, 1972.
74. <u>PZ</u> 4 (1972): 94-96.
75. <u>PZ</u> 16 (1972).
76. <u>RD</u>, April 2, 1972.
77. K. Mishev, <u>Anteni</u>, February 21, 1975, p. 8.
78. <u>RD</u>, March 15, 1975.
79. <u>PZ</u> 11 (1971); <u>RD</u>, August 31, 1971; <u>Trud</u>, February 15, 1972.
80. M. Kalecki, <u>Z zagadnien gospodarczo-spolecznych Polski Ludowej</u> (Warsaw, 1964), pp. 83-90.
81. D. Fidanov, <u>Anteni</u>, February 21, 1975.

CHAPTER 9

PLANNING PHILOSOPHIES AND REFORMS

GROWTH AND REFORM

Until the late 1950s or thereabouts the STEs recorded impressive growth rates of industrial output and national income; these rates, however, fluctuated in time. Moreover, the various composite activities often grew at strikingly disparate rates. The deleteriously unbalanced growth and ensuing dislocations did not seem to concern the c.p. to the extent that a decline in overall growth rates would have, for his performance criterion is the relative dynamics of "short-term" growth rates. In part, the economic reform movement owes its existence to the retardation of growth rates of production and productivity in the 1960s. Growth is propelled by increased inputs and/or a rise in productivity. The STEs relied primarily on increased commitment of resources. The slump in growth rates of output was not accompanied by a corresponding relative decline of inputs but by a deterioration in the growth rates of productivity. This partly accounts for the focus on the determinants of productivity. At first economists and later the leadership became increasingly aware that, with the growing complexity of the economy, proliferating priorities, the increasingly vocal consumer, shifts in factor endowment, and abated possibilities of "extensive growth," the traditional ways of running the economy were becoming increasingly obsolete.

The economic reform blueprints usually referred to increased efficiency derived from economic reform as an additional source of economic growth or as a possible source for redressing some disproportions and increasing living standards. However, the essence of the inverse relationship between growth policy and reform—that is, that reform implementation requires an appropriate slack in the economy—was not readily perceived.* Reform will encounter insurmountable obstacles in an economy overheated by overcommitment of resources to investment.

Whatever the advantages of STEs in manipulating macroproportions, the c.p. is subjected to little immediate scrutiny and the dangers of voluntarism are great. The system is rigid, lacks a microeconomic adaptive mechanism, and does not stimulate technical and organizational dynamism at the production level. Whereas the reform usually focuses on more effective input combinations, improved quality, and wider and better product-mix, together with technical dynamism, a high-pressure economy necessarily stresses quantity, is conducive to wrong factor combinations, and insists on current production at the detriment of innovation. The persisting insistence on high growth rates of output affects the ways of achieving such growth rates. Whereas the reform usually stresses economic calculation, the high-pressure economy invalidates such calculation. The execution of an excessively taut plan involves supply shortages in one branch after another. Under such conditions economic calculation is forgone in favor of "production at all costs," and the c.p. increasingly reverts to central allocation of resources. Whereas the reform involves greater stress on material incentives to motivate the actors, and consequently larger take-home pay, the high-pressure economy encroaches on the volume, quality, and variety of consumer goods and services that are available on the market, reducing the attractiveness of larger pay. The maximization of output pushes quality considerations aside. It is in the makeup of the directors of high-pressure systems to overrate the favorable effects of growth in volume and underrate the adverse effects on the system's dynamic efficiency and welfare. This is partly due to their limited time horizon, for in politics immediate results count, and in the long run the politicians are dead. The valuation scale affects the attitude toward the reform and the trade-off between gains in efficiency and deceleration in growth momentum. While the c.p. largely agrees on the necessity to shift to "intensive growth," he is unwilling to accommodate himself to the accompanying costs in terms of a deterioration of "extensive growth" indexes. Whereas the reform usually involves a set of stable norms to broaden the decision-making horizon of lower units, a taut plan contains the seeds of breakdown and consequent shift that undermine stability and reintroduce the short-term vantage point. Whereas the reform features profit (or a

---

*In this context it is noteworthy that in the final stages of preparation of the Hungarian economic reforms the crucial issue of revision of growth strategy to eliminate planners' tensions was pushed to the background.[1]

variation thereof) at least as a key performance criterion for the plan executants, the high-pressure economy, with its stress on quantity, increasingly reverts to forms of gross value of output as the most important criterion. The reform tends to substitute profitability (achieved through increased output and/or reduced costs) for quantity of output as the important performance criterion of the enterprise. But this is incompatible with physical planning relaying on material balance techniques in plan construction. It is not a matter of indifference whether increased profitability results from increased output or cost reduction and in the latter case whether the reduction is achieved in labor and/or material costs. All these enter into separate balances and are not interchangeable for this type of system.[2] Whereas the reform gives greater leeway to horizontal relationships, the overheated economy requires increasing control over the lower units and consequently strengthens the vertical apparatus.

In final analysis the system of functioning should be consistent and dovetailed with the development strategy. There is an obvious interaction between the plan and the modus operandi. Choice of an appropriate growth rate, structure, and techniques of production, as well as a more effective planning system, would jointly produce better results than could be achieved by attacking the economic riddle on one front only.

THE SOCIALIST ECONOMY

In Bulgaria the traditional growth strategy has not undergone fundamental changes, although it has been subject to periodic flexible interpretation. The principal tenets have persisted, but there is some evidence of greater caution in postulating growth rates, of narrowing the disparity in growth rates of producer and consumer goods, and of taking into account interdependencies—at least at the planning stages in certain periods. However, revealed preferences at the implementation stage still point to the use of the consumer sector as a shock absorber and to sharp discriminatory treatment within the producer sector. The c.p. still displays the industrialization rush psychosis, perhaps with greater recognition of constraints. He still seems to to overrate his ability to overcome the various constraints with a larger dose of the old medicines.

A realistic plan has to be based on investment and foreign trade efficiency calculations. While this is not the place to analyze the principles of these calculations, one should point out that the high-pressure system is not conducive to their application. An indispensable condition for the evaluation of variants is their preparation. Incidentally, this also applies to "scientifically substantiated planning" and evaluation of plan variants. The answer to a taut plan is not the often-prescribed remedy of greater reserves; rather, it is a more realistic plan that would not evince the need for reserves during its implementation. However, such a plan would still require reserves in susceptible or in largely unplanable activities, such as foreign trade and agriculture. The Bulgarian proclamations for "scientifically substantiated planning," forecasting, increased attention to interdependencies through integral planning, enlarged use of computers, and so forth are all methods of improving internal consistency and speed of preparation. But the main precondition for realistic planning—change in plan content—is conspicuously absent. In a nutshell, a shift in planning philosophy is required as a condition for implementation of a radical reform.

VARIETIES OF SOCIALIST ECONOMIES

The STEs have undergone major structural changes, but they have evinced remarkable resistance to fundamental organizational innovations. The command system is only one of the alternative ways of organizing a socialist economy, whose gamut is theoretically large.[3] The alternatives are not necessarily equally efficient, nor are reforms always for the better. The range of alternatives is narrowed down by the constraints of political realities and the center's macropolicy.

The definition of a socialist economy has long been a subject of dispute. Predominant state or collectivist ownership of the means of production is not necessarily a precondition for socialism and planning. Ownership is merely one of the institutional arrangements, and planning in its various forms is merely a technique. Institutions are means, rather than ends, and it is the aims that should distinguish the socialist economy—among which all-round improvement of living and working standards and more equitable income distribution should rank high.[4] But individuals have strong preferences for institutions and techniques, and in some cases they are willing to sacrifice the ends for the means.

Organization of a socialist economy has to resolve, among other things, the manner of determining the c.p.'s aims; the interaction between the c.p.'s and periphery's preferences; the loci of decision making and channels of implementation; the relationship between economic actors and checks and balances; the flow, reliability, and uses of information; and the instruments of influencing economic activity. This is not the place to examine the innumerable alternative organizational patterns, nor even to list the most representative or empirically relevant ones. Three different abstract possibilities—(a) market socialism, (b) indirect centralism, and (c) modified centralism—

PHILOSOPHIES AND REFORMS

are identified in extremely general contours, omitting even the most important details. Each of these alternatives is also open to a number of variations.*

(a) Fundamentally this system differs from b and c in the considerably greater weight accorded to microunits' (producers' and consumers') preferences if they are in conflict with those of the c.p. Contrary to b and c this system could be envisaged with various forms and configurations of ownership of means of production. The government shapes economic activity primarily through investment policy to affect the growth pattern and fiscal and monetary policy to contain fluctuations and inflation. Controls are restricted to a minimum. In influencing the economic actors the government operates mainly through the market, which exists in its institutional sense for both producer and consumer goods. Owing to well-known market flaws, externalities, and social considerations, this is a regulated market. Planning is of an indicative nature, and the relationship among economic units is primarily horizontal rather than vertical. Important functions of the government include the containment of monopolistic power, the influencing of foreign trade, and the shaping of income distribution. The system stresses industrial democracy and workers' participation in management. While this system does not "worship" the market, it leaves it sufficient scope to discharge its allocative functions within a framework of centrally made macroeconomic decisions. Any system tends to constrict the market by imposing various desiderata and to confuse efficiency and equity. Arrow clearly stated the implications of the general competitive equilibrium for the theory of social choice:

> General competitive equilibrium above all teaches the extent to which a social allocation of resources can be achieved by independent private decisions coordinated through the market. We are assured indeed that not only can an allocation

---

*If we were dealing with real economies it would be difficult to establish the precise boundaries when the economy ceases to manifest the dominant features of one system and shifts to another, and to identify the weights of relative factors at work.[5] Planning is a term that can be variously interpreted. Its scope, intensity, and methods differ in concept and application. The best-known existing varieties are imperative and indicative planning, but even this distinction is not clear-cut, and other varieties could be evolved. For obvious reasons, this discussion refers mainly to planning as practiced in STEs and its possible improvements.[6]

be achieved, but the result will be Pareto efficient.
But . . . [as stressed by Bergson and others] there is
nothing in the process which guarantees that the distribution be just. Indeed, the theory teaches us that the final
allocation will depend on the distribution of initial supplies
and of ownership of firms. If we want to rely on the virtues of the market but also to achieve a more just distribution, the theory suggests the strategy of changing the
initial distribution rather than interfering with the allocation process at some later stage.[7]

(b) Under this system central control over economic activity is much greater than under a and more restricted than under c, especially regarding microdecisions, although the scope of key macrodecisions under this system and c is similar. Under this system there is greater tolerance for the periphery's preferences than under c. The system attempts to induce executants to implement the central plan, relying primarily but not exclusively on incentives, rules of behavior, and realistic choice coefficients. In contrast to c, this system relies on outright intervention via specific orders supported by sanctions as an exception rather than as a rule. The system aims at constructing such an end-means structure that, by following his own interests, the economic actor should act in line with the c.p.'s wishes without direct compulsion.[8] Here the c.p. not only makes the major macrodecisions and decides on the methodology of plan construction but also chooses the instruments for plan implementation, such as rules of conduct, essential choice coefficients (including shadow prices, normative rates for calculations of investment and foreign trade efficiency, and exchange rates), reward scales, and planning and financial rules. Certain microdecisions also fall within the c.p.'s purview—for example, important investment projects, choice of techniques, output volume of key goods, and so forth. The major characteristic differentiating b from c is the extent to which the c.p. tolerates sharp deviations from the plan, nonobservance of "plan discipline" and rise of spontaneous forces, and the slow process of microadaptation. Aside from the welfare implication of instruments used, the special characteristic of this system is that the c.p. accepts deviations from the plan as significant correctives and takes into account the informational signals from the periphery, even if these should conflict with his strategy.

(c) This is essentially the traditional STE system, with some organizational changes that do not fundamentally alter it, but aim at removing the inconsistencies accumulated over the years, and generally at streamlining the system. Other changes, however, are again ad hoc measures to deal with this or that manifest shortcoming or desideratum and, as such, may introduce new inconsistencies.

PHILOSOPHIES AND REFORMS

The problem of reforming an STE involves not only the economic merit of the variant chosen but also the restraints imposed by political realities and vested interests and the compatibility of the system with the c.p.'s growth strategy. These are the reasons forcing realistic reformers to look rather to b and c systems.

With emphasis shifting in time and place the c.p. has generally resisted erosion of the following monopoly of party control (which at times of decentralization has been enhanced as a source of countervailing power); preponderance of public ownership, even in domestic trade and services where some private initiative could widen bottlenecks and provide the necessary competitiveness; foreign trade monopoly; central macrodecisions affecting the pace and pattern of growth, consumption, and employment; central determination of output profile and techniques; selection of key investment projects, with control over the bulk of investment and only minor adaptive investments of quick return left to the periphery's decision; central setting of most producers' and consumers' prices and wage scales; distribution of profit between enterprises and the budget and within enterprises among s.p.f.; the vertical organizational structure and superior-subordinate relations; and the directive nature of the plan. Concurrently, fearing that democratization in economic life might lead to democratization in other spheres, the c.p. is strongly opposed to any form of control from below, to checks and balances, to real workers' participation in management, and to genuine representation of workers' interests by trade unions. The c.p. is essentially conservative and would rather retain more than fewer prerogatives (even if these were not absolutely essential for preservation of power). Hence, the paradoxical feature of some of the reforms: Certain decisions are delegated to the periphery, without, however, removing them from the center's orbit.

Even assuming that the reform chosen at the blueprint stage would observe the c.p.'s minimum desiderata (and presumably be a variant of b), it is not at all certain that it could be implemented primarily because it clashes with the high growth rate chosen and the tensions built into the plan. The c.p. not only stands by his macrodecisions, thus undermining the necessary conditions for implementing the reform, but expects immediate improvements of efficiency that should facilitate the implementation of the ambitious plan. As in the case of Bulgaria, reform measures are often introduced experimentally, partially, and inconsistently, with some key changes (prices) emasculated and delayed, so that the initial results are not only disappointing but often worse than under the traditional system. The c.p. then reverts to system c, perhaps with reinforced centralization. If one could advance the broad generalization that a b-type rather than a reform is adopted at the blueprint stage because the c.p. insists on a relatively broad range of decision making, one could also advance the proposition that a type-b reform ends up as type-c at the implementation stage because of the preservation of a high-pressure economy.

PLANOMETRICS

The issue here is not the merit and techniques of optimal planning, but the broader question of planning philosophy.[9] Mathematical techniques and computer technology are merely tools that could be used for various purposes, depending on the preferences of the user. Even in STEs mathematical methods and computers attract individuals of varying economic and political philosophies. There are major differences in the ways in which the existing order could be improved.* A question of some importance is whether at present the optimal plan is feasible. If it were possible, the Bulgarians might be attracted to it. But other problems aside, the present philosophy of the leadership seems to indicate that the c.p. might fail to state explicitly the limiting conditions of an adequate share of consumption in the short and intermediate periods when growth of national income would be set as the optimum for the economy. Given certain assumptions, one could show through mathematical analysis that a high share of productive investment in national income allows for a high rate of growth. This so raises the level of national income in the ensuing years of the long-term plan that consumption throughout the period would be shown to be higher than with a lower share of investment in national income. Within the context of present political regime of STEs, optimal planning has other pitfalls: It curtails the further scope of give and take between economists (planners) and politicians; it enshrouds the plan in spurious precision and develops a complacency about results that may be derived from wholly untenable assumptions; it tends to replace common sense by computomania, with an ensuing loss of benefits from learning by doing. Whatever its drawbacks, however, the method of successive approximations is not only more intelligible and flexible, and permits closer examination of investment efficiency in various sectors, but its redeeming quality is the informal restraints on the c.p. involved in the bargaining process among the various planning levels that, weak though they are, would be irretrievably lost. Of course, all this does not mean that mathematical methods cannot be put to beneficial use in solving many intricate partial problems.**

---

*In order not to diverge too far afield, we shall limit ourselves only to the improvement of the existing order.

**In a similar vein, commenting on the prospects for changes in planning techniques, Janos Kornai saw mathematical methods merely

PHILOSOPHIES AND REFORMS

At this juncture the Bulgarians rely on mathematical techniques and the growing computer network to streamline the traditional system. They appear to be aware of most of the difficulties involved even in such a relatively simple use of computer technology and are attempting to mitigate them. Considerable attention is paid to the necessary redesign of the information system. Yet they seem oblivious to the considerable purposeful distortion of data due to system-made inefficiencies. Computers cannot purify the data fed to them ("garbage in—garbage out"). This may yet largely invalidate this streamlining. The system adopted at present uses computer techniques to permit the c.p. (or the 60 associations closely controlled by him) to regulate and stringently control operative activity at enterprises that have been formally stripped of even the narrow range of prerogatives they enjoyed in the traditional system. It seems to this author that whether or not the present Bulgarian system will be successful in making the traditional system more internally coherent and compatible with growth strategy is not so important as the fact that it is a further encroachment on democracy. This is one of the reasons why the Hungarian reform—even as imperfect and inconsistent as it proved to be in practice—is preferable to the conservative, "technocratic," and autocratic Bulgarian reform.

INCENTIVES

Palpable incentives are required in any system to elicit better performance or compliance. Incentives could be used to support commands, as is preponderantly the case of a c-type system, and to replace commands, as is mostly the case of a b-type system. Incentives may be used to pioneer or initiate new activities or to adapt and imitate superior performance elsewhere. Among the questions that arise are the following: What is expected of incentives? What kind of incentives should be chosen? How can inconsistency between incentives, performance criteria, choice coefficients, and other elements of the system of functioning and specific commands be avoided? Should primarily nonmaterial (political or moral) or material incentives be used? The choice of incentives is further complicated by insufficient knowledge of the strength of incentives required to accomplish a given result.[11]

The traditional STEs and even the more radical reforms (as in Hungary) tend to stress material incentives, whereas the Chinese system relies on moral incentives. In Bulgaria more than in other

---

as a tool to "check, supplement, and correct the plans drawn up by the traditional methods."[10]

STEs nonmaterial incentives have received shifting emphasis, though material incentives have been used considerably. It is difficult to judge to what extent the political incentives have been effective. Even the recent pronouncements that uphold the use of both types of incentives are extremely vague about the combination of the two, and even about the constituents of nonmaterial incentives.

The high-investment and highly politicized Bulgarian economy has relied considerably on campaigns, moral suasion, and exhortations of all sorts to fulfill "mobilizing" plans, together with political advancement and all kinds of orders and decorations. The effectiveness of campaigns to fulfill the "mobilizing" plan (counterplans) is particularly doubtful when the exhortatory targets are not supported by concrete measures, such as increased inputs, for instead of spurring toward greater efforts and improving performance, they merely mobilize executants toward ingenious manipulations. Moreover, campaigns, exhortations, and political appeals are subject to diminishing marginal returns.

The basic problem of financing industrialization and extraction of surplus product cannot be ignored. With a high rate of accumulation, the worker perforce gets considerably less than he produces. There are two questions: How much less, and what will it buy? The rule of thumb has been that not even the entire increment of output attributable to increased productivity should be allocated to consumption.

Economic reforms rely on improved material incentive schemes to relate rewards to actual performance. To be effective such rewards would have to be relatively palpable, substantially increasing the variable portion of the employee's take-home pay. While reforms stress that premiums out of profit must become a much larger share of reward to labor, the prudent planner knows that they tend to be excessive in relation to the flow of consumer goods that will be placed at the consumers' disposal.

Rewards related to productivity call for wider income differentials, and they conflict with egalitarian distribution. This is a major problem that transcends the confines of STEs. One of the answers to the often-posed question "how to make workers work" is to reward them according to their relative productivity, granted that in many cases it is a difficult index to measure. Marginal productivity could only be measured in a fraction of the work force, mainly among blue-collar workers. The relation between the synthetic indexes (for example, profitability and value-added) and performance of specialists and managers is often tenuous, not to speak of the difficulties in measuring productivity in services, especially in health, education, administration, and arts and sciences. But the technical problems of measurement are not the main obstacle to differentiating incomes.

In the postwar period there has been a tendency toward egalitarian distribution of income (with the important exception of high salaries and other privileges of the ruling elite). There is a well-known conflict between equity and efficiency, but established patterns of income distribution are difficult to change. This seems to be one of the fundamental obstacles to a reform linking rewards to performance. A resolute implementation would mean that earnings would fluctuate, and employees are apprehensive of downward earnings flexibility. Moreover, workers (managers) are indifferent and prefer "leisure at work"—a legacy of the period in which efforts were not adequately rewarded—one that has been mitigated but has not yet disappeared. If the higher pay cannot buy high-quality, desired goods, the worker might prefer leisure. As the economy becomes more complex the technocrats play a larger role and have to be adequately rewarded. One of the problems is reconciling their aspirations with the well-ingrained propensity to equalize distribution and preserve the privileges of the stalwart bureaucracy. It seems that the greater attempt to differentiate incomes and increased attention to consumer-goods production (particularly durables) is in response to pressures from the technocracy.

PRICES

In traditional STEs prices are more or less arbitrarily set: They are rigid for prolonged time spans, and price formation is divorced from plan construction. Prices tend to be inconsistent with other instruments of plan execution. Domestic prices are dissociated from world market prices, and within the system, producers' prices are insulated from consumers' prices (the dual-price system). When the logic of physical planning is carried to an extreme, prices should be neutral and producers insensitive to them. As the system developed, producers' prices became more than merely weights to aggregate heterogeneous output, and the planner vacillated between circumventing "wrong" prices by a more direct detailed determination of output and attempting to neutralize sensitive prices or to strengthen their effect by stressing some desideratum or other. Naturally, the planner wants production to be elastic to some prices and inelastic to others, and this is a source of many inconsistencies. Executants will make whatever decisions they can on the basis of existing prices (and will try to influence new prices to their own advantage). It is irrational to ask executants to care or know of the opportunity cost to the economy of inputs that are underpriced or are not entirely accounted for, just as it is to appeal to the consumer to be parsimonious in his consumption of subsidized goods. One of the most expedient measures of reducing the distorting impact of prices is to restrict the periphery's sphere of decision making.

The traditional system requires periodic price revisions that, however, are quite distinct from reform of principles and methodology of pricing. Revision eliminates some distortions and shortcomings and tightens up the system. But here again, elimination of some accumulated inconsistencies tends to create new ones. For example, the revision may set limits on planned profitability, but usually, as a result of cost inflation, ex post profits significantly exceed ex ante. Price reform is a part of overall economic reform and should be consistent with it. Just as it would be counterproductive to widen the scope of decisions at production levels without scarcity prices, provision of such prices without devolution of at least short-run production decisions would not be enough. A shift from crude quantitative indexes necessarily leads to greater reliance on prices as tools of economic calculation. In this respect the issue of congruence between tools of economic calculation at the center and the periphery assumes growing importance. By fixing or manipulating prices, the c.p. can influence the periphery's decisions. The point is to make price setting more consistent with plan construction and with the other instruments of plan execution (such as performance criteria and incentives); and also to allow for more flexibility to reflect changing environment and desiderata.

This raises the question of alternative methods of generating prices. Scarcity (equilibrium) prices do not necessarily have to be formed on the market in its institutional sense. Such prices may be set by a Lange-type CPB that performs the function of the market or by the interaction of electric wires (obtained as the dual solution of a mathematical programming problem).[12] Not every scarcity price is necessarily an efficient price, for it might not account, for example, for externalities, market failures, and social costs, which could be taken into account, however imperfectly, in an optimal plan and in a Lange-type or genuine market by means of taxes, and subsidies.[13]

The legacies of the traditional system are particularly overwhelming in pricing. The touchstones of a price reform are the abolition of the dual-price system, the extent to which producers' prices reflect scarcity, and the correspondence of domestic prices to the price structure on the world market. But prices that more readily reflect dynamic production conditions cannot be equated with efficiency, nor can the administrative locus of price fixing be necessarily equated with arbitrariness. The question is not only of the principles underlying periodic price realignments but of procedures for price mobility. The problem of the traditional system is not so much initially fallacious price but absence of a mechanism of adjustment toward equilibrium.

If prices are to be parameters for the periphery, each actor should be powerless to affect perceptibly the outcome. The c.p. fears that decentralization would strengthen the subunits' opportunities to manipulate prices. As we know, the planning process is characterized by bargaining between participants, whose outcome depends on the relative power of the parties, which, in turn, depends on the access to reliable information. Increased horizontal relations can only subject the seller to the "discipline of the market" if the economy is not overheated. The negotiated price depends on the relative power of negotiators. This depends on the state of disequilibrium, industrial structure, whether there is reciprocal choice of trading partners, and so forth. Often contractual parties merely fill out details in assignments predetermined by superiors. This sort of "adaptation to the specific conditions on the spot" is very restrictive, because, rather than satisfying each other, the actors are interested in satisfying their superiors.

In STEs prices do change, and sometimes for "wrong" reasons, but these are largely ad hoc changes, without permanent price realignment built into the system. Price revisions fail to specify systematic procedures for price changing. The recurring calls for increased price flexibility are not supported by adequate working arrangements for that purpose. In the past price revisions aimed at bringing prices closer to more realistically determined costs and mark-ups. By and large there have been fairly radical changes in approach to costing but rather weak reflection of demand conditions in pricing of producer goods. Not surprisingly, relative prices so formed do not seem to be anywhere near equilibrium levels that would match demand with available supply for each good.

There is a detectable shift in the interpretation of cost elements, in the methodology of cost determination, and in techniques of calculation. In practice the methodological innovations tend to be limited. The introduction of the capital charge—whether treated as a cost element or deducted from profit—is far more important than the ideological and doctrinal connotations. In general, the charge does not reflect opportunity cost of capital at the margin, and discriminatory rates tend to be used. On efficiency grounds, the capital charge (in excess of recovery cost of the investment) should be equal to (uniform) marginal efficiency of investment in all uses throughout the system. And to the extent that the charge is understated, the resulting profit cannot serve as a yardstick for judging comparative efficiency, for extra capital-using producers will show higher profit. Although practice is still riddled with inconsistencies and arbitrariness, there is a noticeable attempt to introduce a measure of determinism into the base to which the profit mark-up is related—a shift away from the traditional mark-up on average costs. There is also an attempt to set more realistic depreciation rates (give greater allowance for obsolescence), with considerable attention given to revaluation of the capital stock. In more or less disguised form various rental charges are made. One can stress either the new elements and direction of change or the timidity and inconsistency of departures from the old.

In STEs prices of producer goods perform very limited allocative function, and the bulk of allocations is centrally made. Reforms attempted to strengthen the allocative (informational) function of prices, but their success was limited. The fairly modest changes introduced vary considerably in time and place. To recall, in Bulgaria even the reform blueprint did not consider bringing producers' prices near clearing levels. Thus, a significant component of redesigning the system (from $\underline{c}$ to $\underline{b}$) was absent. Consistency requires that reform either provide effective prices, or else the system will swing back to direct centralism ($\underline{c}$). To the extent that producers' prices are below clearing levels, excess demand pressures will be exerted, and physical allocation will necessarily remain and flourish. Supply will be allocated to priority uses, while preventing its dissipation among others. Not only will the supply be rationed, but, depending on the payoff, the producer may seek to avoid production of underpriced scarce products, and binding assignments in physical units will proliferate. Thus, partial decentralization, without appropriate price reform leads to creeping and cumulative recentralization.

The pricing system is a rationing method based on the ability to pay. In the STE the function of the market is to allocate consumer goods (made available more or less automatically) by the purse. Thus, relative prices of consumer goods should correspond to microeconomic equilibria in each partial consumer market. But for distributional and social policy reasons the c.p. tends to set and keep prices of certain goods below and others above the demand prices in the consumer market. On welfare grounds a system where consumers can spend their income freely is preferable to one without freedom of consumers' choice.[14] A pricing system should provide a mechanism to guide production by consumer demand (as revealed in demand prices), or a milder version thereof, with production sensitive to "consumer sovereignty," while key decisions on the structure of consumption are made at the center. Even the milder version is largely missed in $\underline{c}$-type modifications. Despite the exhortations to bring production more in line with consumer demand and some progress, the consumer still has little influence on what is produced. There are serious obstacles in adapting production to demand even in spheres that do not encroach on the c.p.'s prerogatives. To be sure, the c.p. is not always interested in overruling consumers' preferences, if for no other reason that because the savings could be used to advance his own objectives if the same "value" of consumption could be produced with smaller resources. Essentially the system still lacks any specific set of rules that would make production more

PHILOSOPHIES AND REFORMS

responsive to market conditions. Such a mechanism should not only ensure that changes in demand invariably lead to changes in production but also that the producer be interested in aggressive innovation and influence demand by providing the market with better products.

In order for the market to influence the pattern of production, producers' prices should be brought into proportion with consumers' prices. The turnover tax, collected as the difference between the retail price and the producers' factory price, as an instrument of the dual-price system, should be replaced with a uniform rate of turnover tax on retail prices, so that producers' price ratios approximate consumers' price ratios and tax collection be shifted closer to final sale. Here also there are serious practical difficulties. Even where more daring (b-type) reforms were introduced (as in Hungary), turnover tax continued to prevail. In Hungary the reform of turnover tax was very limited, partly due to the restricted scope of alterations allowed in the level and pattern of consumers' prices. Even without a radical realignment of consumers' and producers' prices (c-type), correspondence could be achieved. But only an a-type reform is likely to bring consumers' preferences to bear significantly on the structure of production.

The managers (as sellers) prefer a state of seller's market, for a hungry buyer is less choosy and troublesome. Shortages force the consumer to find a solution and adapt himself to the system. As supply of consumer goods increases quantitatively, the buyer becomes more choosy. Overstocking forces the planner to use "downward price flexibility" as a device to get rid of unwanted stocks. But upward price adjustments to choke off excess demand for some goods are unpopular. Interestingly, the Hungarian experience showed that there was "no willingness of producers to test the elasticities of the demand curves facing them."[15] The process of learning to adapt to the signals of the market was inordinately protracted, and drastic price changes, rather than marginal variations, were required to elicit desired response. Since such large price changes are often prohibitive, the workability and elasticity of the mechanism tends to be overrated by market-oriented economists.[16]

In larger perspective, one should keep in mind that the STEs are supply-constrained, and the workability of the market has to be viewed under these conditions. The market works best under conditions of slack. The market method is not only ineffective when massive redirection of resources is required (as in a war economy), but it is also not forceful enough under conditions of strain, and to accomplish large-scale unusual adjustments, without undue time-lags and costs of waiting. Assume, for example, that serious shortages of rolled steel have arisen. What could be expected from an increase in prices? The prices of capital goods and of some consumers' durables (for example, automobiles) would have to be increased. With unchanged funds for investments, the increased prices of capital goods would reduce the physical volume of investment, but it would not be easy to predict to what extent such a reduction in investment would influence the curtailment of demand for rolled steel. The situation would be similar in consumer durables. Price increases would reduce demand for them, but again it would be difficult to predict to what extent and after what lapse of time. In order to equilibrate demand with the reduced supply of rolled steel, prices would have to be raised so that the reduction of demand from these two sources would be sufficient. This can only be achieved by trial and error. Moreover, should there be no considerable reduction in demand of the producer whose output is particularly "steel-intensive" (for example, railroad cars or automobiles), the general reduction of investment and consumption would have to be very high. Such equilibrium would only be achieved with a time-lag during which the use of steel would exceed supply, and stocks would be depleted. It has been suggested that a much simpler and more expedient procedure would be a limitation of investment that would take account of general priorities and their "steel-intensity" and a similar restriction on durables. In case of the latter, prices would also be increased to the level at which demand is more or less equilibrated with supply. As a result, the distribution of steel among the various branches would be reduced, and the production plans would be modified. Even with this method, an increase in prices of rolled steel would be indicated if the shortages were of a recurring nature.[17]

INFLATION

Inflationary forces are generated by the industrialization rush and affected by working arrangements. If the ultimate aim of economic activity were to outlaw inflation (as recorded in conventional statistics), the inflation-prone Western economy should institute tight price-wage controls, isolate domestic from world market prices, and impose comprehensive foreign trade and exchange controls. But the rise in the overall price level is hardly a measure of aggregate economic loss, nor is a stable price level a measure of aggregate economic gain. If the center does not fix equilibrium prices, it cannot expect prices to perform the function of allocating scarce resources by themselves. Whatever the political advantages of a policy of price reduction practiced in Bulgaria, this must be evaluated in view of the persisting disequilibrium on the consumer market. This together with the lack of success in controlling the size of the wage fund, tends

PHILOSOPHIES AND REFORMS

to aggravate the shortage-prone system. A better way to raise living standards would be to increase wages.

Inflationary pressures can surface under various guises. Authorities cannot avoid latent price increases, which take various forms, including:

1. Under conditions of excess demand the producer lowers the quality or use value of goods, so that de facto the consumer gets less per unit of expenditure. Real purchasing power declines, for now a larger expenditure is required to get the same utility.

2. Producers tend to modify slightly the product (without commensurately affecting the quality) and force the hand of the price authorities to raise the relative price. At the same time the production of the cheaper product (which is still in demand) is discontinued. This spurious product differentiation is usually not in response to demand; rather, it is dictated by the producer's interests and often restricts choice. Thus, it is not sufficient to look at the dynamics of overall consumption; its composition must also be examined, for the structure shifts toward more expensive goods.

3. Welfare is reduced when the available options are narrowed down. The consumer is then forced either to buy the goods that are available instead of those that he really wants or to postpone purchases. Thus arises the problem of forced substitution and savings.

4. Shopping becomes a nightmare and the consumer pays a high price in annoyance, mistreatment, and waste of time. Purchasing power then cannot be equated with shopping power. This also gives rise to all sorts of graft and under-the-counter sales and purchases from artisans.

5. Inflation is also introduced through various "private" markets (for example, buying and selling of private homes, antiques, jewelry, and so on) and the black markets (especially in services and repairs). The prices actually paid are not reflected in statistics. Although the center tends to take into account only the price changes it effects, the population is not only fully aware of latent inflation but tends to overrate it.[18]

Market-type reforms tend to increase the risk of open inflation. But this is not an indictment; rather, it is an argument for effective macropolicy. The poor record of Western economies in containing inflation is not sufficient evidence that the problem cannot be resolved more effectively in a socialist economy with a built-in market mechanism. Those who oppose market-type reforms usually point to inflationary upsurge of prices, substantial unemployment, and a deteriorating balance of payments in post-1965 Yugoslavia.*

This argument, however, emphasizes the peculiarities of the decentralized institutional arrangements and underrates mismanagement of macropolicy and structural factors. The critics of the Hungarian "new economic mechanism" have also pointed to the danger of inflation. Those fears circumscribed the implementation of the blueprint.**

In China there is great concern about inflation. On the face of it, China seems to be more successful than the STEs in controlling inflationary pressures—that is, overspending of the wage fund. Some of the reasons include a more realistic investment program (at least in the period after the Great Leap Forward), direct allocation of labor, lack of wage bargaining (which takes place informally in STEs), and lack of competition among producers for scarce labor. But again the outcome might be worse than living with moderate inflation.

THE ASSOCIATION

The organizational structure of Bulgaria now features a high degree of monopolization, somewhat along the lines of the GDR expe-

---

*The record shows an annual rate of inflation between 2 and 3 percent in CMEA countries; this is, however, considerably understated. According to unofficial sources, the estimated cost increase rate in industry and construction in CMEA countries has been somewhere between 4.5 and 9 percent per annum since the mid-1960s.[19] There is also an indication that the producer's pattern of adjustment was in the direction of raising prices, rather than adjusting the flow of output. In Yugoslavia there was a serious deterioration in a number of economic-activity indicators, including a fourfold increase in industrial prices from 1965 to 1972 and a cost-of-living hike by about 150 percent. The average annual growth rate of employment fell from 5.7 percent in 1955-64 to 1.8 percent in 1964-72. Unemployment rose substantially. In 1972 unemployment amounted to 8 percent. Yugoslavia—together with Greece and Ireland—recorded the highest rates in Europe. In addition, 1 million workers sought work abroad.[20]

**In the early 1970s producers were shielded from the increase of world market prices of raw materials. The difference between the administrative domestic price and the actual cost was absorbed by the budget. But with the substantial price increases in 1975 in intra-CMEA trade, the inflationary impact in CMEA countries should be very strong.

rience. The association is presently constituted primarily on a horizontal level, and its scope of decision making has been enlarged not only by emasculating enterprises that effectively have been turned into plants but also by vesting it with some of the prerogatives that were traditionally in the precinct of the industrial ministries, while the control by functional ministries has been enlarged. In essence this is a partial shift of power from the bureaucracy to the technocracy. But it really does not alter the command system, for it even weakens enterprises and forces them into almost total subjection to the association. This is not cohesive but only administrative integration. Should it not work, the c.p. can emasculate the association and once again strengthen the ministries, as implied in some of the Bulgarian rearrangements in the latter part of the Sixth FYP.

Whatever the economic reasons, benefits of scale, propensity for gigantomania, and equating high concentration with progress, there are strong political reasons for such organization. It facilitates central control over a small number of units, shortens and reduces the number of channels of communication and coordination, making the process of aggregation less cumbersome, and provides an opportunity for restricted "decentralization." Such an arrangement avoids some of the excesses of overcentralization but also precludes devolution of decision making to the actual production units—a characteristic of the more daring reforms. Thus, the system cannot be expected to benefit from the initiative and ingenuity of economic actors at the time and place of action. But it seems that the planner has traded off the potential gain in efficiency for maintenance of control, which is necessary in a high-pressure economy. It is also a specific feature of the Bulgarian way, which relies considerably on political compulsion and mistrust of decentralized action, partly due to the relatively low level of training of cadres.

The Bulgarians have recognized one of the problems of the traditional system: Each additional level in the planning process tends to distort information and complicates the bargaining process. The system that relies on the association has been designed to bypass the ministry in the process of aggregating data and disaggregating targets and limits. Such a simplified system is also more suitable for an "automated" management process. However, it means not a weakening but rather a strengthening of the vertical chain of command and subordination.

This development is in line with the process of concentration in the West. The essential difference is that in Bulgaria the monopolization is imposed from above and indiscriminately forced, regardless of the nature of the branch, factor endowment, pattern of investments, and so on. While the degree of centralization differs in many large Western concerns, in most cases key decisions are made at the head office, particularly those decisions regarding consequential investment, and in some cases the subsidiaries work on internal cost accounting. But whatever the other differences and the economic merit of concentration in the West, in Bulgaria the "imitation" has been carried too far, without discernment, and unimaginatively.

One of the reasons for central price setting is the fear of monopolistic or oligopolistic price behavior. But even if central price setting is carried to its extreme, the periphery still can influence prices, especially through the information system. Moreover, monopolistic behavior is not confined to price making; this is particularly due to unequal market power and is strengthened by the tensions built into the plan. Here a seller's market or a state of excess demand prevails, where buyers are forced to compete for the limited supply and to court the sellers, who are in a position to dominate buyers and find no difficulties to sell. In such an economy, an interesting combination of seller's market and monopoly power occurs, with one reinforcing the other. To break this one would need a macropolicy that would remove tensions and effectively contain monopoly power. This is not to be confused with breaking down the existing structure into a large number of small firms so as to satisfy the conditions for perfect competition, if only on grounds of dynamic efficiency. Containment of monopoly power requires a number of measures. One of them is the organization of vertical concerns.[21] Another is providing for competition from abroad, requiring considerable increase and restructuring of imports (more consumer goods and sophisticated machinery, especially from the West).* Above all, some slack should be introduced into the system so as to promote establishment of a buyer's market, and administrative integration, regardless of conditions, should not be forced.

TECHNICAL PROGRESS

Reforms have to cope with moderating the sources of inefficiency. Static inefficiency might be attenuated by a consistent reform, which would be a considerable source of economic growth and improvements in welfare. But the crux of the matter is dynamic efficiency, which requires a reform capable of instilling the spirit of innovation and risk taking at all levels of economic activity. Variants of traditional STEs have not only failed to introduce the necessary stimuli for generating technical progress from below but feature built-in obstacles to implementation of technical progress imposed from above. What-

---

*Again, the Hungarian experience provides interesting insights.

ever the advantages of central decision making for science policy and shaping foreign trade, the major problem remains one of diffusion of technical progress and spillover from growth-promoting and defense activities to the rest of the system. In addition, experience of industrialized countries has shown not only that progress has been propelled by the major inventions, but that the cumulative effect of the minor ones has been equally important.[22] These are the inventions that a tightly centralized system has difficulties in eliciting.

The system-made obstacles to technical progress derive primarily from the disruptive effect of its introduction on current activity, which is at the forefront of attention. These obstacles include preponderant stress on plan fulfillment, the incentive system and performance criteria supporting it, faulty prices, orientation to short-term benefits, mobility of cadres, weak rewards for entrepreneurial and innovational activity, lack of motivation to improve skills, and central restrictions on major output profile and techniques. The underlying factor is again the high-pressure economy that, despite the leadership's rhetoric regarding technical progress and the extension of the planning horizon, stresses quantity at the cost of quality, and current plan fulfillment at the cost of long-term benefits, and results in instability even in the short run, so that executants have to provide larger reserves for current plan fulfillment, which otherwise could have been used for technical-progress measures. In the final analysis the rate of innovation and pattern of adaptation depends not only on the functioning system but, in a most significant sense, on the rate and pattern of utilization of resources in the system, on the content of macropolicy and rates and frequency of shifts, and on the differential response of economic actors under varied conditions and rates of change of pressure or slack.

PRESSURES AND SLACK

Throughout this study arguments have been mustered against the high-pressure economy and for a shift to a lower gear if reform is to be successfully implemented. But how much slack is beneficial and tolerable? There is some truth to the contention that disequilibria, bottlenecks, and shortages emerge in a rapidly developing backward economy, and that they are part and parcel of a development sequence that maximizes pressures and creates palpable incentives (profits) for overcoming bottlenecks and for productivity growth.[23] But a rapidly developing planned economy does not have to be subject to such pressures. This is what realistic planning is all about: It could largely eliminate the pressures while keeping the economy on a maximum sustainable development path. Much depends on the institutions, the predisposition of economic decision makers (trade-off between investment and consumption), and the recent development history (tensions or slack).

It is a valid point that the STEs suffer not from sleepiness but from insomnia. Janos Kornai argues that a certain permanent excess of supply over demand is admissible and even desirable. Such a continuous state of "underemployment disequilibrium" has adverse effects in terms of underutilization of resources and short-term quantitative gaps between actual and potential output, but it has positive effects on technical progress, qualitative performance, and adaptive properties to change.[24] Kornai contends that most of the gains in dynamic efficiency and adaptability to users' needs achieved by the capitalist economy can be only partly explained by institutions and the profit motive. He accords equal, if not higher, rank to the state of relative slack predominant in the capitalist world, which has enabled it to achieve certain qualitative results so far absent in STEs.[25]

The really big question is, To what extent does the overheated economy have to be shifted to a lower gear? There definitely is a danger of overshooting the target. The underutilization of resources prevailing in capitalist economies is a source of waste.[26] In fact, the aim of a rational economy should be to minimize such losses, and a combination of planning and market is advocated as a tool for that end. The danger of advocating too much slack is evident: Some East European economists have suggested that the only way to raise the system's efficiency is to introduce significant unemployment in order to make workers work. The idea of a "reserve army of workers" in a socialist economy is repugnant, whatever its merits in potential gains in efficiency. The basic issue is: for what and for whom? It seems that realistic planning (with due weight to the frustration and technical and organizational barriers) could afford just that minimum amount of slack necessary to implement a meaningful and consistent reform.

ATTITUDES TO REFORM

The strongest opposition to reform is not on grounds of economic merit but on those of the gains that the different interest groups expect to derive from it and the losses they fear. It might seem paradoxical that in view of the recognized needs, consequential reforms are not introduced. But economic solutions are increasingly political questions. By their very nature market-type reforms shift the focus on interaction between contracting parties and checks from below—and divert it from the vertical command system. It is a democratization process that not only weakens the rulers' monopoly over economic decision making but threatens to spread into other domains. A real movement toward political democratization would essentially "expropriate" the "ruling clique," which, though it is not the legal owner of the means of production, possesses all the prerogatives and privileges of a collective owner.[27]

The leadership (Politbureau) cannot be viewed as a homogeneous group. Its decisions probably result from friction and compromise, and in part they are an adjustment to constraints imposed by the environment, with increasing leverage of the heterogeneous, distinct interests and opinions of various pressure groups. Even those among the leadership who favor consequential reform are reluctant to commit themselves to such a course, for it is usually costly, entails some unpopular measures, and does not pay off quickly. Even the original supporters of reform are prompt to withdraw when the results of implementation are not immediately satisfactory and unforeseen problems have to be solved. Better results require a simultaneous attack on all fronts and a strong take-off, whereas the inbred caution of the leadership inclines it to piecemeal, timid experimentation. When the leadership proclaims its pursuit of technical progress and improvement in living standards, it is not merely acting for propaganda purposes, it might be convinced of the desirability of such action. But when it comes to implementation, these intentions prove to be in conflict with the leadership's overall aim of "maintaining its monopoly of uncontrolled power over all areas of social activity."[28]

Industrialization of STEs relies heavily on "mobilizing" planning and the political apparatus, and a growing bureaucracy plays an important role. The aparatchiks are compensated by relatively very high earnings (which grow apace, even when salaries and wages in production units are stagnant or fall) and a broad and growing network of privileges that include special stores where they can acquire at special prices high-quality goods and luxuries unavailable elsewhere, their own hospitals and resorts, private villas, expensive gifts, proliferation of jobs at full pay for one individual, special luxury holidays abroad, and so forth.[29] Moreover, the supreme privilege and key to the bureaucrats' power is the network of favors they can wield to influence their subordinates. Concurrently, there is a tendency to tolerate only those "experts" who do not aspire to decision making and to staff the advisory bodies with those who are politically servile rather than competent. To allow experts any essential role in the decision-making process would be to reduce the rulers' (and bureaucrats') power. Therefore, inefficiency, waste of resources, and the rule of incompetence are in a way built into the administrative process and cannot be viewed as periodic aberrations that can be remedied in the future.[30]

An expanding industrialized economy has to rely increasingly on the technocracy to keep the wheels turning. There is understandable friction between the old-time privileged aparatchik and the increasing aspirations of the technocrat for a share in power, prestige, position, and income. While the forces at work cannot be analyzed here, it seems that the increasing attention accorded to the consumer-goods sector is a response to these demands. The line of demarcation between the bureaucracy and the technocracy is difficult to draw because some individuals belong to both and their attitudes are influenced by their positions in the capital or the provinces, the necessary room at the top, and so on. Essentially, the technocracy is characterized by a pragmatic and deterministic (nondogmatic) approach, is inclined to stress material incentives and technical solutions, and promotes the running of the economic machinery by technicians rather than by politicians. The bureaucracy seeks the preservation of the voluntaristic approach and the administrative planning methods with which it is familiar, together with the maintenance of its command over resources and the distribution of favors, which perpetuates its power, and thus it opposes the goals of the technocrats. Although the system is more inclined to listen to the arguments for greater technical efficiency, for it means less loss of control and power than the introduction of fundamental reforms (the market type, with checks and balances), the technocratic reform still usurps some of the bureaucrats' powers, or it necessitates a sharing of the bureaucrats' power with the technocracy.[31] However, the conflict between the bureaucracy and the technocracy should not be overrated, because most bureaucrats and technocrats are united in opposing fundamental overhaul of the system. Most of them are products and beneficiaries of this system and fear a loss of power and/or benefits already achieved. The struggle between them is then confined to the technocrats' demands for a readjustment of the power structure (so that they could achieve a better place in the sun) and the bureaucrats' resistance. Thus, the differences concentrate mainly on the nature of the modifications of the existing system and are a contributing factor to the dynamics of perpetuation.

Surely, in the long run, the worker would stand to benefit from a reform that would provide a great availability, variety, and quality of consumer goods. However, the stress on efficiency entails widening of income differentials and spur on exertion of effort. While a strong argument could be made in favor of the effort-releasing effect of the market, the point is that the worker is not particularly interested in exerting the necessary efforts, because he fears that the system would be designed to make him work harder for the benefits of the bureaucracy and the tehnocracy, rather than for his own benefit. This inherent distrust is a well-founded legacy of the past, for, despite its stress on equity, the system has always distributed income and other benefits in favor of "privileged classes."

In the final analysis meaningful reform cannot be implemented by individuals who fear its consequences and resent it. Neither can it be introduced by those who have been brought up in the bureaucratic way, for sooner or later their well-ingrained habits will predominate and overrule the attempted reform. The vertical hierarchic chain of command is staunchly defneded by the bureaucracy. Such a system is incompatible with market-type reforms and perpetuates the salient features of the traditional system. In particular, despite pronouncements about differentiation between guiding and directive indexes, the executants depend on the source from which these instructions emanate and thus superiors can de facto turn guiding indexes into directive ones, and reduction of the number of binding indexes is not a measure of decentralization. On the whole it is difficult to judge to what extent the bureaucracy has deliberately exerted its energies on thwarting reform. At any rate, one should remember that the Zhivkov regime is cast in the traditional conservative mold, with strong allegiance to the USSR. Although it has shown some pragmatism and flexibility, it lacks imagination and daring and is particularly skillful in the centralistic and autocratic ways.

WELFARE AND DEMOCRACY

In the eyes of the leadership technical advancement tends to overshadow the human factor in increasing productivity (including the motivation to produce and to improve skills). The worker is not only a producer but also a consumer, and this dual function seems to be underestimated by the leadership. Probably one of the most enduring consequences of the industrialization rush is the adverse effect that sacrifice of the present for the future has had on the quality of performance, standards of morality, attitude to work, and social consciousness. The slow improvements in living standards and inequitable distribution of sacrifice have alienated the worker so that he has seen his interests as sharply divergent from those of the power elite. The bureaucratization of life deepened the alienation, and the system lacked even the ideology of the worker as a participant in the decision-making process (for example, the one created in Yugoslavia).

The imposition of central preferences is only a part of the story; the manifold ways of resistance and adaptation to the rules of the game and the development of the entire mechanism of manipulations and illegal activities are all equally important. Management's pattern of adaptation is well-known in the literature.[32] But the effect of depressed living standards on the rise of economic crimes needs to be emphasized. It was the worker-consumer who bore the brunt of unrealistic plans. It is the neglect of the worker-consumer from the standpoint of better working and living conditions that is one of the strongest indictments against the system.

A socialist system cannot be judged by economic performance alone. In the final analysis it is the system that should serve the individual, and not vice versa. The choice of an appropriate organization depends not so much on economic efficiency for its own sake as on the reason for efficiency and its beneficiaries, on the freedom and dignity of the individual, and on his ability and opportunity to use his talents to the full advantage. To paraphrase Schumpeter, the reformed socialist order should command the moral allegiance that is being increasingly refused to the system of authoritarian discipline: "This, it need hardly be emphasized, will give the workman a healthier attitude toward his duties than he possibly can have under a system he has come to disapprove."[33] The system should elicit from the worker loyalty and pride in good performance. A successful reform has to go beyond the economic sphere and to introduce fundamental changes in the fabric of society that profoundly affect the willingness to produce, the quality of life, the spiritual values, and the basic relationships between individuals as producers, consumers, and citizens. It is the very danger of the consequences of economic reform as it permeates the system that fuels resistance and limits its scope.

Success of reforms may be evaluated in terms of the preferences of the leadership and of the various interest groups. The essence of the improvement depends on the frame of reference adopted and the value scale. The aim might be difficult to evaluate exactly, but the distance away from the traditional system, even if it is not very far, might be significant, for the participants are learning by doing. Even if the lessons often appear to be forgotten, they introduce elements of change. Although undoubtedly the political framework limits the scope of reforms, one cannot assume a priori that without political reform there is no scope for economic reform. Realistically, less radical reforms are better than none at all, but not every reform is an improvement.

At the beginning of the 1930s Carl Landauer wrote: "Because Socialism concentrates economic power, and because this concentration necessarily gives an extraordinary degree of power to some individuals, regardless of how much one might decentralize the socialist economic system, democracy is more necessary to socialism than to any other type of economic system."[34] Another socialist, Oskar Lange, stressed democratic socialism and management by rules, warned of the danger of bureaucratic arbitrariness and degeneration, and advocated diffusion of power as the principal bulwark of industrial democracy, requiring, among other things, the separation of economic management from the organized political force of the state, and the participation of workers in the process of management. Only by subjecting the c.p. "to democratic control and operation can social functions be prevented from being distorted in the interest of a privileged minority."[35] It is almost four decades since Lange set down these thoughts, and many changes have intervened, but his prescriptions did not lose their relevance.

NOTES

1. Compare J. Kornai, Anti-Equilibrium (New York, 1971), p. 328.
2. Compare E. Mateev, Balans na narodnoto stopanstvo (Sofia, 1972).
3. Compare A. Bergson, JPE, October 1967, pp. 655-73; B. Ward, The Socialist Economy (New York, 1967); G. R. Feiwel, KES 1 (1971): 67-100.
4. Compare J. Tinbergen, in On Political Economy and Econometrics (Warsaw, 1964), pp. 593-94; O. Lange, The Working Principles of the Soviet Economy (New York, 1944); M. H. Dobb, Welfare Economics and the Economics of Socialism (Cambridge, Mass., 1969).
5. Compare J. G. March and H. A. Simon, Organizations (New York, 1958), p. 1.
6. Compare G. R. Feiwel, KES 1 (1971). For the interesting Chinese case, see J. Robinson, Economic Management in China 1972 (London, 1972); D. H. Perkins, Market Control and Planning in Communist China (Cambridge, Mass., 1966); A. Donnithorne, China's Economic System (New York, 1967). On planning under market socialism, see Bergson, JPE; Dobb, Welfare Economics; G. R. Feiwel, The Economics of a Socialist Enterprise (New York, 1965); Kyklos 3 (1972); Ward, The Socialist Economy; A. Zauberman, Aspects of Planometrics (New Haven, Conn., 1967).
7. K. J. Arrow, AER, June 1974, p. 269.
8. Compare O. Lange, Political Economy, vol. 1 (New York, 1963), pp. 177-81; G. R. Feiwel, The Soviet Quest for Economic Efficiency (New York, 1972), Chapter 4.
9. On the merit and techniques of optimal planning, compare Feiwel, The Soviet Quest, Chapter 4; Zauberman, Aspects of Planometrics; M. Ellman, Soviet Planning Today (Cambridge, Mass., 1971).
10. J. Kornai, Mathematical Planning of Structural Decisions (Amsterdam, 1967), p. 44. Compare G. R. Feiwel, JFL, December 1974.
11. Compare H. Leibenstein, AER, June 1966; H. A. Simon, in Surveys of Economic Theory, vol. 3 (New York, 1968), p. 10.

12. O. Lange, in B. E. Lippincott, ed., On the Economic Theory of Socialism (Minneapolis, 1938); O. Lange, ZG 43 (1965); Feiwel, Kyklos; V. V. Novozhilov, EMM 5 (1965).

13. For a discussion of the economic merit of various ways of generating prices, see, among others, Feiwel, The Soviet Quest, Chapter 4; G. R. Feiwel, The Economics of a Socialist Enterprise (New York, 1965), Chapter 2; see also the references therein.

14. Compare Lange in Lippincott, ed., Economic Theory, pp. 98-104.

15. R. Portes, AER, May 1970, p. 312.

16. Compare Kornai, Anti-Equilibrium, passim.

17. M. Kalecki, Z zagadnien gospodarczo-spolecznych Polski Ludowej (Warsaw, 1964), pp. 17-19.

18. Compare RD, November 20, 1973; B. Csikos-Nagy, Socialist Economic Policy (Budapest, 1973), pp. 170-75.

19. K. Engel, FAZ, February 21, 1975.

20. B. Horvat, Ekonomist (Belgrade) 1-2 (1974).

21. Compare Kalecki, Z. zagadnien, pp. 24-27.

22. Compare W. E. G. Salter, Productivity and Technical Change (Cambridge, Eng., 1969), p. 5.

23. A. Hershman, The Strategy of Economic Development (New Haven, Conn., 1958).

24. Compare J. A. Schumpeter, Capitalism, Socialism and Democracy (New York, 1947), p. 85.

25. Kornai, Anti-Equilibrium.

26. M. Kalecki, Theory of Economic Dynamics (London, 1965).

27. L. Kolakowski, Kultura (Paris) 6 (1971): 4.

28. Ibid., p. 15.

29. For the illuminating Soviet experience, see, in particular, R. Medvedev, Let History Judge (New York, 1971).

30. Kolakowski, Kultura, p. 5.

31. Ibid., p. 14.

32. Compare J. S. Berliner, Factory and Manager in the USSR (Cambridge, Mass., 1957); Feiwel, Economics of a Socialist Enterprise.

33. Schumpeter, Capitalism, Socialism, and Democracy, p. 211.

34. Quoted after G. Grossman, ed., Essays in Socialism and Planning in Honor of Carl Landauer (Englewood Cliffs, N.J., 1970), p. 184.

35. Lange, in Lippincott, Economic Theory, pp. 121-29.

APPENDIX A
APPENDIX TO CHAPTER 6:
REVENUE GENERATION AND DISTRIBUTION

Profit Generation and Distribution at Associations

1. Revenue from sales (less turnover tax)
2. Other revenues and disbursement
   a. Of subsidiaries
   b. Of the headquarters (including subsidies for subsidiaries)
3. Budget subsidies for current activity
4. Total (1 ± 2 + 3)
5. Costs of output sold
6. Profit from domestic activity (4 - 5)
7. Profit or loss from foreign trade
   a. On exports
   b. On activity of f.t.e.
8. Export subsidies
9. Profit from total activity (6 ± 7 + 8)
10. WF
11. AMIF
12. EAF
13. Remaining profit (9 - 10 - 11 - 12)
14. Profit tax
15. Profit tax rebate
16. Profit for distribution (13 - 14 + 15)
17. Interest on loans for working capital
18. Social security payments
19. Remaining profit for distribution (16 - 17 - 18), allocated to IFTEF, MAF, specific funds, DF

Computation of the Association's Wage Fund

1. Profit from total activity
2. Wage fund included in costs of goods sold
3. Gross income (1 + 2)
4. Reduction of gross income by
   a. Interest charges on past-due loans for working capital
   b. Others
5. Remaining gross income (3 - 4)
6. Wage fund normative
7. Wage fund computed according to normative $[(5 \times 6)/100]$
8. AMIF of subsidiaries
9. Contributions to EAF (7 - 2 - 8)

Distribution of MAF

1. Funds derived from profit
2. Expenditures on materials
3. Contributions to WF
4. Wage fund
    a. Basic wages
    b. AMIF
5. Balance to the EAF

Results of Foreign Trade Activity

1. Equivalent of foreign currency obtained from export
2. Exports according to enterprise sale prices
3. Subsidies according to normative
4. Markups and commissions paid
5. Other expenditures
    a. Directly tied into exports
    b. Overhead
6. Result of foreign trade activity (1 - 2 + 3 - 4 - 5)

Contributions to AMIF

1. From the wage fund
    a. The wage fund computed according to the normative
    b. The wage fund according to the counterplan
    c. The wage fund included in costs of sales
    d. Savings (a - c)
    e. Contribution to the AMIF, including:
        (i) From planned savings (a - c): 100 percent
        (ii) From above-plan savings (b - c): 50 percent
        (iii) Less 100 percent of overexpenditures (c - a)
    f. Total contribution to the AMIF
2. From material costs
    a. Costs according to preset standards
    b. Costs according to counterplan standards
    c. Reported costs
    d. Savings (a - c)
    e. Contribution to the AMIF, including:
        (i) From planned savings (a - b): 40 percent
        (ii) From above-plan savings (b - c): 20 percent
        (iii) Less 40 percent of cost excesses (c - a)
    f. Total contribution to the AMIF

APPENDIX A

3. For quality and technical standards
   a. Revenue, including:
      (i) Sales revenue with the reported quality structure
      (ii) Sales revenue with the planned quality structure
      (iii) Revenue from the sale of goods with specific quality structure, according to confirmed normatives
   b. The amount of penalties for sale of second quality goods
   c. The result of improvement or deterioration of quality and technical standards

APPENDIX B
STATISTICAL APPENDIX

TABLE B.1

Area and Population of CMEA Countries

Country	Area		Census data as of	Population		
	Thousands of sq. kms.	Share in CMEA		Thousands of people	Share in CMEA	Share in the world
Bulgaria	111	0.4	1965	8,228	2.5	0.2
Czechoslovakia	128	0.5	1970	14,362	4.2	0.5
GDR	108	0.4	1971	17,068	5.0	0.7
Hungary	93	0.4	1970	10,322	3.0	0.4
Poland	313	1.3	1970	32,642	9.4	1.0
Rumania	238	1.0	1966	19,103	5.7	0.6
USSR	22402	89.6	1970	241,720	69.3	6.9

Source: Statisticheski Yezhegodnik Stran-Chlenov Soveta Ekonomicheskoy Vzaimopomoshchi 1973. Moscow, 1973, p. 5; Rozwoj gospodarczy krajow RWPG 1950-1968. Warsaw, 1969, pp. 8-13.

TABLE B.2

GNP per Capita of Selected European Countries in U.S. Dollars, 1955
(estimates derived by the use of purchasing power parities)

Countries	GNP per capita	
	1	2
Bulgaria	515	
Czechoslovakia	1,163	
Hungary	835	
Poland	755	
Rumania	551	
Greece		336
Ireland		704
Norway		1,394
Spain		516
Yugoslavia	444	427

1 - derived from Pryor and Staller estimates.
2 - derived from Balassa estimates.

Source: Balassa B. and T.J. Bertrand: AER, May, 1970, p. 314.

TABLE B.3

Estimates of GDP (Gross Domestic Product) per Capita in CMEA and Some Western Countries, 1950-67

Countries	a 1953-55	In dollars 1963 "average" prices		Regional average=100		Average annual geometric growth rates %
		1950	1967	1950	1967	1950-1960
East Europe		431	1,093	100	100	5.6
Bulgaria	391	312	917	72	84	6.5
Czechoslovakia	828	698	1,421	162	130	4.3
GDR	716	553	1,476	128	135	6.0
Hungary	564	475	984	110	90	4.4
Poland	542	449	961	104	88	4.6
Rumania	318	243	723	56	66	6.6
USSR	528	417	1,104	97	101	5.9

		In dollars b				Average annual growth rates %
	1953-55	1959-61		1965-67		1953-67 c
Austria	608	853		1,070		5.7
France	1,069	1,316		1,654		4.5
GFR	954	1,296		1,589		4.7
Italy	549	726		926		6.0
Sweden	1,434	1,718		2,165		4.1
UK	1,183	1,329		1,516		4.3
Greece	317	408		580		n.a.
Spain	340	384		576		n.a.
Yugoslavia	172	268		384		n.a.
Japan	264	n.a.		868 d		9.4
USA	2,542	2,658		3,250		3.5

Notes: a - "adjusted GDP per capita at 1965 market prices
  b - at 1963 factor cost
  c - estimated as a function of per capita GDP
  d - in 1965 at official exchange rate

Source: _____ ; Economic Survey of Europe in 1969, Part I. New York, 1967, pp. 9, 60-61, 82, 99-100, 150.

TABLE B.4

Per Capita GDP, Investment and Consumption in Eastern Europe and Some Western Countries in 1965

Countries	GDP 1	GDP 2	Fixed capital formation 1	Fixed capital formation 2	Fixed capital formation 3	Consumption 1	Consumption 2	Consumption 3
Bulgaria	n.a.	877	n.a.	274	31.24	n.a.	573	65.34
Czechoslovakia	n.a.	1,427	n.a.	437	30.62	n.a.	935	65.52
GDP	n.a.	1,437	n.a.	422	29.37	n.a.	959	66.74
Hungary	n.a.	1,015	n.a.	282	27.78	n.a.	715	70.44
Poland	n.a.	989	n.a.	306	30.94	n.a.	646	65.32
Rumania	n.a.	697	n.a.	223	31.99	n.a.	464	66.57
USSR	n.a.	1,053	n.a.	353	33.52	n.a.	655	62.20
Yugoslavia	n.a.	692	n.a.	208	30.05	n.a.	479	69.22
Austria	1,273	1,459	326	381	26.11	944	1,061	72.72
France	1,922	1,616	417	393	24.32	1,482	1,198	74.13
GFR	1,913	1,854	503	490	26.43	1,375	1,319	71.14
Italy	1,021	1,190	208	323	27.14	851	860	72.27
Sweden	2,536	2,171	588	543	25.01	1,944	1,579	72.73
UK	1,802	1,929	324	417	21.62	1,474	1,477	76.57
Greece	677	758	163	181	23.88	574	582	76.78
Spain	680	939	155	268	28.54	526	680	72.42
Japan	868	1,293	274	332	25.68	566	955	73.86
USA	3,553	2,597	616	535	20.60	2,281	2,023	77.90
Norway	1,910	1,668	542	462	27.70	1,346	1,285	77.04

1 - in dollars at official exchange rates in 1965.
2 - ECE physical indicators estimates in "average prices".
3 - GDP in column 2 = 100.

Source: Economic Survey of Europe in 1969, Part I. New York, 1970, pp. 150-52.

TABLE B.5

Comparisons of Estimated per Capita Level of Industrial Output
in Eastern Europe, Selected Years, 1955-67
(Czechoslovakia = 100)

	Bulgaria	GDR	Hungary	Poland	Rumania
Pryor-Staller, 1955	26	99	64	52	34
M. Ernst, 1961	30	100	49	55	36
U.N., 1963	52	126	61	60	34
Z. Roman, 1964	39	103	58	51	32
Projected for 1967 Pryor-Staller	54	96	66	52	54
M. Ernst	48	99	55	59	52
U.N.	67	116	62	61	41
Z. Roman	47	111	57	52	37

Source: U. S. Congress. Joint Economic Committee: *Economic Developments in Countries of Eastern Europe.* Washington, 1970, p. 436.

TABLE B.6

Alton-Project Estimates of Levels of per Capita Agricultural Output: CMEA and United States, 1959, 1966, and 1970
(USSR = 100)

	Year	Total Output	Crop Output	Livestock Output
Bulgaria:				
	1959	104	137	71
	1966	114	143	85
	1970	105	128	86
Czechoslovakia:				
	1959	98	94	101
	1966	89	78	100
	1970	92	77	104
GDR:				
	1959	107	99	113
	1966	114	98	126
	1970	108	92	118
Hungary:				
	1959	142	160	124
	1966	131	141	121
	1970	122	127	118
Poland:				
	1959	125	134	116
	1966	123	128	119
	1970	114	126	105
Rumania:				
	1959	92	128	64
	1966	94	115	76
	1970	74	80	68
East Europe:				
	1959	112	124	101
	1966	112	117	107
	1970	102	106	100
USA:				
	1959	156	140	172
	1966	138	121	153
	1970	135	118	147
USSR:				
	1959	100	100	100
	1966	100	100	100
	1970	100	100	100

Source: Alton, T. P. (ed.): OP-46, Lazarcik. <u>Comparative Levels of Agricultural Output and Purchasing Powers of Currencies for Agricultural Products of Eastern European Countries, USSR, and USA, 1959, 1966, and 1970.</u> New York. 1975, p. 16.

TABLE B.7

Average Annual Growth Rates of National Income in CMEA Countries, 1951-74

Countries				Periods			
	1951-67	1961-72	1951-55	1956-60	1961-65	1966-70	1971-74
BULGARIA	9.6	7.6	12.2	9.6	6.6	8.7	7.4
CZECHOSLOVAKIA	6.0	4.5	8.1	7.0	1.9	6.9	5.1
GDR	7.7	4.5	13.2	7.4	3.4	5.2	5.5
HUNGARY	5.7	5.5	5.7	6.0	4.1	6.8	6.4
POLAND	7.1	6.6	8.6	6.6	6.2	6.0	10.2
RUMANIA	9.8	8.9	14.2	6.6	9.2	7.6	11.6
USSR	8.8	6.7	11.3	9.2	6.5	7.8	6.0

Source: Rozwoj gospodarczy krajow RWPG 1950-1968. Warsaw, 1969, p. 44; Rocznik statystyki miedzynarodowej 1973. Warsaw, 1973, p. 97; Rocznik statystyczny 1975. Warsaw, 1975, p. 563.

TABLE B.8

Fluctuating Annual Growth Rates of National Income in CMEA Countries, 1951-74

Years	Bulgaria	Czechoslovakia	GDR	Hungary	Poland	Rumania	USSR
1951	41	10	22	16	8	31	12
1952	-1	10	14	-2	6	5	12
1953	21	7	5	12	10	15	9
1954	0	4	9	-5	11	0	13
1955	5	11	9	-8	8	22	12
1956	1	5	4	-11	7	-7	11
1957	13	7	7	23	11	16	7
1958	7	8	11	6	6	4	12
1959	22	6	9	7	5	12	8
1960	7	8	4	9	4	11	8
1961	3	7	4	6	8	11	7
1962	6	1	2	5	2	4	6
1963	7	-2	3	6	7	10	4
1964	10	1	5	5	7	12	9
1965	7	3	5	1	7	10	7
1966	11	10	6	8	7	8	8
1967	9	7	5	9	6	8	7
1968	6	7	5	5	9	7	8
1969	10	7	5	8	3	7	5
1970	7	5	6	5	5	7	9
1971	7	5	5	7	8	13	6
1972	8	6	6	5	10	10	4
1973	8	5	6	7	11	11	8
1974	8	6	6	7	11	13	6

Source: Rozwoj gospodarczy krajów RWPG 1950-1968. Warsaw, 1969, p. 15-41; Rocznik statystyki międzynarodowej 1973. Warsaw, 1973, p. 96; Rocznik statystyczny 1975. Warsaw, 1975, p. 563.

TABLE B.9

Growth and Fluctuations of per Capita National Income in CMEA Countries, 1951-68

Years	Bulgaria		Czechoslovakia		GDR		Hungary		Poland		Rumania		USSR	
	1	2	1	2	1	2	1	2	1	2	1	2	1	2
1951	41	141	8	108	n.a.	n.a.	15	115	6	106	n.a.	n.a.	10	110
1952	-1	139	10	119	n.a.	n.a.	-3	112	4	110	n.a.	n.a.	9	120
1953	21	168	5	125	n.a.	n.a.	11	125	8	119	n.a.	n.a.	8	130
1954	-2	165	2	128	n.a.	n.a.	-6	117	8	129	n.a.	n.a.	10	143
1955	4	172	9	140	n.a.	191	7	126	6	137	n.a.	182	10	157
1956	0	172	4	146	n.a.	n.a.	-12	110	5	144	-8	167	10	172
1957	12	192	6	155	n.a.	n.a.	30	144	9	157	14	191	5	181
1958	6	204	7	166	n.a.	n.a.	0	144	4	163	2	195	10	199
1959	20	245	6	176	n.a.	n.a.	6	152	4	169	11	216	6	211
1960	6	260	7	188	n.a.	283	10	166	3	173	9	236	6	223
1961	2	265	6	199	4	297	6	174	7	186	10	259	5	234
1962	5	279	1	201	2	303	4	182	1	188	4	269	4	243
1963	7	298	-3	195	3	311	5	193	6	198	9	293	3	250
1964	9	325	0	194	6	328	4	201	5	208	11	325	8	269
1965	6	345	3	200	4	342	1	203	6	221	9	354	6	284
1966	10	381	10	218	5	360	8	219	6	235	8	381	7	304
1967	9	414	7	233	5	377	9	238	5	246	7	405	7	326
1968	6	438	6	247	5	396	5	249	8	265	5	426	7	348

1 - annual growth rates
2 - index 1950=100

Source: Rozwoj gospodarczy krajow RWPG 1950-1968. Warsaw, 1969, pp. 14-21, 24-43.

TABLE B.10

Alton-Project (Provisional) Estimates of GNP Dynamics in Eastern Europe, 1950-72
(1965 =100)

Date	Bulgaria	Czechoslovakia	GDR	Hungary	Poland	Rumania
1950	36.7	55.3	45.5	52.5	51.0	42.3
1955	49.9	65.4	68.0	68.3	64.2	59.7
1960	73.2	88.9	86.5	82.5	80.3	77.0
1961	76.7	92.4	87.5	86.5	86.7	82.9
1962	82.4	93.7	89.8	90.2	85.6	84.7
1963	85.6	91.9	92.5	95.2	90.7	88.7
1964	93.7	96.4	95.4	100.7	94.8	94.9
1965	100.0	100.0	100.0	100.0	100.0	100.0
1966	108.2	105.3	103.3	106.3	106.3	111.0
1967	116.2	110.4	106.8	112.3	110.6	117.9
1968	122.5	116.3	111.1	115.1	118.1	122.9
1969	130.7	120.7	114.8	119.7	118.8	130.0
1970	140.5	127.3	118.5	124.9	123.5	137.3
1971	147.7	133.9	122.4	133.1	131.2	153.7
1972	158.0	138.1	127.6	138.6	142.6	166.5

Source: _____ : Reorientation and Commercial Relations of the Economies of Eastern Europe. Washington, 1974, p. 270.

TABLE B.11

Dynamics of National Income in Eastern Europe, 1950-72
(1965 = 100)

	1950	1955	1960	1965	1970	1971	1972
Bulgaria	36[a]	46	72	100	152	163	175
Czechoslovakia	44	64	91	100	139	146	154
GDR	32	60	84	100	129	135	143
Hungary	46	61	82	100	139	148	156
Poland	36	54	74	100	134	144	159
Rumania	24	46	65	100	145	164	180

a) 1952

Source: Reorientation and Commercial Relations of the Economies of Eastern Europe. Washington, 1974, p. 271.

TABLE B.12

Comparison of Official and Alton-Project Revised Estimates of
Growth of Bulgarian National Product, 1950, 1955, and 1960-74
(1968 = 100)

	Official NMP	Alton-Project Indexes	
		NMP Coverage	GNP
1950	20	27	31
1955	35	39	43
1960	56	59	62
1961	58	61	65
1962	61	68	70
1963	66	71	73
1964	72	77	80
1965	78	86	86
1966	86	93	93
1967	94	98	98
1968	100	100	100
1969	110	104	104
1970	118	110	110
1971	126	113	113
1972	136	118	118
1973	146	123[a]	123[a]
1974	157[a]	125[a]	125[a]

a) Provisional figures

Source: Alton, T. P. (ed.): OP-44, Bass. <u>Bulgarian GNP by Sectors of Origin, 1950-1974.</u> New York, p. 3.

TABLE B.13

Shares of Industry (1), Construction (2), and Agriculture (3) in National Income Produced in CMEA Countries, 1950–72

Countries		1950	1955	1960	1965	1970	1972
Bulgaria	1	36.8	34.3	45.6	45.0	49.1	50.8
	2	6.6	8.0	7.1	7.3	8.7	8.8
	3	42.1	29.2	31.5	32.5	21.9	23.5
Czechoslovakia	1	62.5	63.3	63.4	66.4	62.1	62.0
	2	8.7	10.4	10.7	9.3	11.4	12.3
	3	16.2	14.7	14.7	12.0	10.1	10.5
GDR	1	47.0	52.1	56.4	59.2	60.9	61.1
	2	6.1	5.8	7.0	7.4	8.2	8.2
	3	28.4	20.2	16.4	13.8	11.7	11.1
Hungary	1	48.6	54.2	60.1	61.3	43.6	41.2
	2	6.8	5.9	10.4	9.0	12.0	12.5
	3	24.9	32.8	22.5	19.9	18.2	18.3
Poland	1	37.1	43.6	47.0	53.4	57.5	50.1
	2	7.9	8.9	9.3	9.0	9.8	9.1
	3	40.1	28.2	23.3	19.2	13.1	16.4
Rumania	1	43.4	39.3	42.1	48.6	60.1	57.6
	2	6.2	5.8	8.9	8.4	10.8	9.1
	3	27.3	36.8	34.8	28.2	20.1	22.6
USSR	1	57.5	54.1	52.3	51.7	51.2	52.0
	2	6.1	8.5	10.0	9.3	10.3	11.4
	3	21.8	22.8	20.5	22.5	21.8	18.8

Source: Statisticheski Yezhegodnik Stran-Chlenov Soveta Ekonomicheskoy Vzaimopomoshchi 1971. Moscow, 1971, p. 46; Statisticheski Yezhegodnik Stran-Chlenov Soveta Ekonomicheskoy Vzaimopomoshchi 1973. Moscow, 1973, p. 48.

TABLE B.14

Average Annual Rates of Growth of Value Added
in Key Sectors of CMEA Countries, 1951-65

Countries	Periods								
	1951-55			1956-60			1961-65		
	1	2	3	1	2	3	1	2	3
Bulgaria	11.6	15.8	6.8	18.5	9.8	3.7	9.0	7.7	2.0
Czechoslovakia	9.0	16.7	-1.5	8.8	11.6	-2.1	3.9	1.2	-5.6
GDR	14.3	15.1	6.8	8.7	7.9	7.1	4.5	2.8	9.2
Hungary	11.1	6.6	1.9	7.8	10.6	-1.7	7.5	2.9	-1.6
Poland	12.2	11.2	1.2	8.2	7.5	2.6	8.9	5.5	2.1
Rumania	16.9	17.2	11.2	11.2	13.8	0.5	13.7	7.0	0.3
USSR	14.5	12.6	2.3	9.9	11.6	5.3	9.1	4.1	1.2

1 - Industry
2 - Construction
3 - Agriculture

Source: Rozwoj gospodarczy krajow RWPG 1950-1968. Warsaw, 1969, p. 44.

TABLE B.15

Dynamics of Value Added in Key Sectors of CMEA Countries, 1950-71
(1960 = 100)

Years	Bulgaria			Czechoslovakia			GDR			Hungary			Poland			Rumania			USSR		
	1	2	3	1	2	3	1	2	3	1	2	3	1	2	3	1	2	3	1	2	3
1950	25	30	60	43	28	120	36	37	74	41	44	99	38	41	83	27	24	60	32	32	69
1965	154	145	110	121	106	75	126	125	101	140	114	97	153	131	111	190	140	101	154	122	106
1966	171	168	126	129	126	90	132	134	106	153	120	107	164	142	116	209	152	117	169	129	116
1967	194	196	121	134	136	98	140	143	113	167	137	108	176	159	114	237	176	116	187	143	116
1968	224	224	100	141	142	107	148	159	112	176	148	107	192	174	124	266	199	108	207	156	117
1969	253	236	106	150	143	112	157	172	104	184	161	120	208	185	99	299	213	110	227	162	106
1970	277	258	105	164	151	101	167	182	110	199	177	100	222	191	102	346	242	98	246	176	118
1971	302	269	103	174	166	102	176	192	105	212	190	108	241	201	111	386	262	123	264	192	112

1 - Industry
2 - Construction
3 - Agriculture

Source: Rocznik statystyki międzynarodowej 1973, pp. 98-99.

TABLE B.16

Alton-Project Estimates (Revised) of Dynamics of GNP by Sectors of Origin in Eastern Europe,
1965-74
(1965 = 100)

Countries		1966	1967	1968	1969	1970	1971	1972	1973	1974
Bulgaria[a]	1	108.0	113.9	116.2	121.3	128.0	131.1	137.2	143.0	145.6
	2	110.8	121.9	133.3	141.7	148.5	153.1	157.2	165.6	166.7
	3	106.8	106.4	96.4	97.4	101.9	100.8	107.8	108.3	105.0
	4	111.5	122.2	127.4	129.2	138.2	140.0	141.8	147.6	146.6
	5	103.4	107.1	110.9	115.2	119.8	124.4	129.6	134.8	140.1
	6	104.9	108.5	111.5	115.9	120.8	125.9	132.5	139.7	146.4
Czechoslovakia[b]	1	103.8	108.0	113.2	114.9	116.5	120.1	123.9	126.8	130.0
	2	101.5	105.5	109.8	111.7	116.8	118.7	122.4	125.8	128.9
	3	112.6	119.5	129.3	127.6	117.2	121.3	124.6	125.1	127.0
	4	105.4	110.7	112.6	109.7	117.5	125.7	129.4	132.0	136.2
	5	100.9	101.5	101.5	103.5	104.5	105.9	107.5	109.1	110.8
	6	100.9	103.6	106.6	111.5	109.8	113.8	116.9	118.8	120.6
GDR[a]	1	103.1	107.4	111.2	114.2	116.9	119.6	122.9	125.6	131.9
	2	103.0	106.5	110.5	115.3	119.5	122.5	122.1	123.3	131.8
	3	104.5	109.5	111.5	105.9	102.0	98.3	110.8	110.3	115.3
	4	106.8	114.6	127.3	137.6	145.4	152.2	157.6	164.9	175.1
	5	101.0	101.7	102.9	103.8	104.2	104.4	105.6	107.0	108.2
	6	100.8	101.5	103.8	105.4	106.9	108.9	110.2	112.9	115.6
Hungary[c]	1	106.3	112.6	114.3	118.2	117.1	123.0	125.6	132.4	139.3
	2	105.0	110.0	114.9	116.9	122.5	124.9	127.0	133.2	141.9
	3	111.8	122.3	115.3	122.4	100.2	110.9	113.4	122.7	127.7
	4	107.7	115.5	133.1	134.3	145.6	156.1	156.3	159.3	165.3
	5	102.7	104.3	104.6	104.9	105.3	107.0	108.8	111.5	113.7
	6	102.3	105.2	111.9	114.4	117.0	123.3	129.8	136.2	142.9
Poland[c]	1	106.3	110.5	117.8	117.0	121.1	131.4	141.8	152.7	163.5
	2	105.3	112.6	121.0	129.5	133.3	147.9	160.1	173.1	189.7
	3	108.9	107.9	114.0	95.2	99.3	107.3	112.8	117.9	120.0
	4	108.9	121.9	133.2	141.9	146.5	153.9	181.5	211.3	238.8
	5	104.3	108.2	112.0	116.0	119.9	124.9	128.0	131.5	135.4
	6	102.5	106.8	109.5	112.1	114.8	119.1	124.1	127.8	133.5
Rumania[a]	1	111.4	116.4	118.8	124.6	126.8	144.5	153.7	160.0	168.5
	2	112.2	124.8	136.8	153.4	168.7	183.4	198.2	217.5	241.5
	3	114.6	113.3	105.5	104.1	94.7	120.5	129.0	124.4	120.6
	4	105.7	111.9	123.0	134.1	142.4	160.8	162.7	186.0	201.8
	5	101.8	104.4	107.2	109.9	113.0	116.7	120.2	123.5	127.1
	6	102.3	103.6	103.1	104.6	101.9	102.3	105.9	106.1	100.5

1 - Total GNP; 2 - Industry; 3 - Agriculture; 4 - Construction; 5 - Housing;
6 - Government services

a) at 1968 adjusted factor cost
b) at 1967 adjusted factor cost
c) at 1969 adjusted factor cost

Source: Alton, T. P. (ed.): OP-48, P. Alton, M. Bass, Czirjak and Lazarcik.
Statistics on East European Economic Structure and Growth. New York, 1975, pp. 52-57.

TABLE B.17

Alton-Project Estimates (Revised) of Dynamics of GNP by Sectors Origin in Bulgaria, 1950-74
(1968 = 100; weights in percent of total GNP)[a]

	Industry	Agriculture & Forestry	Construction	Transport & Communications	Trade	Housing	Government & Other Services	GNP
1968 weights (%)	33.35	29.72	6.99	7.37	5.90	6.45	10.22	100.00
1950	13.0	50.4	19.0	18.2	17.0	62.0	42.0	31.3
1955	21.6	66.5	27.7	27.3	29.3	66.4	62.6	43.3
1960	43.1	84.5	50.5	45.7	54.2	77.6	69.1	61.6
1961	46.9	84.6	57.1	49.7	59.0	79.9	74.0	64.6
1962	52.5	91.9	60.8	54.1	61.6	82.0	77.1	69.9
1963	58.3	89.2	67.4	59.8	67.3	84.6	80.5	72.7
1964	65.6	100.9	70.8	67.4	71.2	87.4	84.7	80.3
1965	75.0	103.5	78.5	71.8	76.9	90.2	89.7	86.1
1966	83.1	110.5	87.5	78.8	83.4	93.3	93.4	93.0
1967	91.4	110.2	95.9	89.6	93.0	96.6	96.2	98.1
1968	100.0	100.0	100.0	100.0	100.0	100.0	100.0	100.0
1969	106.3	101.1	101.4	111.5	107.8	103.9	103.8	104.5
1970	111.4	105.7	108.5	124.4	116.2	108.1	108.3	110.2
1971	114.8	104.6	109.9	133.5	123.8	112.2	111.6	112.8
1972	117.9	111.8	111.3	142.5	131.9	116.9	117.5	118.2
1973*	124.2	112.3	115.9	154.4	143.5	121.6	122.4	123.1
1974	125.0	109.0	115.1	172.9	157.4	126.4	127.3	125.3

*Provisional figures

a) Weights - Based on 1968 adjusted factor cost.

Source: Alton, T. P. (ed.): OP-44, Bass. Bulgarian GNP by Sectors of Origin, 1950-1974. New York, p. 2.

TABLE B.18

Distribution of National Income into Consumption (1) and Accumulation (2) in Some CMEA Countries,
1950-73
(in "constant" prices)

Years	Bulgaria		Czechoslovakia		GDR[a]		Hungary		Poland	
	1	2	1	2	1	2	1	2	1	2
1950	n.a.	n.a.	94.1[b]	5.9[b]	90.4	9.6	77.0	23.0	78.9	21.1
1951	n.a.	n.a.	88.2	11.8	89.3	10.7	70.4	29.6	79.3	20.7
1952[c]	76.2	23.8	85.1	14.9	87.6	12.4	74.7	25.3	76.8	23.2
1953	69.7	30.3	83.3	16.7	86.6	13.4	71.9	28.1	71.5	28.5
1954	79.7	20.3	90.6	9.4	89.7	10.3	78.5	21.5	76.4	23.6
1955[d]	79.3	20.7	85.6	14.4	88.3	11.7	77.0	23.0	77.3	22.7
1956	85.7	14.3	86.9	13.1	85.5	14.5	91.2	8.8	79.2	20.8
1957	79.9	20.1	85.1	14.9	84.3	15.7	85.6	14.4	77.3	22.7
1958	80.5	19.4	80.9	19.1	81.4	18.6	80.2	19.8	77.4	22.6
1959	69.9	30.1	79.7	20.3	81.0	19.0	78.3	21.7	76.9	23.1
1960[e]	72.5	27.5	80.5	19.5	82.9	17.1	75.0	25.0	75.8	24.2
1961	77.7	22.3	78.2	21.8	83.4	16.6	74.4	25.6	75.0	25.0
1962	74.6	25.4	79.8	20.1	80.6	19.4	73.0	27.0	75.7	24.3
1963	71.0	29.0	84.8	15.2	81.3	18.7	72.1	27.9	74.5	25.5
1964[f]	69.0	31.0	85.9	14.1	80.9	19.1	72.0	28.0	74.3	25.7
1965[g]	71.7	28.3	86.5	13.5	79.8	20.2	75.6	24.4	72.9	27.1
1966	65.8	34.2	83.7	16.3	79.6	20.4	74.9	25.1	72.0	28.0
1967[h]	66.9	33.1	78.2	21.8	75.7	24.3	75.0	25.0	72.5	27.5
1968[i]	67.8	32.2	77.4	22.6	80.3	19.7	76.0	24.0	71.2	28.8
1969	68.6	31.4	74.0	26.0	78.4	21.6	76.2	23.8	72.3	27.7
1970	70.8	29.2	72.9	27.1	77.2	22.8	72.8	27.2	74.7	25.3
1971	76.0	24.0	74.0	26.0	78.0	22.0	69.0	31.0	71.0	29.0
1972	73.0	27.0	75.0	25.0	78.0	22.0	74.0	26.0	68.0	32.0
1973	72.0	28.0	73.0	27.0	78.0	22.0	74.0	26.0	65.0	35.0

a) 1950-54 in current prices
b) in 1955 prices for Czechoslovakia
c) in 1957 prices for Bulgaria
d) for Bulgaria 1955 in current prices, for Czechoslovakia 1955 in 1962 prices
e) all in current prices, except Poland in 1961 prices
f) for Bulgaria 1964 in 1962 prices, for Czechoslovakia 1964 in 1960 prices
g) for Bulgaria 1965 at current prices for Czechoslovakia 1965 in 1960 prices
h) all in current prices, except Poland in 1961 prices
i) for Bulgaria 1968 in 1962 prices, for Czechoslovakia 1968 in 1967 prices, for Hungary 1968 in 1967 prices

Source: Rozwoj gospodarczy krajow RWPG 1950-1968. Warsaw, 1969, p. 62; Polska wsrod krajow europejskich 1950-1970. Warsaw, 1971, p. 34; Rocznik statystyczny 1966. Warsaw, 1966, p. 602; Rocznik statystyczny 1967. Warsaw, 1967, p. 630; Rocznik statystyczny 1970. Warsaw, 1970, p. 600; Rocznik statystyczny 1971. Warsaw, 1971, p. 660; Rocznik statystyczny 1972. Warsaw, 1972, p. 626; Rocznik statystyczny 1973. Warsaw, 1973, p. 653; Rocznik statystyczny 1974. Warsaw, 1974, p. 652; Rocznik statystyczny 1975. Warsaw, 1975, p. 566; Statistical Yearbook of Bulgaria 1971. Sofia, 1971, p. 58.

TABLE B.19

Average Annual Rates of Growth of Consumption and Accumulation in Some CMEA Countries, 1951-67

Countries	1951-1967		1951-1955		1956-1960		1961-1965		1966-1967	
	1	2	1	2	1	2	1	2	1	2
Bulgaria	7.9[a]	11.8[a]	8.6[a]	2.3	8.7	17.2	6.6	8.3	7.7	22.8
Czechoslovakia	5.0	12.5	5.9	29.2	5.7	13.7	3.4	-4.3	5.8	17.0
GDR	n.a.	n.a.	n.a.	n.a.	6.6	16.5	2.5	6.8	4.1	6.3
Poland	6.5	8.7	8.3	.0.4	6.5	8.2	4.9	8.3	5.8	6.3
Hungary	4.9	8.9	5.1	6.8	5.6	12.6	4.1	3.0	5.2	20.8

1 - consumption; 2 - accumulation

a) in 1953-67 and in 1953-55

Source: Rozwoj gospodarczy krajow RWPG 1950-1968. Warsaw, 1969, pp. 45-46.

TABLE B.20

Rates of Growth of Consumption (1) and Accumulation (2) in Some CMEA Countries, 1955–73 (in "constant" prices)

Years	Bulgaria		Czechoslovakia		GDR		Hungary		Poland	
	1	2	1	2	1	2	1	2	1	2
1955	-22	-7	n.a.	n.a.	n.a.	n.a.	5	27	10	-2
1956	11	-29	6	-5	n.a.	n.a.	7	65	11	24
1957	5	58	8	25	n.a.	n.a.	3	71	3	3
1958	10	5	1	36	n.a.	n.a.	6	22	6	10
1959	12	100	5	12	n.a.	n.a.	6	39	2	8
1960	7	-6	9	4	n.a.	n.a.	2	5	6	11
1961	8	-18	4	26	4	1	4	12	4	0
1962	5	24	3	-8	n.a.	n.a.	6	11	5	12
1963	6	32		-18	4	16	6	6	5	6
1964	7	18	3	-5	4	11	3	-17	6	14
1965	7	-6	5	0	4	13	4	16	6	11
1966	6	40	5	26	4	6	6	27	5	2
1967	10	8	5	5	4	-10	6	-2	7	14
1968	9	1	9	10	5	21	6	-3	5	-1
1969	5	10	6	7	4	11	4	7	9	21
1970	6	-1	2	17	4	-2	6	27	8	16
1971	7	-11	6	-1	6	1	3	-25	9	22
1972	5	13	6	4	6	6	4	-1	11	27
1973	7	16	5	14						

Source: Rocznik statystyczny 1967. Warsaw, 1967, p. 630; Rocznik statystyczny 1970. Warsaw, 1970, p. 600; Rocznik statystyczny 1971. Warsaw, 1971, p. 660; Rocznik statystyczny 1972. Warsaw, 1972. p. 626; Rocznik statystyczny 1973. Warsaw, 1973, p. 653; Rocznik statystyczny 1974. Warsaw, 1974, p. 652; Rocznik statystyny 1975. Warsaw, 1975, p. 566; Statistical Yearbook of Bulgaria 1971. Sofia, 1971, p. 58.

TABLE B.21

Growth of National Income, Consumption, and Accumulation in Bulgaria, 1939–70
(1952 = 100)

Years	National Income	Per Capita National Income	Consumption	Accumulation
1939	72	83	n.a.	n.a.
1948	73	74	n.a.	n.a.
1955	127	124	128	107
1956	128	123	142	76
1957	145	138	149	120
1958	155	146	164	126
1959	189	176	183	252
1960	202	187	195	237
1961	208	190	211	194
1962	221	201	221	241
1963	237	214	233	317
1964	261	233	251	374
1965	279	248	269	353
1966	310	273	286	495
1967	339	296	313	533
1968	360	312	341	539
1969	396	341	360	594
1970	424	363	380	590

Source: Statisticheski godishnik na narodna republika Bulgaryi
1968. Sofia, 1968, pp. 100 and 102; Statistical Yearbook of Bulgaria
1971. Sofia, 1971, pp. 56 and 58.

TABLE B.22

Percentage Distribution of Accumulation in Some East European Countries, 1960–64

Countries		1960	1961	1962	1963	1964
BULGARIA	1	52.5	64.6	54.1	59.2	61.7
	2	47.5	35.4	45.9	40.8	38.8
HUNGARY	1	72.8	63.2	62.9	63.9	64.3
	2	27.2	36.8	37.1	36.1	35.7
POLAND	1	69.3	67.7	79.0	71.5	71.0
	2	30.7	32.3	21.0	28.5	29.0
USSR	1	66.1	59.0	63.1	66.9	58.8
	2	33.9	41.0	36.9	33.1	41.2
YUGOSLAVIA	1	87.4	94.5	98.9	86.9	76.2
	2	12.6	5.5	1.1	13.1	23.8

1 - investment expenditures
2 - increase of inventories

Source: _____, G. Michajlow, and S. Milewski: *Gospodarka zapasami w Krajach socjalistycznych*. Warsaw, 1967, p. 82.

TABLE B.23

Growth of Investments in CMEA and Some Western Countries, 1950-71
(in "constant" prices)

Countries	1950	1955	1965	1966	1967	1968	1969	1970	1971
				Index 1960=100					
Bulgaria	23	43	146	175	216	236	238	263	268
Czechoslovakia	34	53	111	123	127	141	160	170	165
GDR	22	51	127	137	149	165	189	202	205
Hungary	51	54	126	138	164	168	181	203	247
Poland	39	66	140	152	170	184	200	211	220
Rumania	23	53	171	188	219	244	258	281	321
USSR	30	54	136	146	158	171	178	194	209
Belgium	73[a]	83	141	150	152	150	161	177	167
France	58	73	162	176	186	196	216	231	246
Greece	40	41	130	150	143	180	217	213	n.a.
GFR	38	69	142	142	132	143	161	175	180
Italy	43[b]	68	114	118	130	141	153	165	155
Norway	72[b]	94	139	148	168	148	144	169	186
Spain	n.a.	n.a.	195	216	220	227	254	268	260
Sweden	61	76	137	145	153	154	160	163	161
U.K.	60	76	135	138	147	155	149	149	156

a) in 1953
b) in 1951

Source: Polska wsrod krajow europejskich 1950-1970. Warsaw, 1971, p. 35; Rocznik statystyczny 1975. Warsaw, 1975, p. 570; Rocznik statystyki miedzynarodowej 1973. Warsaw, 1973, p. 106.

TABLE B.24

Average Annual Rates of Growth of Investments in CMEA Countries, 1951-67
(in "constant" prices)

Countries	1951-67		1951-55		1956-60		1961-65		1966-67	
	1	2	1	2	1	2	1	2	1	2
Bulgaria	14.6	15.5	13.3	14.4	18.3	15.3	8.7	13.9	23.5	23.3
Czechoslovakia	8.1	8.1[a]	9.4	4.4	13.5	16.3	2.2	3.4	7.0	n.a.
GDR[b]	11.9	14.3[c]	18.2	19.2[c]	14.6	19.5[c]	5.0	7.6[c]	7.7	7.2[c]
Hungary	7.1	8.8	1.2	5.2	13.1	11.5	4.7	5.9	14.5	7.4
Poland	9.1	9.5	11.2	13.6	8.7	7.4	7.0	8.1	10.1	33.0
Rumania	14.2	15.0	18.2	22.4	13.5	8.7	11.3	13.7	12.7	16.2
USSR	10.2	9.4	12.4	12.3	12.9	11.2	6.3	7.0	7.6	4.6

1 - total investments
2 - investments in industry

a - in 1951-66
b - in current prices
c - in state owned

Source: Rozwoj gospodarczy krajow RWPG 1950-1968. Warsaw, 1969, p. 46.

TABLE B.25

Annual Rates of Growth of Investment in CMEA Countries, 1956-73
(in "constant" prices)

Years	Bulgaria	Czechoslovakia	GDR	Hungary	Poland	Rumania	USSR
1951	27	21	25[a]	n.a.	12	37	14
1952	22	18	29[a]	n.a.	19	32	11
1953	9	4	21[a]	n.a.	16	27	5
1954	6	-2	3[a]	n.a.	7	-10	18
1955	5	8	15[a]	n.a.	4	12	13
1956	2	13	26	n.a.	4	15	15
1957	-3	9	4	n.a.	7	-2	13
1958	22	13	14	n.a.	10	11	16
1959	63	19	21	n.a.	17	16	13
1960	18	12	10	n.a.	6	30	7
1961	4	8	1	-5	7	18	4
1962	7	-3	2	10	11	13	5
1963	15	-11	2	14	3	8	5
1964	10	11	10	4	4	10	9
1965	8	8	9	1	10	9	8
1966	22	10	7	9	9	10	7
1967	25	3	9	22	11	17	8
1968	9	10	10	2	9	11	8
1969	1	13	15	8	9	6	3
1970	11	6	7	15	4	12	11
1971	2	5	1	11	8	10	7
1972	10	9	4	-2	24	10	7
1973	7	9	8	-1	25	8	4

a) in current prices

Source: Rozwoj gospodarczy krajow RWPG 1950-1968. Warsaw, 1969, pp. 14-40; Rocznik statystyczny 1971. Warsaw, 1971, p. 661; Rocznik statystyczny 1972. Warsaw, 1972, p. 628; Rocznik statystyczny 1975. Warsaw, 1975, p. 570.

TABLE B. 26

Sectoral Distribution of Investments in Some CMEA and Western Countries, 1951-73 (in "constant" prices)

Countries	Years	Industry & Construction	Agriculture & Forestry	Transport & Communications	Housing
Czechoslovakia	1951-55	43.1	11.2	12.7	18.0
	1956-60	43.2	16.2	10.1	15.3
	1961-65	47.2	15.0	10.7	13.4
	1966-70	43.0	11.9	13.8	14.2
	1973	41.8	11.7	12.3	15.9
Hungary	1951-55	43.3	13.3	11.0	14.4
	1956-60	39.3	12.8	9.7	15.4
	1961-65	39.1	16.8	11.3	18.7
	1973a	36.5	14.1	10.9	18.6
Poland	1951-55	47.4	10.0	12.3	12.3
	1956-60	43.5	12.6	9.5	19.4
	1961-65	45.2	13.7	10.9	16.1
	1966-70	43.4	16.5	12.3	15.5
	1974	49I9	13.3	12.2	12.2
Rumania	1951-55	n.a.	11.3	n.a.	10.1
	1956-60	46.6	17.3	9.8	15.7
	1961-65	49.7	19.5	9.8	11.5
	1966-70	53.9	16.2	11.2	9.0
	1973	56.6	14.1	10.6	9.1
USSR	1951-55	43.4	15.1	9.7	20.0
	1956-60	38.2	15.1	9.3	23.6
	1961-65	38.9	11.6	10.3	18.2
	1966-70	39.0	17.0	9.6	17.0
	1973	39.1	20.2	10.7	15.3
Belgium	1956-60	32.8	4.3	12.1	26.5
	1961-65	37.5	3.0	10.6	23.6
	1966-70	36.2	2.8	10.4	25.3
	1972	34.3	2.5	10.9	21.6

Greece	1951-55	28.2[b]	7.7	19.2	29.1
	1956-60	17.7	11.7	30.2	29.9
	1961-64	16.4	12.5	26.9	31.4
	1966-70	22.7[b]	11.7	21.8	29.1
	1972	23.5[b]	10.6	21.0	32.2
GFR	1956-60	36.5	6.6	15.2	22.0
	1961-65	36.9	5.9	15.0	18.7
Italy	1956-60	31.4	11.2	10.2	28.8
	1961-65	32.2	9.1	10.2	29.2
	1966-70	27.8	7.8	8.6	33.2
	1973	34.2	6.6	11.3	28.3
Norway	1952-55	28.2	9.5	n.a.	17.8
	1956-60	25.4	9.4	36.9	14.2
	1961-65	29.3	7.4	34.9	13.0
	1966-70[c]	27.0	6.9	28.7	17.1
	1973[d]	31.1	5.9	18.0	n.a.
Portugal	1952-55	33.2	12.3	17.1	20.7
	1956-60	38.2	10.5	17.0	18.9
	1961-65	40.5	8.0	16.9	18.2
	1966-70	40.8	7.3	16.0	19.8
	1972	39.9	4.9	22.5	16.6
Sweden	1956-60	33.5	4.0	23.9	24.8
	1961-65	34.2	4.0	21.6	23.8
	1966-70	31.1	4.8	15.5	n.a.
	1972	33.9	5.0	15.9	n.a.
U.K.	1951-55	42.2	4.2	10.6	22.7
	1956-60	41.4	3.7	13.1	18.1
	1961-65	38.7	3.4	9.2	19.3
	1966-70	37.7	2.9	11.7	19.1
	1973	30.5	3.6	14.3	17.9

a) in current prices
b) without construction (included mainly in transport)
c) in 1963 prices
d) in 1970 prices

Source: Rocznik statystyczny 1967. Warsaw, 1967, p. 632; Rocznik statystyczny 1975. Warsaw, 1975, p. 571.

TABLE B.27

Sectoral Distribution of Investments in Bulgaria, 1949–73

Years	Industry	Construction	Agriculture	Transport	Housing
1949	31.4	2.2	11.7	16.5	20.0
1952	34.9	1.7	13.7	10.2	15.9
1956	36.8	0.5	21.6	6.2	18.6
1957	37.4	0.3	19.6	5.9	21.1
1960	34.2	1.6	27.9	5.4	14.1
1965	44.8	2.7	18.8	6.1	12.0
1967	45.8	3.8	16.2	8.9	9.1
1968	45.8	3.0	16.9	8.4	8.7
1969	47.8	2.3	15.4	8.6	9.6
1970	45.2	2.9	14.9	7.8	9.7
1973	45.8[a]	n.a.	15.7[b]	10.6[c]	11.8

a) including construction
b) including forestry
c) including communications

Source: Statistical Yearbook of Bulgaria 1971. Sofia, 1971, p. 51; Rocznik statystyczny 1975. Warsaw, 1975, p. 571.

TABLE B.28

Branch Distribution of Investment in Industry in CMEA Countries, 1961-67
(in "constant" prices)

Branches	Years	Bulgaria[a]	Czechoslovakia[b]	GDR[a,c]	Hungary	Poland	Rumania	USSR
Electric power	1961-65	15.3	12.3	14.9	14.0	12.1	13.5	11.9
	1966	15.9	14.0	12.4	11.8	10.5	18.2	12.3
	1967	15.1	12.9	11.3	11.2	10.2	14.7	12.1
Fuels	1961-65	16.9	14.4	23.2	19.3	21.8	20.4	19.4
	1966	11.1	14.4	22.2	19.3	18.5	21.2	21.5
	1967	9.5	11.0	17.4	15.7	16.9	16.8	20.9
Ferrous metallurgy	1961-65	12.7	14.6	9.1[d]	8.7	8.3	10.4	8.8
	1966	11.0	12.3	7.5[d]	6.7	7.2	10.9	7.7
	1967	9.3	9.8	7.9[d]	8.2	7.4	10.8	8.1
Machine building	1961-65	10.8	17.6	17.9	14.2	15.5	8.5	15.9
	1966	13.3	15.2	20.1	15.7	17.8	8.0	16.6
	1967	15.4	16.0	21.8	17.1	18.6	9.5	16.7
Chemicals	1961-65	8.8	11.1	15.7	13.9	12.1	14.3	10.2
	1966	14.4	12.0	16.0	14.4	16.4	11.0	10.2
	1967	18.6	13.0	17.8	13.8	16.0	17.1	9.5
Building materials	1961-65	4.0	5.0	4.6	5.0	6.5	4.1	5.5
	1966	5.3	4.4	3.7	4.1	5.0	4.4	4.6
	1967	5.2	6.5	3.5	6.7	5.7	4.1	4.5
Paper	1961-65	2.2	1.7	1.3	1.9	2.0	4.5	1.7
	1966	1.1	2.1	1.9	1.6	3.0	2.2	1.6
	1967	0.9	1.9	1.7	0.7	1.9	1.7	1.5
Textiles	1961-65	4.7	3.0	3.0	5.2	3.9	2.8	2.5
	1966	3.3	5.0	2.7	4.4	3.8	3.7	3.0
	1967	3.2	2.8	3.6	5.6	4.1	4.7	3.9
Clothing	1961-65	0.3	0.4	0.2	0.4	0.3	0.4	0.2
	1966	0.4	0.1	0.2	0.4	0.4	0.7	0.4
	1967	0.5	0.9	0.3	0.5	0.6	0.6	0.4
Food	1961-65	7.1	5.1	4.1	8.3	8.9	5.2	8.2
	1966	9.0	5.9	5.7	9.2	8.8	5.9	8.8
	1967	9.5	5.9	5.9	9.3	8.5	6.0	8.8

a) in current prices
b) in 1967 prices
c) investment expenditures in state-owned industry
d) includes non-ferrous metallurgy

Source: Rozwoj gospodarczy krajów RWPG 1950-1968. Warsaw, 1969, p. 64.

TABLE B.29

Branch Distribution of Fixed Assets in Industry in Bulgaria, 1952–70

Branches	1952	1956	1960	1965	1968	1969	1970
Electric power	19.48	20.29	21.00	16.70	16.02	15.02	14.40
Fuels	5.16	9.04	8.40	12.51	13.17	13.51	13.55
Ferrous metallurgy	0.29	2.55	4.26	9.56	11.75	11.68	11.02
Non-ferrous metallurgy	4.95	9.32	9.33	9.31	7.20	6.57	6.33
Machine building	10.79	10.57	11.39	11.39	11.74	12.19	12.84
Chemicals	6.28	7.24	6.80	8.54	9.39	10.69	11.97
Building materials	3.14	3.49	5.24	5.17	4.97	4.72	4.87
Timber & woodworking	5.20	4.49	4.14	3.05	2.89	2.85	2.75
Paper	1.05	1.66	1.17	2.03	1.71	1.64	1.46
Glassware & china	0.90	0.95	1.07	1.19	1.30	1.23	1.20
Textiles	12.72	9.68	7.78	6.37	5.76	5.83	5.79
Clothing	0.10	0.13	0.21	0.26	0.33	0.38	0.38
Leather & footwear	1.25	0.81	0.70	0.57	0.61	0.61	0.60
Printing	1.21	1.23	0.89	0.55	0.49	0.45	0.45
Food	24.00	14.84	15.00	11.24	10.99	10.88	10.68
Others	3.44	3.76	2.62	1.57	1.67	1.75	1.69

Source: Statistical Yearbook of Bulgaria 1971. Sofia, 1971, p. 89.

TABLE B.30

Indexes of Production and Inventory Dynamics in Some CMEA Countries, 1961-63
(1960 = 100)

	1961		1962		1963	
	Output	Inventory	Output	Inventory	Output	Inventory
Bulgaria	112.0	102.1	124.3	118.8	138.0	132.5
Czechoslovakia	108.8	106.3	115.4	117.8	115.1	124.0
GDR	104.9	120.4	108.9	128.6	111.3	135.5
Poland	110.3	114.3	119.7	124.2	126.2	133.4
USSR	109.0	107.5	119.6	117.5	129.6	127.3

Source: Cholinski T.: <u>Zapasy w prezedsiebiorstwie przemyslowym</u>. Warsaw, 1969, p. 31.

TABLE B.31

Share of Increase of Inventories in Distribution of National Income
in Some East European and Western Countries, 1960-64

Countries	1960	1961	1962	1963	1964
Bulgaria	13.0	8.2	11.4	11.5	11.8
Czechoslovakia	1.4	5.1	5.4	3.4	n.a.
Hungary	6.8	9.4	10.0	10.1	9.8
Poland	7.4	8.1	5.1	7.4	7.4
USSR	9.1	11.7	10.2	8.4	11.4
Yugoslavia	4.6	1.9	0.4	5.1	10.2
Austria	2.8	3.7	1.2	1.6	2.8
France	2.5	0.9	1.6	1.2	1.9
GFR	3.2	2.0	1.1	0.7	1.4
UK	2.3	1.2	0.3	0.7	1.7
USA	0.7	0.4	1.2	1.0	0.6

Source: Cholinski T.: <u>Zapasy w przedsiebiorstwie przemyslowym</u>. Warsaw, 1969, p. 18.

TABLE B.32

Average Annual Growth Rates of Gross Industrial Output in CMEA Countries, 1951-74

Countries	Periods						
	1951-67	1961-72	1951-55	1956-60	1961-65	1966-70	1971-74
Bulgaria	13.6	10.9	13.7	15.9	11.7	10.9	9.0
Czechoslovakia	8.6	6.1	10.9	10.5	5.2	6.7	6.8
GDR	9.1	6.1	13.8	9.2	6.0	6.5	6.6
Hungary	9.2	6.8	13.2	7.6	7.5	6.2	6.8
Poland	11.0	8.5	16.2	9.9	8.5	8.3	10.3
Rumania	13.2	12.6	15.1	10.9	13.8	11.9	13.2
USSR	10.5	8.3	13.2	10.4	8.6	8.5	7.4

Source: Rozwoj gospodarczy krajow RWPG 1950-1968. Warsaw, 1969, p. 47; Rocznik statystyki miedzynarodowej 1973. Warsaw, 1973, p. 118; Rocznik statystyczny 1975. Warsaw, 1975, p. 572.

TABLE B.33

Fluctuating Annual Growth Rates of Gross Industrial Output in CMEA Countries, 1951-74

Years	Bulgaria	Czechoslovakia	GDR	Hungary	Poland	Rumania	USSR
1951	19	14	23	24	19	24	16
1952	16	18	16	21	17	17	12
1953	15	9	12	12	19	15	12
1954	11	4	10	2	11	6	13
1955	8	11	8	9	8	14	12
1956	15	9	6	-8	9	11	11
1957	16	10	8	16	10	8	10
1958	15	11	11	11	10	10	10
1959	20	11	12	10	9	10	11
1960	12	12	8	12	11	16	10
1061	11	9	6	10	10	15	9
1962	10	6	6	8	8	14	10
1963	10	-1	4	7	5	12	8
1964	10	4	6	9	9	14	7
1965	15	8	6	5	9	13	9
1966	12	7	7	7	8	11	9
1967	13	7	6	9	8	14	10
1968	12	5	7	5	9	12	8
1969	9	5	7	3	9	11	7
1970	10	9	6	9	8	12	9
1971	9	7	6	7	8	12	8
1972	9	7	6	5	11	12	7
1973	9	7	7	7	11	15	7
1974	9	6	7	8	11	15	8

Source: Rozwoj gospodarczy krajow RWPG 1950-1968. Warsaw, 1969, pp. 15-41; Rocznik statystyki miedzynarodowej 1973. Warsaw, 1973, p. 117; Rocznik statystyczny 1975. Warsaw, 1975, p. 572.

TABLE B.34

Growth and Fluctuations of per Capita Gross Industrial Output in CMEA Countries, 1951-68

Years	Bulgaria		Czechoslovakia		GDR		Hungary		Poland		Rumania		USSR	
	1	2	1	2	1	2	1	2	1	2	1	2	1	2
1951	19	119	13	113	23	123	23	123	20	120	23	123	14	114
1952	16	139	17	132	16	142	19	146	17	140	16	143	11	126
1953	13	157	8	142	14	162	11	162	15	161	14	163	10	138
1954	10	172	3	146	10	179	1	164	9	176	4	170	11	153
1955	7	184	9	159	9	195	8	177	9	193	12	190	11	170
1956	14	209	8	172	8	210	-10	160	7	206	9	208	9	185
1957	15	240	9	187	9	228	23	196	8	222	7	222	8	200
1958	14	273	10	206	12	255	5	205	8	240	8	240	9	217
1959	19	326	10	226	13	288	9	224	7	258	9	262	9	237
1960	12	366	11	251	8	311	12	250	9	282	15	301	8	255
1961	11	406	8	270	7	334	10	274	9	309	15	345	7	273
1962	10	449	5	284	6	355	7	294	7	331	13	390	8	295
1963	10	492	-1	281	4	369	7	314	4	345	12	435	7	315
1964	9	537	4	291	8	397	8	340	8	371	14	495	6	333
1965	14	611	7	312	6	421	4	352	8	400	14	565	8	358
1966	11	680	6	332	6	445	7	375	7	427	11	616	7	384
1967	13	767	7	354	7	476	8	407	7	456	13	622	9	419
1968	11	853	5	371	6	502	4	424	8	494	10	759	7	448

1 - Annual rates of growth
2 - Index 1950=100

Source: Rozwoj gospodarczy krajow RWPG 1950-1968. Warsaw, 1969, pp. 14-43.

TABLE B.35

Shares of Producer Goods (A) and Consumer Goods (B) in Industrial Output in Some CMEA Countries, 1950-72

(in "constant" prices)

Countries	1950		1960		1965		1967		1970		1972	
	A	B	A	B	A	B	A	B	A	B	A	B
Bulgaria	38.2	61.8	47.2	52.8	52.3	47.7	53.0	47.0	54.7	45.3	58.1	41.9
Czechoslovakia	48.6[a]	51.4[a]	58.4	41.6	60.6	39.4	60.0	40.0	61.6	38.4	62.6	37.4
GDR	66.6[a]	33.4[a]	66.5	33.5	68.1	31.9	68.1	31.9	70.2	29.8	70.9	29.0
Hungary	65.0	35.0	66.0	34.0	65.4	34.6	65.2	34.8	65.1	34.9	64.7	35.3
Poland	52.6	47.4	59.4	40.6	60.9[b]	39.1[b]	64.4	35.6	65.0	35.0	64.7	35.3
Rumania	52.9	47.1	62.8	37.2	69.9	30.1	67.8	32.2	70.4	29.6	70.7	29.3
USSR	68.8	31.2	72.5	27.5	74.1	25.9	74.4	25.6	73.4	26.6	74.0	26.0

[a] 1955
[b] in 1963 prices

Source: Rozwoj gospodarczy krajow RWPG 1950-1968. Warsaw, 1969, p. 66; Statisticheski Yezhegodnik Stran-Chlenov Soveta Ekonomicheskoy Vzaimopomoshchi 1971. Moscow, 1971, pp. 59-60; Statisticheski Yezhegodnik Stran-Chlenov Soveta Ekonomicheskoy Vzaimopomoshchi 1973. Moscow, 1973, pp. 67-68.

TABLE B.36

Fluctuating Annual Growth Rates of Producer (A) and Consumer (B) Goods Output in Some CMEA Countries, 1951-68

Years	Bulgaria		Czechoslovakia		Hungary		Poland		Rumania		USSR	
	A	B	A	B	A	B	A	B	A	B	A	B
1951	17	21	18	0	n.a.	n.a.	25	19	27	22	17	16
1952	34	6	25	22	n.a.	n.a.	23	15	23	9	12	10
1953	13	16	12	4	n.a.	n.a.	19	15	20	8	11	13
1954	20	4	5	5	n.a.	n.a.	13	9	3	12	14	13
1955	7	8	6	13	n.a.	n.a.	11	12	12	15	15	8
1956	17	14	10	8	n.a.	n.a.	11	7	14	7	11	10
1957	13	19	9	10	n.a.	n.a.	8	12	10	7	11	8
1958	17	13	12	11	n.a.	n.a.	9	11	10	9	11	8
1959	25	16	13	8	12	9	13	5	14	4	12	10
1960	19	8	12	10	14	12	14	8	16	16	11	7
1961	10	13	9	9	11	11	12	8	16	14	10	7
1962	15	9	7	4	8	9	10	6	18	8	11	7
1963	14	7	-1	0	7	6	7	2	14	9	10	5
1964	14	7	6	3	8	10	10	8	17	10	9	4
1965	17	12	6	7	6	4	10	8	14	11	9	8
1966	13	12	7	6	8	6	8	6	12	10	9	7
1967	14	12	12	5	7	11	9	5	14	13	10	9
1968	n.a.	n.a.	n.a.	n.a.	n.a.	n.a.	10	8	n.a.	n.a.	n.a.	n.a.
1966-70[a]	12	10	7	6	6	7	9	6	13	9	8	8
1971	12	6	7	6	4	6	8	8	12	10	8	8
1972	11	6	7	6	4	7	10	13	11	12	7	6

[a] Average annual

Source: Rozwoj gospodarczy krajow RWPG 1950-1968. Warsaw, 1969, pp. 14-41; _____: GP, No. 4, 1973, p. 238; _____: Economic Survey of Europe in 1973. New York, 1974, p. 108.

TABLE B.37

Structure of Socialized Industry in Some CMEA Countries, 1950-67

Country	Years	Fuels	Machine-building	Chemicals	Timber & Paper	Textiles & Clothing	Food
Bulgaria	1950	3.8	9.1	2.9	11.0	17.6	41.0
	1960	2.8	12.4	3.8	7.2	18.7	33.7
	1967	3.7	18.8	5.4	5.0	14.7	28.9
Czechoslovakia	1950	9.0	13.8	2.5	10.1	11.6	30.9
	1960	7.7	24.5	4.2	6.7	9.3	21.6
	1967	7.0	27.8	6.1	6.2	8.1	18.3
Hungary	1950	6.7	24.5	4.9	3.5	21.2	21.2
	1960	9.9	25.1	4.4	3.9	10.9	18.8
	1967	8.2	27.9	7.6	3.9	9.7	17.1
Poland	1950	16.8	8.1	4.6	5.4	13.1	31.7
	1960	9.6	19.7	7.4	5.8	12.6	24.3
	1967	7.6	27.6	10.1	5.0	11.0	18.4
Rumania	1950	12.1	12.9	3.7	11.2	16.8	25.1
	1960	9.0	24.3	6.6	8.5	13.1	19.0
	1967	5.8	22.5	8.3	8.7	11.8	20.6

Source: Rozwoj gospodarczy krajow RWPG 1950-1968. Warsaw, 1969, p. 67.

TABLE B.38

Fluctuating Growth Rates of Output of Various Branches of Socialized Industry in Bulgaria, 1951-70

Years	Electric Power	Fuels	Machine Building	Chemicals	Textiles	Food
1951	32	6	26	-2	12	22
1952	35	13	21	50	30	11
1953	20	16	22	22	12	9
1954	14	27	17	17	13	1
1955	17	16	11	26	3	-2
1956	15	15	14	23	16	23
1957	11	-3	25	19	6	8
1958	14	4	24	17	9	16
1959	30	15	45	48	20	7
1960	21	12	22	21	9	13
1961	15	4	13	12	-3	18
1962	13	8	20	16	8	8
1963	13	19	21	16	6	9
1964	15	57	16	22	4	8
1965	18	13	21	19	14	11
1966	13	11	22	20	8	7
1967	16	9	19	21	11	8
1968	16	11	19	26	7	5
1969	11	24	14	18	9	13
1970	10	21	13	23	7	4

Source: Rozwoj gospodarczy krajow RWPG 1950-1968. Warsaw, 1969, pp. 14-15; Statistical Yearbook of Bulgaria 1971. Sofia, 1971, p. 79.

TABLE B.39

Alton-Project Preliminary Estimates of Industrial Growth in Eastern Europe
Compared to Some Western Countries, Prewar to 1967
(Eastern Europe: 1955 or 1956 = 100; Western Countries: 1958 = 100)

	Prewar	1950	1960	1965	1967
Bulgaria	24	60	200	348	444
Czechoslovakia	60	77	157	190	213
GDR	95	58	140	168	177
Hungary	43	61	132	186	205
Poland	60	62	147	205	230
Rumania	43	63	161	276	342
Austria	44	64	133	166	174
Belgium	64	80	112	148	153
France	64	76	133	166	183
West Germany	60	56	140	185	184
Greece	63	63	131	187	220
Italy	51	64	152	212	252
Netherlands	53	74	133	178	198
Norway	51	75	125	167	186

Source: Rocznik statystyczny 1973. Warsaw, 1973, p. 243.

TABLE B.40

Indexes of Industrial Production in Eastern Europe in 1973
(1965 = 100)

Countries		1973	Ratios of Indexes (Project = 100 in 1973)
Bulgaria	1	166[a]	1.00
	2	230	1.39
	3	218	1.31
Czechoslovakia	1	126	1.00
	2	156	1.24
	3	161	1.28
GDR	1	123	1.00
	2	156	1.27
	3	163	1.32
Hungary	1	133	1.00
	2	174	1.31
	3	162	1.22
Poland	1	173	1.00
	2	195	1.13
	3	198	1.14
Rumania	1	218	1.00
	2	268	1.23
	3	251	1.15

1 - Alton-Project (revised) estimates
2 - Official value-added series
3 - Official gross value series

a) Provisional figure

Source: Alton, T. P. (ed.): OP-48, P. Alton, M. Bass, Czirjak and Lazarcik. Statistics on East European Economic Structure and Growth. New York, 1975, p. 67.

TABLE B.41

Comparisons of Alton-Project (Provisional) Independently Constructed Indexes of Growth of Bulgarian Industrial Output with Official Indexes, 1939 and 1948-65
(1956 = 100)

	Alton-Project Index	Official Gross Value Output Index			Official Value-Added Index
		A	B	C	C
1939	22.2	14.8		15.1	25.4
1948	36.0	28.2		29.9	40.3
1949	44.5	38.0		39.6	48.1
1950	55.3	45.6		45.9	56.4
1951	62.4	54.6		54.9	55.1
1952	69.4	63.9		64.4	67.8
1953	74.6	73.5	78.0	73.6	74.8
1954	82.8	80.8		81.1	82.0
1955	91.9	86.5		88.2	91.1
1956	100.0	100.0		100.0	100.0
1957	111.3	116.1		n.a.	n.a.
1958	128.6	133.8		n.a.	n.a.
1959	160.1	162.3	161.0	n.a.	n.a.
1960	183.4	184.2	182.0	n.a.	n.a.
1961	199.6	206.5	203.0	n.a.	n.a.
1962	223.5	229.9	227.0	n.a.	n.a.
1963	248.3	254.4	250.0	n.a.	n.a.
1964	279.5	281.1	276.0	n.a.	n.a.
1965	319.4	322.1	317.0	n.a.	n.a.

A - According to Alton-Project in 1956 prices
B - According to Statistical Yearbook of Bulgaria 1971 prior to 1958 in 1956 prices, from 1960 in 1962 prices. Figures reproduced only when consequential differences from A occur.
C - in 1939 prices

Source: Alton, T. P. (ed.): op. 27, Lazarcik and Wynnyczuk. Bulgaria: Growth of Industrial Output, 1939 and 1948-1965. New York. 1968, p. 4; Statistical Yearbook of Bulgaria 1971. Sofia, 1971, p. 72.

TABLE B.42

Comparison of Official and Alton-Project Estimates of Growth of Production by Branches of Industry in Bulgaria, 1967-72
(1963 = 100)

Sectors	1967			1972		
	Official	A-P	Ratio O ÷ A-P	Official	A-P	Ratio O ÷ A-P
Power	178.9	189.7	0.943	285.7	310.0	0.922
Fuels	216.4	112.3	1.927	582.8	111.3	5.236
Ferrous Metals	249.0	242.9	1.025	500.0	456.7	1.095
Non-ferrous Metals	143.2	120.3	1.190	n.a.	120.8	--
Machinery & Metalworking	204.9	196.1	1.045	387.8	246.1	1.576
Chemicals	209.3	184.9	1.132	463.6	301.1	1.540
Construction Materials	163.5	137.5	1.189	233.5	149.9	1.558
Woodworking	118.5	139.3	0.851	154.6	190.5	0.812
Paper	198.5	172.6	1.150	311.9	211.6	1.474
Glass & Ceramics	177.4	121.3	1.462	280.5	121.9	2.301
Textiles	141.4	123.9	1.141	198.2	147.7	1.342
Clothing	179.5	n.a.	--	239.3	n.a.	--
Leather & Shoes	131.9	114.0	1.157	200.7	131.1	1.531
Printing	149.1	118.7	1.256	203.6	139.5	1.459
Food	139.6	140.8	0.991	178.4	166.8	1.070
Other	164.1	255.6	0.642	240.6	425.2	0.566
TOTAL	162.2	161.5	1.004	256.3	216.8	1.182

Source: Alton, T. P. (ed.): OP-47, J. Staller. Bulgaria: A New Industrial Production Index, 1963-1972 with Extension for 1973 and 1974. New York, 1975, p. 12.

TABLE B.43

Alton-Project Estimated Index of Industrial Production in Bulgaria, 1963-74
(1963 = 100)

Sector	Relative Weight	1964	1965	1966	1967	1968	1969	1970	1971	1972	1973	1974
Power	7.8	121.1	142.6	163.7	189.7	215.1	239.8	271.6	292.5	310.0	307.1	319.0
Fuels	8.7	107.8	109.0	109.9	112.3	117.6	114.7	118.0	110.9	111.3	108.9	97.9
Ferrous Metals	3.2	103.9	155.0	183.8	242.9	307.1	335.0	387.0	429.3	456.7	483.7	471.2
Non-ferrous Metals	8.5	109.6	115.0	116.7	120.3	119.8	119.5	123.0	120.8	120.8	n.a.	n.a.
Machinery & Metalworking	16.6	111.0	130.4	169.1	196.1	227.8	248.9	249.1	250.8	246.1	266.7	268.4
Chemicals	5.9	130.8	152.3	170.6	184.9	221.3	245.9	278.4	296.7	301.1	319.7	386.4
Construction Materials	5.5	107.1	118.5	124.2	137.5	140.6	143.3	147.0	145.2	149.9	187.8	193.2
Woodworking	7.1	100.7	113.6	127.0	139.3	152.0	163.2	171.3	174.8	190.5	194.1	196.4
Paper	1.3	125.3	134.6	159.7	172.6	198.2	201.5	199.7	209.0	211.6	233.5	286.4
Glass & Ceramics	1.6	100.9	114.4	108.4	121.3	128.5	120.3	125.2	125.1	121.9	130.6	109.8
Textiles & Clothing	10.7	103.7	113.2	119.2	123.9	130.2	139.5	138.9	142.3	147.7	155.1	158.5
Leather & Shoes	1.6	94.5	98.4	107.4	114.0	126.7	123.6	113.1	125.6	131.1	n.a.	n.a.
Printing	0.6	94.7	107.4	110.5	118.7	120.0	121.1	132.0	144.0	139.5	n.a.	n.a.
Food	13.4	109.1	122.8	132.9	140.8	142.0	149.4	156.2	162.3	166.8	171.8	157.7
Other	7.5	134.0	189.3	216.8	255.6	293.9	334.2	363.3	391.2	425.2	n.a.	n.a.
TOTAL	100.0	111.7	129.1	145.1	161.5	178.8	192.1	202.5	209.9	216.8	227.2	247.6

Source: Alton, T. P. (ed.): OP-47, J. Staller. Bulgaria: A New Industrial Production Index, 1963-1972, with Extension for 1973 and 1974. New York, 1975, p. 6.

TABLE B.44

Shares of Various Countries in the Output of Some Industrial and Agricultural Products in CMEA, 1950-67

Products	Bulgaria 1950	Bulgaria 1967	Czechoslovakia 1950	Czechoslovakia 1967	GDR 1950	GDR 1967	Hungary 1950	Hungary 1967	Poland 1950	Poland 1967	Rumania 1950	Rumania 1967	USSR 1950	USSR 1967
Industrial														
Electric Power	0.6	1.7	6.8	4.9	14.4	7.6	2.2	1.6	7.0	6.5	1.6	3.1	67.4	74.5
Lignite	2.2	5.0	10.4	13.2	51.9	45.1	4.5	4.3	1.8	4.5	n.a.	n.a.	28.7	26.3
Pit coal	n.a.	n.a.	6.1	4.3	1.0	0.3	0	0	27.2	20.2	n.a.	n.a.	64.5	73.6
Petroleum	n.a.	n.a.	n.a.	n.a.	n.a.	n.a.	1.2	0.6	n.a.	n.a.	11.5	4.4	86.8	94.5
Czude Steel	0	0.9	8.7	7.4	3.6	3.4	2.9	2.0	7.0	7.8	1.5	3.0	76.3	75.5
Tin ore	0	10.2	n.a.	n.a.	n.a.	n.a.	n.a.	n.a.	46.3	19.7	n.a.	n.a.	52.4	63.8
Lead ore	0	15.3	n.a.	n.a.	n.a.	n.a.	n.a.	n.a.	n.a.	n.a.	n.a.	n.a.	77.4	69.2
Sulfuric acid	0	2.5	8.2	7.0	9.7	6.9	2.2	3.0	9.2	8.4	1.7	4.7	69.0	67.5
Cement	3.2	2.8	10.8	5.3	7.6	5.9	4.3	2.2	13.6	9.1	5.5	5.2	55.0	69.5
Cellulose	0	1.3	16.6	9.6	13.9	7.7	n.a.	n.a.	8.0	8.8	3.6	7.2	56.9	64.0
Automobiles	n.a.	n.a.	25.4	21.9	7.5	21.9	0.8	0.7	n.a.	n.a.	0	n.a.	67.1	49.3
Trucks	n.a.	n.a.	2.0	3.7	0.3	4.1	3.5	3.3	9.9	9.8	3.1	4.3	96.6	81.8
Cotton yarn	1.7	3.4	8.2	5.7	2.6	3.9	3.5	3.3	9.9	9.8	3.1	4.3	71.0	69.6
Wool yarn	3.2	3.9	15.5	8.6	4.5	6.3	5.5	3.6	19.1	15.2	5.9	5.7	46.4	56.7
Agricultural														
Wheat	4.2	2.8	3.6	2.2	3.0	1.7	4.6	2.5	4.4	3.8	5.9	5.5	74.3	81.1
Corn	7.1	6.5	n.a.	n.a.	n.a.	n.a.	18.6	13.6	n.a.	n.a.	21.4	23.8	50.1	53.4
Sugar beet	1.4	1.6	13.2	6.9	13.1	6.0	4.9	3.4	14.3	12.8	2.3	3.2	50.8	66.0
Tobacco	15.7	28.3	n.a.	n.a.	n.a.	n.a.	n.a.	n.a.	n.a.	n.a.	6.0	8.3	84.5	35.6
Meat	n.a.	2.7	n.a.	5.5	6.4	6.0	n.a.	5.1	n.a.	11.8	n.a.	4.2	n.a.	62.7
Eggs	2.8	3.0	7.8	6.5	6.0	7.3	4.9	5.0	16.9	11.3	4.9	5.4	56.3	60.9

Source: Rozwoj gospodarczy krajow RWPG 1950-1968. Warsaw, 1969, pp. 8-13.

TABLE B.45

Index of Agricultural Output of CMEA Countries, 1950-68
(1950 = 100)

Years	Bulgaria	Czechoslovakia	GDR	Hungary	Poland	Rumania	USSR
1951	140	101	119	n.a.	93	125	93
1952	117	98	123	n.a	94	116	101
1953	142	98	133	103	97	136	104
1954	125	97	138	104	103	137	110
1955	137	108	140	117	105	162	122
1956	128	112	136	103	113	131	138
1957	149	111	145	117	118	163	143
1958	148	114	151	123	121	141	158
1959	175	113	146	129	120	168	159
1960	181	120	159	120	126	169	163
1961	176	120	141	121	140	179	167
1962	183	111	139	123	128	164	169
1963	186	119	151	130	133	170	157
1964	208	122	158	135	135	181	175
1965	212	117	170	130	145	192	183
1966	243	130	175	140	153	219	199
1967	250	137	186	146	157	222	202
1968	228	142	190	148	164	213	209

Source: Rozwoj gospodarczy krajow RWPG 1950-1968. Warsaw, 1969, pp. 16-43.

TABLE B.46

Index of Agricultural Output of CMEA Countries, 1960-73
(1970 = 100)

Countries	1960			1965			1971			1972			1973
	1	2	3	1	2	3	1	2	3	1	2	3	1
Bulgaria	72	75	67	84	84	85	102	99	106	108	108	108	109
Czechoslovakia	81	91	74	79	77	81	103	103	104	107	107	108	112
GDR	89	102	79	93	97	90	100	95	105	111	113	110	112
Hungary	82	90	73	87	91	83	109	112	106	114	121	107	118
Poland	80	78	83	91	91	93	104	101	107	112	108	117	120
Rumania	81	87	72	91	98	81	119	126	109	130	136	122	131
USSR	72	72	73	82	79	84	103	103	104	96	90	103	111

1 - Total agricultural
2 - Plant production
3 - Livestock production

Source: Statisticheski Yezhegodnik Stran-Chlenov Soveta Ekonomicheskoy Vzaimopomoshchi 1973. Moscow, 1973, p. 187; Rocznik statystyczny 1975. Warsaw, 1975, p. 596.

TABLE B.47

Fluctuating Growth Rates of Agricultural Output in CMEA Countries, 1951-73

Years	Bulgaria			Czechoslovakia			GDR			Hungary			Poland			Rumania			USSR		
	1	2	3	1	2	3	1	2	3	1	2	3	1	2	3	1	2	3	1	2	3
1951	40	54	15	1	6	-3	19	13	25	n.a.	n.a.	n.a.	-7	-10	-3	25	37	4	-7	-12	6
1952	-16	-26	7	-3	-10	4	3	-6	16	n.a.	n.a.	n.a.	2	3	0	-7	-10	1	9	11	3
1953	22	33	2	0	20	-18	8	16	1	18	n.a.	n.a.	3	0	7	17	26	-1	3	0	9
1954	-12	-18	1	-1	-11	11	4	4	2	2	n.a.	n.a.	6	8	3	1	-3	13	6	4	8
1955	9	10	7	11	13	10	1	-8	11	13	n.a.	n.a.	3	1	6	18	21	12	11	14	11
1956	-7	-8	-2	4	0	7	-3	-6	0	-12	n.a.	n.a.	7	8	8	-19	-25	-6	13	15	11
1957	17	22	6	-1	-3	4	7	5	8	13	n.a.	n.a.	4	2	8	24	35	5	4	-2	11
1958	-1	-6	11	3	5	1	4	3	5	4	n.a.	n.a.	3	2	4	-13	-21	3	10	15	5
1959	18	26	5	-1	-4	3	-3	-11	4	5	n.a.	n.a.	-1	-1	0	19	26	7	1	-5	8
1960	3	2	7	6	10	1	9	23	-2	-6	-8	-5	5	8	2	2	-1	8	3	2	-1
1961	-3	-9	8	0	-3	3	-11	-25	0	0	-4	6	8	12	9	5	2	10	2	2	4
1962	4	9	-4	-7	-12	-3	-1	16	-12	3	3	0	-8	-14	1	-8	-9	-7	1	-1	3
1963	2	4	-1	7	15	0	8	0	16	5	8	2	4	12	-6	4	10	-7	-7	-9	-6
1964	12	10	15	3	-2	8	4	3	5	4	2	12	1	1	2	6	3	14	11	29	-2
1965	2	-2	8	-4	-11	3	8	8	8	-4	-6	-4	8	8	7	6	7	5	5	-8	17
1966	15	19	6	11	19	3	3	1	5	8	11	4	5	6	5	14	16	11	9	14	4
1967	3	2	6	5	6	6	5	11	3	4	4	4	3	4	0	-1	-1	8	1	0	3
1968	-9	-13	-2	4	3	4	2	2	5	1	-1	5	4	5	3	-4	n.a.	n.a.	4	5	2
1969	4	n.a.	n.a.	1	n.a.	n.a.	-7	n.a.	n.a.	7	n.a.	n.a.	-5	n.a.	n.a.	3	n.a.	n.a.	-3	n.a.	n.a.
1970	4	n.a.	n.a.	1	n.a.	n.a.	4	n.a.	n.a.	-6	n.a.	n.a.	2	n.a.	n.a.	-5	n.a.	n.a.	10	n.a.	n.a.
1971	2	n.a.	n.a.	3	n.a.	n.a.	-1	n.a.	n.a.	9	n.a.	n.a.	4	n.a.	n.a.	19	n.a.	n.a.	1	n.a.	n.a.
1972	6	n.a.	n.a.	4	n.a.	n.a.	10	n.a.	n.a.	3	n.a.	n.a.	8	n.a.	n.a.	9	n.a.	n.a.	-4	n.a.	n.a.
1973	1	n.a.	n.a.	5	n.a.	n.a.	1	n.a.	n.a.	5	n.a.	n.a.	7	n.a.	n.a.	1	n.a.	n.a.	14	n.a.	n.a.

1 - total agricultural output
2 - plant production
3 - livestock production

Source: Rozwoj gospodarczy krajow RWPG 1950-1968. Warsaw, 1969, pp. 14-39; Rocznik statystyki miedzynarodowej 1973. Warsaw, 1973, p. 195; Rocznik statystyczny 1975. Warsaw, 1975, p. 596.

TABLE B.48

Alton-Project Estimates of Levels of Agricultural
Output per Agricultural Worker and per Hectare of
Agricultural Land in CMEA and United States, 1959, 1966, and 1970
(USSR=100)

	Year	Agricultural Output Per Person Employed	Agricultural Output Per Hectare	Crop Output Per Hectare	Livestock Output Per Hectare
Bulgaria:					
	1959	63	344	454	236
	1966	88	378	472	281
	1970	88	334	406	274
Czechoslovakia:					
	1959	163	428	411	441
	1966	172	413	361	461
	1970	173	419	351	472
GDR:					
	1959	265	688	634	728
	1966	285	709	609	788
	1970	297	659	564	720
Hungary:					
	1959	150	470	530	410
	1966	210	450	484	415
	1970	196	414	430	399
Poland:					
	1959	136	428	459	397
	1966	127	463	481	449
	1970	112	426	473	392
Rumania:					
	1959	53	276	384	192
	1966	57	284	346	227
	1970	48	224	243	208
East Europe:					
	1959	112	416	462	375
	1966	120	431	451	412
	1970	111	390	404	379
USA:					
	1959	745	138	123	152
	1966	885	138	121	153
	1970	944	139	122	152
USSR:					
	1959	100	100	100	100
	1966	100	100	100	100
	1970	100	100	100	100

Source: Alton, T. P. (ed.): OP-46, Lazarcik. Comparative Levels of Agricultural Output and Purchasing Powers of Currencies for Agricultural Products of Eastern European Countries, USSR, and USA, 1959, 1966, and 1970. New York, 1975, p. 18.

TABLE B.49

Alton-Project Estimates of Average Annual Growth Rates of Outputs, Inputs, and Selected Productivity Measures in Bulgarian Agriculture, 1948-70

	1948-1954	1954-1960	1960-1965	1965-1970	1948-1970
TOTAL PRODUCTION	4.2	5.9	4.3	1.7	4.3
Crop	5.4	5.9	4.8	0.9	4.1
Livestock	2.1	5.9	3.4	3.1	4.7
INTERMEDIATE PRODUCE	5.0	4.0	-0.4	4.9	3.3
Crop	6.0	4.4	-0.4	5.4	3.9
Livestock	-1.2	-0.6	-0.7	-1.8	-0.5
AGRICULTURAL OUTPUT	3.7	6.5	5.4	1.0	4.6
Crop	5.2	6.6	7.1	-0.8	4.3
Livestock	2.3	6.3	3.6	3.3	5.0
OPERATING EXPENSES	5.8	10.4	7.9	8.1	9.0
GROSS PRODUCT	3.5	5.8	4.9	-1.0	3.8
DEPRECIATION	2.2	5.8	5.7	6.9	5.3
NET PRODUCT	3.5	5.8	4.9	-1.6	3.6
CROP YIELDS					
Total production per ha. of agricultural land (1948 prices)	5.2	5.9	4.3	0.0	3.8
Per hectare sown:					
Wheat	7.7	7.7	5.4	0.6	4.6
Rye	3.0	2.9	2.3	0.1	0.8
Barley	10.5	7.9	1.9	1.0	3.8
Corn	4.2	6.3	2.2	8.2	7.0
Cotton	2.2	8.2	7.5	-4.2	4.8
Tobacco	-4.2	4.1	11.4	-2.1	3.1
Grapes	-0.7	0.8	13.2	-3.0	1.9
LIVESTOCK YIELDS					
Meat per pig	1.6	-8.5	5.6	0.4	0.2
Milk per cow	3.9	18.1	3.8	4.5	9.4
Mil per ewe	0.9	5.6	-7.4	3.4	0.2
Eggs per hen	0.6	3.3	1.6	3.0	4.4

Source: Alton T. P. (ed.): OP-35, Lazarcik. Bulgarian Agricultural Production, Output, Expenses, Gross and Net Product, and Productivity, at 1968 Prices, 1939, and 1948-1970. New York, 1973, p. 39.

TABLE B.50

Birthrate in CMEA Countries, 1950-74
(per 1,000 population)

Years	Bulgaria	Czechoslovakia	GDR	Hungary	Poland	Rumania	USSR
1950	25.2	23.3	16.5	20.9	30.7	26.2	26.7
1955	20.1	20.3	16.3	21.4	29.1	25.6	25.7
1960	17.8	15.9	17.0	14.7	22.6	19.1	24.9
1965	15.3	16.4	16.5	13.1	17.4	14.6	18.4
1969	17.0	15.5	14.0	15.0	16.3	23.3	17.0
1970	16.3	15.9	13.9	14.7	16.6	21.1	17.4
1974	17.2	19.8	10.6	17.8	18.4	20.3	18.2

Source: Polska wsrod krajow europejskick 1950-1970. Warsaw, 1971, p. 5; Rocznik statystyczny 1975. Warsaw, 1975, p. 557.

TABLE B.51

Rates of Birth and Natural Increase in Total, Urban, and Rural Population
in Bulgaria, 1939-70
(per 1,000 population)

Years	Birth Rates			Natural Increase		
	Total	Urban	Rural	Total	Urban	Rural
1939	21.4	16.9	22.7	8.0	4.7	9.0
1945	24.0	21.6	24.8	9.1	6.2	10.0
1950	25.2	24.9	25.3	15.0	15.5	14.8
1953	20.9	20.8	20.9	11.6	12.6	11.2
1955	20.1	18.7	20.8	11.1	11.5	10.9
1960	17.8	16.1	18.8	9.7	9.6	9.8
1961	17.4	15.8	18.3	9.5	9.6	9.4
1962	16.7	15.3	17.7	8.0	8.7	7.6
1963	16.4	15.5	17.0	8.2	9.3	7.5
1964	16.1	15.6	16.4	8.2	9.2	7.3
1965	15.3	14.7	15.9	7.2	8.5	6.1
1966	14.9	15.5	14.4	6.6	9.0	4.6
1967	15.0	15.9	14.2	6.0	8.9	3.3
1968	16.9	18.0	15.9	8.3	11.2	5.5
1969	17.0	18.7	15.2	7.5	11.3	3.5
1970	16.3	18.0	14.6	7.2	11.0	3.2

Source: Statistical Yearbook of Bulgaria 1971. Sofia, 1971, pp. 9-10.

TABLE B.52

Sectoral Distribution of Employment in Bulgaria, 1948-70

Sectors	1948	1956	1960	1965	1968	1970
Industry	7.9	12.9	21.9	26.3	29.0	30.4
Construction	2.0	3.3	5.2	7.0	7.9	8.4
Agriculture[a]	82.1	70.5	55.5	45.3	39.6	35.8
Transport[b]	1.5	3.0	4.1	5.1	5.7	5.7
Supply and Procurement	2.2	3.0	4.0	5.2	5.7	6.1
"Non-productive" sector	4.3	7.2	9.2	10.8	11.8	13.4

[a] Including forestry
[b] Including communications

Source: Statistical Yearbook of Bulgaria 1971. Sofia, 1971, p. 38.

TABLE B.53

Average Annual Growth Rates of Industrial Employment (1) and Labor Productivity (2) in CMEA Countries, 1951-67

Countries		1951-67	1951-55	1955-60	1961-65	1966-67
Bulgaria	1	6.9	5.2	11.5	4.4	6.8
	2	6.6	8.1	5.2	6.8	5.8
Czechoslovakia	1	2.8	3.7	3.2	1.8	1.9
	2	5.9	7.1	7.3	3.5	5.4
GDR	1	n.a.	n.a.	0.7	0.4	0.3
	2	n.a.	n.a.	7.8	5.6	5.4
Hungary	1	5.1	10.1	3.5	2.8	2.4
	2	4.4	3.9	4.1	4.9	5.4
Poland	1	3.9	7.0	2.1	3.2	3.4
	2	6.9	9.2	7.8	5.1	3.8
Rumania	1	4.8	5.9	3.0	6.0	3.6
	2	8.6	9.9	8.2	7.8	8.6
USSR	1	3.8	4.2	3.4	3.9	3.5
	2	6.4	8.3	6.5	4.6	6.3

Source: Rozwoj gospodarczy krajow RWPG 1950-1968. Warsaw, 1969, pp. 48-49.

TABLE B.54

Dynamics of Employment in CMEA Countries, 1951-67
(1950 = 100)

Years	Bulgaria		Czechoslovakia		GDR		Hungary		Poland		Rumania		USSR	
	1	2	1	2	1	2	1	2	1	2	1	2	1	2
1951	112	n.a.	n.a.	107	106	n.a.	113	108	108	108	114	110	105	106
1952	126	112	n.a.	111	110	n.a.	126	119	114	115	120	119	109	110
1953	135	116	117	114	114	n.a.	137	131	122	123	131	127	112	114
1954	146	124	121	117	121	129	140	140	126	129	135	131	122	121
1955	153	129	124	120	122	129	138	143	131	135	139	133	124	123
1956	157	136	127	123	121	128	140	145	136	140	141	138	130	126
1957	166	156	131	127	123	132	137	148	139	146	137	137	137	130
1958	175	168	133	130	123	135	141	153	138	147	139	142	140	134
1959	196	200	137	134	125	139	149	159	141	149	144	147	145	139
1960	220	222	141	140	124	139	157	167	143	149	153	154	153	144
1961	228	227	147	145	123	138	160	171	147	153	164	167	163	150
1962	234	233	151	149	123	137	165	175	153	159	176	179	169	155
1963	247	248	154	149	123	137	170	181	158	163	185	188	174	160
1964	258	257	157	150	124	138	176	187	161	166	194	195	181	166
1965	273	275	161	153	125	140	176	188	168	174	203	206	190	171
1966	298	297	166	157	126	140	178	191	174	180	212	213	197	176
1967	312	313	168	159	128	142	182	197	181	188	220	221	204	181
1968	321	322	171	161	129	143	189	203	187	194	224	230	212	186

1 - total
2 - in industry

Source: Rozwoj gospodarczy krajow RWPG 1950-1968. Warsaw, 1969, pp. 16-43.

TABLE B. 55

Sectoral Distribution of Employment in Some CMEA and Western Countries for Selected Years, Prewar-1972

Countries	Years	Industry and Construction	Agriculture and Forestry	Services a)
Bulgaria	1934	8.0	80.0	12.0
	1956	18.7	64.2	17.1
	1965	33.3	45.3	21.4
	1972	40.1	32.8	27.1
Czechoslovakia	1930	36.1	38.3	25.6
	1950	36.3	38.6	25.1
	1965	47.1	21.4	31.5
	1972	47.3	16.7	36.0
GDR	1946	41.3	29.2	29.5
	1960	48.3	17.3	34.4
	1965	48.1	15.5	36.4
	1972	49.8	12.0	38.2
Hungary	1930	24.1	53.0	22.9
	1950	23.7	50.1	26.2
	1964	37.4	31.7	30.9
	1972	43.3	25.0	31.7
Poland	1931	13.2	70.3	16.5
	1950	23.0	57.2	19.8
	1960	29.0	47.1	23.9
	1970	34.2	38.6	27.2
Rumania	1930	7.2	78.2	14.6
	1950	14.2	74.3	11.5
	1965	25.8	56.7	17.5
	1972	34.5	44.2	21.3
Austria	1951	37.2	32.3	30.5
	1961	41.0	22.8	36.2
	1972	40.7	16.4	42.9
Belgium	1930	47.8	17.0	35.2
	1947	48.7 b)	12.1	39.2
	1964	44.9	5.8	49.3
	1972	41.4	4.0	54.6
France	1931	33.7	35.6	30.7
	1954	36.3	26.7	37.0
	1962	37.6	19.8	42.6
	1972	37.0	12.3	50.7
Greece	1951	18.9	48.2	32.9
	1961	19.1	53.9	27.0
	1971	25.6	40.4	34.0
Italy	1936	29.3	48.2	22.5
	1951	30.4	40.0	29.6
	1965	40.4	25.1	34.5
	1972	43.0	17.5	39.5
Norway	1930	26.5	35.3	38.2
	1950	36.6	25.9	37.5
	1960	36.5	19.5	44.0
GFR	1950	42.9	23.2	33.9
	1971	48.9	8.2	42.9
Spain	1940	23.9	51.7	24.4
	1950	24.9	48.8	26.3
	1965	34.4	34.5	31.1
	1970	37.2	24.9	37.9
USA	1931	31.7	22.0	46.3
	1950	36.0	'2.2	51.8
	1965	34.1	6.2	59.7
	1972	32.0	4.2	63.8

a) As a residual, by deducting industry, construction, and agriculture from total.
b) Without electricity and gas producing industries.

Source: Rocznik statystyczny 1967. Warsaw, 1967, p. 626; Rocznik statystyczny 1971. Warsaw, 1971, p. 657; Rocznik statystyczny 1974. Warsaw, 1974, p. 647.

TABLE B.56

Index of Growth of Labor Productivity in Industry in CMEA Countries, 1965–72
(1960 = 100)

Years	Bulgaria	Czechoslovakia	GDR	Hungary	Poland	Rumania	USSR
1965	140	117	129	125	128	143	125
1966	143	122	137	132	133	154	130
1967	155	130	144	139	137	169	138
1968	171	136	152	139	145	181	144
1969	182	141	163	138	151	189	151
1970	195	152	174	153	163	202	162
1971	207	161	182	165	171	213	173
1972	222	167	185	173	181	225	183

Source: Rocznik statystyki międzynarodowej 1973. Warsaw, 1973, p. 122.

TABLE B.57

Growth of Employment and Labor Productivity as Factors in Growth of Industrial Output of CMEA Countries, 1961-72
(in percentages of growth of output)

	Bulgaria	Czechoslovakia	GDR	Hungary	Poland	Rumania	USSR
Growth of Employment							
1961-1972	23	21	10	23	29	27	27
1961-1965	32	34	27	34	34	37	41
1966-1970	31	18	5	33	37	31	30
1971-1972	25	26	52	0	37	48	18
Growth of Labor Productivity							
1961-1972	77	79	90	77	71	73	73
1961-1965	68	66	73	66	66	63	59
1966-1970	69	82	95	67	63	69	70
1971-1972	75	74	48	100	63	52	82

Source: Rocznik statystyki międzynarodowej 1973. Warsaw, 1973, p. 122.

TABLE B.58

Changes in Labor Productivity (1), Capital Productivity (2), and Capital Intensity (3) in CMEA Countries, 1961-69
(annual cumulative percentage rates of change)

Countries	Years	Productive Sectors			Industry		
		1	2	3	1	2	3
Bulgaria	1961-65	6.7	-3.0	10.2	7.2	-2.3	9.7
	1966-69	8.4	-1.9	10.5	6.5	-0.9	7.4
Czechoslovakia	1961-65	1.4	-2.5	4.1	3.3	-0.1	3.4
	1966-69	6.1	2.7	3.3	4.6	1.7	2.9
GDR	1961-65	3.7	-2.3	6.2	5.5	-0.8	6.4
	1966-69	5.1	0.7	4.4	5.8	1.3	4.4
Hungary	1961-65	4.8	0.0	4.9	5.1	-0.7	5.7
	1966-69	5.8	2.9	2.9	2.3	-0.5	2.5
Poland	1961-65	3.4	1.6	1.8	5.3	1.7	3.6
	1966-69	3.4	-0.1	3.6	4.6	0.5	4.2
Rumania	1961-65	9.2	1.0	8.2	7.5	3.8	3.5
	1966-69	7.9	-0.4	8.3	7.3	-1.1	8.5
USSR	1961-65	4.3	-3.0	7.5	4.5	-2.3	6.9
	1966-69	6.2	-0.4	6.7	5.1	0.0	5.1

Source: United Nation: Economic Survey of Europe in 1970. Part II. New York, 1971, pp. 72 and 117.

TABLE B.59

Incremental Gross Capital-Output Ratios in Some CMEA Countries, 1955-75

Years	Bulgaria	Czechoslovakia	GDR	Hungary	Poland	USSR
1955-60	1.92	3.14	n.a.	2.84	3.68	2.53
1960-65	3.89	14.28	n.a.	5.65	4.62	3.83
1965-70 Plan	3.8	6.7	n.a.	6.9	4.7	3.7
1966-70	4.0	4.5	5.3	4.8	4.7	3.7
1971	4.8	6.2	6.4	5.8	3.5	5.2
1972	5.1	5.5	4.9	6.9	3.2	7.7
1973 Plan	4.1	6.5	5.1	7.7	4.3	4.9
1973	4.0	6.4	5.3	4.9	3.6	3.6
1974 Plan	n.a.	n.a.	5.9	6.7	3.9	4.1
1971-73	4.7	6.0	5.5	5.8	3.5	5.1
1971-75 Plan	3.9	6.3	5.7	6.2	4.3	4.0

Source: United Nations: Economic Survey of Europe 1966. New York, 1967, Chapter II, p. 45; United Nations: Economic Survey of Europe in 1973. New York, 1974, pp. 122-23.

TABLE B.60

Alton-Project Estimated Annual Growth Rates of Industrial Output per Unit of Labor, Capital, and Combined Inputs in Some East European Countries, 1950-67[a]

	Bulgaria	Czechoslovakia	GDR	Hungary	Poland	Rumania
Output per unit of labor:						
1950-67	5.1	3.4	4.2	2.8	3.5	5.1
1950-60	4.1	4.2	5.0	2.2	4.0	4.7
1960-67	6.6	2.1	3.6	3.8	2.7	5.6
Output per unit of capital:						
1950-67	.6	.3	n.a.	.8	n.a.	n.a.
1950-60	2.5	1.7	n.a.	1.9	n.a.	n.a.
1960-67	-2.0	-2.3	-2.5	-.9	-.5	.7
Output per unit of combined inputs:						
1950-67	3.6	2.0	n.a.	1.9	n.a.	n.a.
1950-60	3.9	3.2	n.a.	2.1	n.a.	n.a.
1960-67	3.3	0	1.8	1.7	1.3	3.3

[a] The initial year for the periods differs for some countries from 1950, as follows: Czechoslovakia, 1948; East Germany, 1955; Hungary, 1949. For Bulgaria, the capital input begins with 1952.

Source: U.S. Congress. Joint Economic Committee: *Economic Developments in Countries of Eastern Europe.* Washington, 1970, p. 438.

TABLE B.61

Alton-Project Estimated Average Annual Rates of Growth of Fixed Capital Inputs and Capital Productivity in East European Countries, 1960-72[a]
(in constant prices; percent)

	1960-65	1965-70	1967-72	1960-72
Bulgaria:				
Capital inputs:				
Industry	13.6	12.7	11.0	12.3
Agriculture	9.0	7.1	6.6	7.7
Construction	16.8	16.0	10.9	15.7
Capital Productivity:				
Industry	-5.1	-.5	-.9	-2.0
Agriculture	-4.5	-7.9	-5.9	-5.6
Construction	-9.2	-4.9	-4.3	-6.2
Czechoslovakia:				
Capital inputs:				
Industry	5.2	4.2	4.9	4.6
Agriculture	4.9	4.3	5.0	4.5
Construction	7.5	7.7	8.6	7.3
Capital Productivity:				
Industry	-2.	1.9	1.1	.4
Agriculture	-8.5	.6	-1.5	-2.4
Construction	-8.3	-5.0	-5.0	-4.9
GDR				
Capital inputs:				
Industry	6.4	4.9	5.5	5.7
Agriculture	6.6	5.6	5.0	5.7
Construction	9.9	10.1	10.4	9.2
Capital Productivity:				
Industry	-2.8	-1.1	-1.5	-2.1
Agriculture	-5.2	-5.1	-6.2	-3.8
Construction	-5.8	-2.2	-4.2	-2.8
Hungary:				
Capital inputs:				
Industry	5.6	6.0	6.4	6.0
Agriculture	5.2	7.0	8.6	6.7
Construction	11.2	10.9	15.0	11.9
Capital Productivity:				
Industry	.8	-1.3	-1.5	-.5
Agriculture	-3.9	-4.5	-7.2	-5.1
Construction	-7.2	-2.1	-8.5	-5.5
Poland:				
Capital inputs:				
Industry	6.5	7.5	7.6	7.1
Agriculture	2.1	3.8	3.9	3.2
Construction	9.4	11.4	9.3	9.9
Capital Productivity:				
Industry	0	-1.3	-1.1	-.7
Agriculture	-1.2	-4.8	-4.2	-2.0
Construction	-6.1	-3.4	-1.9	-3.3
Rumania:				
Capital inputs:				
Industry	9.5	12.3	12.4	11.2
Agriculture	4.2	5.8	7.7	6.0
Construction	10.4	10.5	11.3	11.1
Capital Productivity				
Industry	.5	-1.8	-2.0	-.9
Agriculture	-4.3	-7.6	-4.6	-4.3
Construction	-4.4	.7	-1.4	-2.3

[a] The capital productivity indexes are the ratios of the sectoral output (GNP) indexes to the capital input indexes.

Source: United Nations: Reorientation and Commercial Relations of the Economies of Eastern Europe. Washington, 1974, p. 281.

TABLE B.62

Index of Nominal Nonfarm Wages in Some CMEA Countries, 1955-74
(1960 = 100)

Years	Bulgaria	Czechoslovakia	Hungary	Poland	Rumania	USSR
1955	80	88	73	65	63	90
1965	118	108	112	120	131	120
1966	123	111	117	125	138	124
1967	137	118	121	130	142	130
1968	145	128	124	136	146	140
1969	150	137	127	141	152	145
1970	158	142	139	144	168	151
1974	183	164	171	205	194	175

Source: Polska wsrod krajow europejskich 1950-1970. Warsaw, 1971, p. 119; Rocznik statystyczny 1975. Warsaw, 1975, p. 567.

TABLE B.63

Sectoral Indexes of Monthly Wages in Some CMEA Countries, 1960-72
(industry = 100)

Countries	1960			1965			1970			1972		
	1	2	3	1	2	3	1	2	3	1	2	3
Bulgaria	98	120	112	97	118	108	100	121	112	99	117	110
Czechoslovakia	95	105	101	95	108	103	98	113	114	99	112	114[a]
Hungary	97	103	92	99	104	100	103	112	107	102	111	108
Poland	91	105	90	92	108	95	93	113	98	94	115	100
USSR	88	101	95	92	108	102	92	113	103	92	112	106

1 - national economy without agriculture and forestry

2 - construction

3 - transport

[a] in 1971

Source: Rocznik statystyki międzynarodowej 1973. Warsaw, 1973, p. 342.

TABLE B.64

Index of Growth of Real Wages in CMEA Countries, 1951-67
(1950 = 100)

Year	Bulgaria	Czechoslovakia	GDR	Hungary	Poland[a]	Rumania[b]	USSR[c]
1951	95	n.a.	n.a.	89			
1952	104	n.a.	n.a.	81			
1953	117	n.a.	n.a.	86			
1954	137	104	n.a.	101			
1955	150	109	211	105	100		
1956	171	116	229	117	112		
1957	178	121	244	138	121		
1958	180	124	260	144	125		100
1959	188	129	287	151	131		103
1960	202	136	302	154	129	100	105
1961	210	140	313	154	132	103	109
1962	210	139	314	157	133	108	111
1963	212	140	319	163	136	113	112
1964	216	142	328	168	139	115	115
1965	222	143	341	168	139	122	123
1966	234	147	350	171	144	129	129
1967	257	152	353	179	147	132	134

[a] 1955 = 100

[b] 1960 = 100

[c] 1958 = 100

Source: Rozwoj gospodarczy krajow RWPG 1950-1968. Warsaw, 1969, pp. 16-43.

TABLE B.65

Index of Growth of Real Wages in Some CMEA and Western Countries, 1965-74
(1960 = 100)

Countries	1965	1970	1974
Bulgaria	110	143	159[a]
Czechoslovakia	106	127	142[a]
Hungary	109	129	147
Poland	108	120	163
Rumania	122	147	157[a]
USSR	109	134	149[a]
Austria[b]	129	168	201[a]
Belgium[b]	129	161	219
France[c]	121	148	197
GFR[bde]	139	173	197[a]
Italy[bd]	132	168	207
Japan[f]	120	174	230
Spain[fdg]	108	159	195[a]
UK[hi]	116	136	156[a]
USA	109	118	121
Yugoslavia[fg]	116	163	174

[a] in 1973

[b] without the trade sector

[c] without the mining industry

[d] without transport

[e] including family allowances

[f] including salaries of white-collar workers

[g] 1963 = 100

[h] wages of men

[i] excluding wages of workers in coal mining, trade, and railway transport

Source: Rocznik statystyczny 1975. Warsaw, 1975, p. 568.

TABLE B.66

Index of Growth of Average Annual Wages and Salaries by Sectors of Bulgarian Economy, 1948-70
(1960 = 100)

	1948	1952	1956	1960	1965	1967	1968	1969	1970
Total	52	69	83	100	118	137	145	150	158
Industry	48	69	84	100	119	133	142	146	154
Construction	55	70	82	100	117	136	142	146	156
State-owned farms	43	53	73	100	111	128	133	137	144
Cooperative farms	n.a.	n.a.	n.a.	100	161	180	189	200	207
Transport	60	68	77	100	115	131	141	146	156
Trade & distribution	54	75	84	100	122	147	160	162	169
Housing & public utilities	47	77	83	100	118	137	150	158	166
Science	80	72	96	100	134	169	176	176	190
Education & culture	56	71	92	100	118	145	155	157	165
Health	49	70	87	100	113	140	150	150	156
Administration	49	65	81	100	125	157	167	174	179

Source: *Statistical Yearbook of Bulgaria 1971.* Sofia, 1971, p. 280.

TABLE B.67

Index of Average Annual Wages and Salaries by Sectors of Bulgarian Economy, 1948-70
(average economy - wide annual wage = 100)

	1948	1960	1970
Industry	94	102	100
Construction	132	123	121
State-owned farms	79	95	87
Cooperative farms	n.a.	76	99
Transport	132	114	113
Trade & distribution	90	85	91
Housing & public utilities	77	84	88
Science	144	93	112
Education & culture	99	90	94
Health	83	88	86
Administration	96	101	115

Source: Statistical Yearbook of Bulgaria 1971. Sofia, 1971, p. 280.

TABLE B.68

Per Capita Consumption of Some Foodstuffs in CMEA and Some Western Countries, 1960-73
(in kilograms)

Countries	Years[a]	Grain[b]	Potatoes	Vegetables	Fruit	Meat	Fish	Fats[c]	Milk	Eggs	Sugar
Bulgaria	1960	190	35	122	70	29	2.0	14[d]	95	4.7	18
	1973	170	27	116	89	50	5.7	17[d]	134	7.5	33
Czechoslovakia	1960	126	100	63	46	57	4.7	19	173	9.9	36
	1973	109	105	79	40	77	5.5	20	206	16	37
GDR	1960	102	174	61	80	55	n.a.	27	95[e]	11	29
	1973	95	143	100	69	74	8.3	27	102[e]	14	36
Hungary	1960	136	98	84	55	48	1.5	24[a]	114	8.9	27
	1973	125	67	85	74	63	2.8	28[d]	117	15	37
Poland	1960	147	223	n.a.	n.a.	43	4.5	14	234	7.9	28
	1973	127	183	93	20	62	7.2	19	271	11	42
Rumania	1964-66	183	66	60	35	38	1.5	9	116	5.1	18
USSR	1960	164	143	70	22	36	9.9	12[d]	153	6.6	28
	1973	145	124	85	40	48	16	17[d]	188	11	41
Austria	1960-61	107	88	69	n.a.	60	2.1	17	211[f]	12	35
	1972-73	87	60	73	105[f]	81[g]	3.7[f]	25[g]	199[f]	15	37
France	1960	99	114	140	66	71	7.3	21	190	11	31
	1971-72	76	96	122	87[f]	96	8.0[f]	26	230[f]	13	37
Greece	1960-62	157	39	135	110	26	9.1	18	126	6.9	16
	1967	121	59	139	145	41	10.0	19	164	11	20
GFR	1960-61	82	132	49	105	60	6.6	25	198	13	32
	1972-73	68	94	67	110	79	4.0	26	207[f]	17	34
Italy	1960	144	44	127	81	29	5.1	15	126	8.8	20
	1971-72	134	39	159	108[i]	59	6.2[f]	25	144[f]	11	28
Norway	1960-62	78	99	36	64	40	20	23	243	8.8	42
	1971	70	83	36	74[f]	44	18[f]	25	261[f]	10	43
Spain	1960-62	116	115	130	87	21	14	20	77	7.3	19
	1971	82	108	134	92[f]	44	14[f]	18	108[f]	12	27
Sweden	1960	71	88	30	72	49	18	21	262	12	41
	1971	61	84	41	96	52	21	20	264	13	42
USA	1960	66	49	111	97	95	4.7	21	246	19	46
	1971	64	54	100	101[j]	114	6.6[j]	24	251[j]	18	51
Yugoslavia	1960	186	70	57	41	30	0.5	10	109	3.6	15
	1971	174	65	71	60	36	0.9[k]	18	102[k]	7.6	27

[a] during periods, average annual.
[b] recomputed by processed foods.
[c] recomputed for 100% fat.
[d] without recomputing for 100% fat.
[e] without processed foods.
[f] 1969-70.
[g] 1971.
[i] 1970-71.
[j] 1970.
[k] 1969.

Source: Rocznik statystyczny 1975. Warsaw, 1975, p. 569.

TABLE B.69

Retail Sales of Some Durables in CMEA Countries, 1960-72
(numbers per 1,000 population)

Countries	Years	Radios	TV sets	Motorcycles	Washing Machines	Refrigerators & Freezers	Vacuum Cleaners
Bulgaria	1960	19.8	0.5	4.2	4.9	0.4	n.a.
	1965	17.3	9.4	5.8	13.8	5.7	n.a.
	1970	20.7	25.3	5.3	14.2	22.2	8.2
	1972	20.3	17.4	5.7	8.9	21.4	8.5
Czechoslo-vakia	1960	22.1	22.1	11.1	18.9	8.0	n.a.
	1965	23.2	18.2	6.4	16.5	16.1	n.a.
	1970	31.7	18.9	4.0	20.9	15.6	n.a.
	1972	38.3	19.5	3.1	25.6	17.8	n.a.
GDR	1960	37.2	26.6	13.0	9.2	8.4	24.3
	1965	28.3	28.2	7.8	18.7	19.5	13.6
	1970	48.2	16.2	9.7	16.9	22.4	13.9
	1972	56.8	21.9	9.9	16.1	24.1	17.4
Hungary	1960	18.0	5.3	4.0	12.2	1.3	2.9
	1965	14.0	15.4	3.0	16.1	7.9	10.2
	1970	62.0	22.6	8.4	15.7	22.0	13.4
	1972	52.3	20.7	8.7	13.8	23.8	13.2
Poland	1960	20.2	6.8	6.2	18.1	1.8	4.4
	1965	16.6	13.7	6.4	14.0	7.0	8.2
	1970	25.6	16.7	6.2	11.6	10.9	10.2
	1972	32.9	24.4	7.0	14.3	15.8	12.1
Rumania	1960	8.6	1.4	2.0	1.7	0.4	0.3
	1965	14.1	8.4	1.0	4.4	6.2	3.2
	1970	13.9	12.0	0.3	5.4	7.1	3.2
	1972	18.6	14.1	0.5	5.8	8.0	2.9
USSR	1960	19.5	6.9	2.3	4.2	2.4	1.9
	1965	21.6	14.5	3.0	13.6	6.3	2.8
	1970	24.2	23.0	3.4	17.0	15.5	4.7
	1972	24.0	27.5	3.6	12.7	18.2	6.1

Source: Rocznik statystyki miedzynarodowej 1973. Warsaw, 1973, pp. 355-57.

TABLE B.70

Finished Housing per 1,000 Population in CMEA and Some Western Countries, 1955–73

Countries	1955	1960	1964	1965	1966	1968	1969	1970	1971	1972	1973
Bulgaria	5.7	6.3	5.8	5.5	5.3	5.1	5.6	n.a.	5.7	5.4	6.3
Czechoslovakia	3.9	5.4	6.2	6.1	5.3	5.9	6.1	7.4	7.5	8.0	8.4
GDR	1.8	4.7	4.5	4.0	3.8	4.4	4.4	4.5	4.5	5.0	5.6
Hungary	4.4	5.8	5.3	5.4	5.5	6.5	6.0	7.7	7.3	8.7	8.2
Poland	3.5	4.8	5.3	5.6	5.6	5.9	6.1	5.9	5.8	6.2	6.8
Rumania	3.2	7.3	6.5	6.4	6.1	5.8	7.2	8.0	7.3	6.8	7.4
USSR	7.3	12.1	9.6	9.7	9.7	9.5	9.4	9.5	9.4	9.0	9.2
Austria	5.9	6.0	6.2	6.3	n.a.	7.0[b]	6.6	6.1	6.0	6.7	5.8
France	5.0	6.9	7.6	8.4	8.4	8.2	8.5	8.9	9.3	10.6	9.6
GFR	10.7[a]	10.3	11.1	10.0	10.2	9.6	8.1	7.0	9.1	11.0	11.5
Italy	4.5	5.9	8.8	7.5	5.6	5.1[b]	5.5	5.5	6.7	4.4	4.8
Spain	3.9	4.2	8.2	9.0	8.5	6.4[b]	8.2	9.3	9.3	9.8	10.0
Sweden	7.9	9.1	11.4	12.5	11.4	13.4	13.7	13.6	13.2	12.8	12.0
UK	6.4	5.9	7.2	7.3	7.3	7.8[b]	6.8	6.5	6.7	6.1	5.4
Yugoslavia	1.7	4.1	6.3	6.3	6.5	6.4	6.4	6.3	6.1	6.4	6.4

[a]Without the Saar
[b]In 1967

Source: Rocznik statystyczny 1966. Warsaw, 1966, p. 659; Rocznik statystyczny 1967. Warsaw, 1967, p. 692; Rocznik statystyczny 1968. Warsaw, 1968, p. 695; Rocznik statystyczny 1969. Warsaw, 1969, p. 690; Rocznik statystyczny 1970. Warsaw, 1970, p. 664; Rocznik statystyczny 1971. Warsaw, 1971, p. 724; Rocznik statystyczny 1972. Warsaw, 1972, p. 694; Rocznik statystyczny 1973. Warsaw, 1973, p. 715; Rocznik statystyczny 1974. Warsaw, 1974, p. 714; Rocznik statystyczny 1975. Warsaw, 1975, p. 626.

TABLE B.71

Share of Individual Countries in CMEA Trade, 1960–72[a]

Countries	1960			1965			1970			1972		
	1	2	3	1	2	3	1	2	3	1	2	3
Bulgaria	4.5	4.3	4.7	5.9	5.9	6.0	6.3	6.5	6.1	6.3	6.4	6.2
Czechoslovakia	14.1	14.6	13.6	13.5	13.4	13.5	12.2	12.2	12.2	11.6	12.0	11.2
GDR	16.6	16.7	16.4	14.8	15.3	14.2	15.4	14.8	16.0	14.7	15.1	14.2
Hungary	6.9	6.6	7.1	7.5	7.5	7.5	7.8	7.5	8.1	7.7	8.1	7.4
Poland	10.6	10.1	11.2	11.5	11.1	11.9	11.7	11.4	11.9	12.4	12.1	12.8
Rumania	5.1	5.5	4.8	5.5	5.5	5.5	6.2	6.0	6.5	6.3	6.4	6.3
USSR	42.2	42.2	42.2	40.8	40.8	40.9	40.1	41.3	38.8	38.1	37.6	38.5

1 - total turnover

2 - export

3 - import

[a]Cuba and Mongolia are omitted, hence from 1965 the shares do not total up to 100.

Source: Statisticheski Yezhegodnik Stran-Chlenov Soveta Ekonomicheskoy Vzaimopomoshchi 1973. Moscow, 1973, p. 354.

TABLE B.72

Growth of Export and Import in CMEA and Some Western Countries, 1965-72
(1960 = 100)

Countries	1965		1967		1970		1972	
	Export	Import	Export	Import	Export	Import	Export	Import
Bulgaria	206	186	255	248	350	289	457	403
Czechoslovakia	139	147	148	148	197	203	255	257
GDR	139	128	157	149	208	221	280	269
Hungary	173	156	195	187	265	259	377	325
Poland	168	157	191	177	268	241	372	356
Rumania	153	166	194	239	258	302	363	404
USSR	147	143	173	152	230	209	276	285
Austria	143	148	162	163	255	251	347	368
Belgium	169	164	185	184	306	286	424	393
France	146	165	166	197	259	301	380	426
CFR	157	174	196	173	300	296	408	399
Greece	162	162	244	169	317	279	429	334
Italy	197	156	239	208	362	316	508	408
Japan	208	182	257	260	476	420	707	523
Norway	164	151	197	188	279	253	369	296
Portugal	176	169	214	194	289	290	392	400
Spain	133	417	189	479	329	654	524	937
Sweden	155	151	177	162	264	242	341	278
UK	134	127	141	140	189	172	238	220
USA	133	142	153	178	209	264	240	367
Yugoslavia	193	156	221	207	297	348	395	391

Source: Rocznik statystyki międzynarodowej 1973. Warsaw, 1973, p. 292.

TABLE B.73

Structure of Export and Import According to Product Groups in CMEA Countries, 1960-72

Countries	Years	Product Groups									
		1		2		3		4		5	
		Export	Import	Export	Import	Export	Import	Export	Import	Export	Import
Bulgaria	1960	12.9	43.9	9.2	24.3	56.4	16.7	17.9	7.6	3.6	7.5
	1965	24.8	43.7	7.6	26.7	49.9	17.3	13.6	5.1	4.1	7.2
	1967	25.5	49.1	7.2	24.0	48.7	13.7	14.8	4.5	3.8	8.7
	1970	29.0	40.6	8.1	29.1	43.4	15.9	14.7	5.7	4.8	8.7
	1972	35.0	46.4	8.3	27.3	40.4	12.2	12.4	6.0	3.9	8.1
Czechoslovakia	1960	45.7	21.7	19.1	27.9	10.4	37.1	20.4	3.4	4.4	9.9
	1965	49.1	30.0	19.7	27.4	9.1	28.4	16.6	5.3	5.5	8.9
	1967	48.6	30.6	17.9	25.6	9.1	29.2	18.3	5.5	6.1	9.1
	1970	50.4	33.4	18.6	23.5	7.3	24.1	16.6	8.5	7.2	10.5
	1972	49.4	33.8	17.2	26.0	7.6	22.4	18.7	7.8	7.1	10.0
GDR	1960	49.0	12.7	15.7	38.5	5.9	39.2	15.1	5.3	14.3	4.3
	1965	49.8	18.0	12.9	39.1	6.0	33.1	19.1	4.0	12.2	5.8
	1970	51.7	34.2	10.1	27.6	7.4	28.1	20.8	4.5	10.6	5.6
	1972	51.3	32.1	10.0	28.1	8.0	25.4	18.8	5.6	11.9	8.8
Hungary	1960	38.6	28.5	12.8	27.7	27.4	29.2	17.8	5.1	3.4	9.5
	1965	33.2	28.8	14.1	27.9	26.9	27.8	21.3	5.3	4.5	10.2
	1967	31.1	32.8	13.5	22.5	27.5	26.6	23.1	6.4	4.8	11.7
	1970	32.6	30.9	14.4	23.6	26.7	24.4	21.3	7.7	5.0	13.4
	1972	33.5	35.3	12.1	24.0	26.4	20.9	22.8	6.7	5.2	13.1
Poland	1960	28.3	27.1	34.0	25.3	23.0	33.9	10.1	5.5	4.6	8.2
	1965	34.8	32.9	24.7	24.7	22.6	27.4	12.4	6.8	5.5	8.2
	1967	36.1	37.0	23.7	25.2	19.3	23.4	15.4	5.6	5.5	8.8
	1970	38.5	36.2	23.9	26.6	15.9	21.4	16.1	6.4	5.6	9.4
	1972	39.1	38.9	22.7	24.2	15.0	20.1	16.0	7.9	7.2	8.9

	Year	1	2	3	4	5		
Rumania	1960	16.7	33.6	36.9	34.3	35.9	18.1	5.8	4.7	8.8
	1965	18.8	39.9	25.1	31.4	35.3	14.3	11.0	9.8	7.7
	1967	19.0	48.8	20.5	24.9	40.7	11.6	11.1	8.7	7.8
	1970	22.8	40.3	22.7	30.4	26.8	15.4	18.1	9.6	8.4
	1972	24.9	46.1	16.5	26.6	28.6	15.4	18.9	11.1	6.8
USSR	1960	20.7	31.0	37.6	20.0	27.3	23.7	2.9	11.5	8.4
	1965	20.0	34.0	40.1	11.3	22.1	30.2	2.4	16.9	10.4
	1967	21.1	34.2	35.9	11.1	24.6	25.7	2.6	15.4	9.4
	1970	21.5	35.5	38.2	11.8	19.5	24.8	2.7	15.8	9.6
	1972	23.6	34.6	39.1	12.1	16.8	25.6	3.1	18.1	9.1

1 — Machines, equipment and transportation equipment
2 — Fuels, minerals, and metals
3 — Agricultural products (including agricultural raw materials)
4 — Industrial consumer goods
5 — Chemicals, fertilizers, rubber, and building materials

Source: Statisticheski Yezhegodnik Stran-Chlenov Soveta Ekonomicheskoy Vzaimopomoshchi 1973. Moscow, 1973, pp. 355–60; Rozwoj gospodarczy krajow RWPG 1950–1968. Warsaw, 1969, p. 114.

TABLE B.74

Structure of Export and Import of CMEA Countries by Geographical Areas, 1950-67

Countries	Partners	1950 Import	1950 Export	1955 Import	1955 Export	1960 Import	1960 Export	1965 Import	1965 Export	1967 Import	1967 Export	1972 Import	1972 Export
Bulgaria	Socialist	85.8	91.8	89.3	89.7	83.9	84.0	74.2	79.4	73.9	77.3	79.8	80.6
	Others	14.2	8.2	10.7	10.3	16.1	16.0	25.8	20.6	26.1	22.7	20.2	19.4
Czechoslovalia	Socialist	56.5	54.8	71.5	68.8	71.3	72.3	73.4	73.1	72.0	71.6	70.7	71.9
	Others	43.5	45.2	28.5	31.2	28.7	27.7	26.6	26.9	28.0	28.4	29.3	28.1
GDR	Socialist	75.9	68.2	70.7	73.5	73.8	75.7	72.8	74.8	73.0	75.2	66.5	75.4
	Others	24.1	31.8	29.3	26.5	26.2	24.3	27.2	25.2	27.0	24.8	33.5	24.6
Hungary	Socialist	56.6	66.0	54.9	66.5	70.9	71.5	67.5	70.1	67.0	68.6	66.4	69.8
	Others	43.4	34.0	45.1	33.5	29.1	28.5	32.5	29.9	33.0	31.4	33.6	30.2
Poland	Socialist	61.1	56.9	64.9	62.9	63.5	62.6	66.1	63.2	65.7	63.8	61.2	63.6
	Others	38.9	43.1	35.1	37.1	36.5	37.4	33.9	36.8	34.3	36.2	38.8	36.4
Rumania	Socialist	78.1	89.2	81.8	81.7	73.1	73.0	61.2	68.6	48.8	56.5	51.6	55.8
	Others	21.9	10.8	18.2	18.3	26.9	27.0	38.8	31.4	51.2	43.5	48.4	44.2
USSR	Socialist	78.0	83.6	79.0	79.6	70.7	75.7	69.6	68.0	69.6	66.1	64.0	65.1
	Others	22.0	16.4	21.0	20.4	29.3	24.3	30.4	32.0	30.4	33.9	36.4	34.9

Source: Rozwoj gospodarczy krajow RWPG 1950-1968. Warsaw, 1969, p. 115-18; Statisticheski Yezhegodnik Stran-Chlenov Soveta Ekonomicheskoy Vzaimopomoshchi 1973. Moscow, 1973, p. 353.

TABLE B.75

Geographical Distribution of Bulgarian Exports and Imports by CMEA and Some Western Countries, 1939-70

(total = 100)

Countries	1939 Exports	1939 Imports	1956 Exports	1956 Imports	1960 Exports	1960 Imports	1965 Exports	1965 Imports	1970 Exports	1970 Imports
Czechoslovakia	3.6	4.3	10.6	14.4	9.6	9.7	7.8	6.5	4.4	5.2
GDR	67.8[a]	65.5[a]	13.1	13.2	9.8	11.1	9.2	7.2	8.7	8.6
Hungary	1.9	0.9	4.3	3.5	2.0	1.8	1.2	1.6	2.5	1.4
Poland	3.8	5.6	3.6	2.8	3.6	3.4	3.4	3.9	3.9	3.5
Rumania	0.2	3.9	3.3	3.5	1.5	1.4	1.2	0.8	2.1	1.5
USSR	0	0	48.7	41.3	53.8	52.5	52.2	49.9	53.8	52.1
Austria	n.a.	n.a.	1.8	2.7	2.0	1.5	1.1	2.6	0.8	2.0
Cuba	0	0	0	0	0.1	0	1.4	1.9	1.2	1.7
France	0.9	1.13	1.2	1.7	0.8	1.2	0.6	2.1	1.9	2.4
GFR	[a]	[a]	2.6	3.1	3.3	5.9	3.5	5.8	2.6	2.6
Greece	1.1	0.4	0.7	0.5	0.6	0.2	1.4	0.8	0.7	0.9
India	0	0	0.1	0.06	0.1	0.05	0.7	0.6	1.0	0.5
Italy	6.0	6.9	0.5	0.7	1.6	1.0	3.3	2.7	2.8	3.1
Japan	0	0	0	0	0.4	0.2	0.6	1.2	0.4	1.1
Pakistan	0	0	0.1	0	0	0	0.02	0	0.5	0.6
UAR	1.1	0.2	1.1	1.8	0.6	1.1	0.3	0.4	1.2	0.5
UK	3.2	2.8	1.0	2.9	1.1	1.7	1.5	1.5	1.2	2.4
USA	3.5	2.3	0.06	0	0.3	0.04	0.2	0.06	0.1	0.3
Yugoslavia	0.3	0.4	0.7	0.8	1.4	1.3	1.9	1.7	1.7	1.1

[a] Includes GDR & GFR.

Source: Statistical Yearbook of Bulgaria 1971. Sofia, 1971, p. 196.

TABLE B.76

Average Annual Growth Rates of National Income (1),
Industrial Output (2), Agricultural Output (3),
Investment Expenditures (4), and Labor Productivity (5) in CMEA
Countries, 1956-75

Years	Categories	Bulgaria	Czechoslovakia	GDR	Hungary	Poland	Rumania	USSR
1966-70P	1	8.5	5.6	5.1-5.7	4.0	6.0	8.1	6.6-7
	2	10.5-11.2	6.5	6.5-7	5.7-6.3	7.5	11.5	8.9
	3	4.6-5.4	2.7	2.5-2.8	2.5-2.8	2.3-2.5	4.7-5.7	4.6-5.4
	4	14.9	3.7	8.2-8.7	5.3	6.6	10.7	7.4
	5	n.a.	4.1	7.0	4.5	4.6	7.6	5.9
1966-70F	1	8.7	7.0	5.2	6.8	6.0	7.7	7.6
	2	10.9	6.7	6.5	6.2	8.4	11.9	8.5
	3	3.5	4.9	1.5	2.8	1.8	1.9	3.9
	4	12.5	7.3	9.8	10.5	8.3	10.8	7.3
	5	6.9	5.3	6.5	3.5	4.9	7.3	5.8
1971-75P	1	7.8-8.5	4.7-5.4	4.9	5.4-5.7	7.0	11-12	6.7
	2	9.8	6.0-6.3	6.0	5.7-6.0	8.5	11-12	8.1
	3	3.2-4.0	2.7	2.4	2.6-2.8	3.5	6.3-8.3	3.7-4.1
	4	5.9-7.0	6.2-6.5	5.2	5.9-6.2	7.8	8.7	7.3
	5	8.1	5.4-5.7	6.2	4.1-4.7	5.4	6.8-7.6	6.7
1971-75F[a]	1	7.8	5.1	5.6	6.3	9.8	11.8	5.9
	2	8.7	6.6	6.5	7.0	10.7	13.5	7.3
	3	3.5	3.5	2.8	4.8	5.1	7.0	3.9
	4	6.4	7.9	4.2	5.3	17.3	13.6	6.4
	5	6.5	5.8	5.4	6.5	7.9	7.0	6.0

P - Plan
F - Fulfillment
[a] fulfilment expected as of early 1975

Source: Golebiowski M. and B. Zielinska: GP, No. 4, 1968, pp. 1-6; Golebiowski M. and B. Zielinska: GP, No. 5, 1972, pp. 261-65; Golebiowski M. and B. Zielinska: GP, No. 4, 1975, pp. 262-71.

TABLE B.77

Annual Plan and Fulfillment of Growth Rates of National Income in CMEA Countries, 1966-74

Years	Bulgaria		Czechoslovakia		GDR		Hungary		Poland		Rumania		USSR	
	P	F	P	F	P	F	P	F	P	F	P	F	P	F
1966	9.5	11.0	3.8	7.0	5.1	4.5	4.0	6-6.5	3.7	6.0	7.0	7.9	6.4	7.5
1967	n.a.	9.0	n.a.	8.0	5.1	5.1	3.5-4.0	7.0	3.4	5.6	8.9	7.5	6.6	6.7
1968	10.5	6.5	6.0	7.0	5.4	5.3	5-6	5.0	4.8	8.0	8.6	7.0	6.8	7.2
1969	8.9	7.7	7.0	6.5	6.0	5.0	n.a.	5-6	5.0	3.5	9.0	7.3	6.5	6.0
1970	9-10	7.0	6.3	5.0	6.3	5.2	5-6	5.0	6.2	6.0	12.0	7.0	6.0	8.5
1971	9.0	9.4	5.2	5.3	4.9	4.5	7.0	7-8	5.8	7.3	13.9	12.5	6.1	6.0
1972	n.a.	7.0	5.0	5.9	4.6	5.4	5-6	5.0	6.1	9+	11-12	10.0	6.2	3.8
1973	9.0	9.0	5.1	5.2	5.7	5.5	4-5	6.5-7.0	7.9	11.0	14.0	10.8	6.0	6.8
1974	10.0	7.5	5.2	5.2	5.4	6.3	5.0	7.0	9.5	10.0	14.6	12.0	6.5	5.0

P - Plan
F - Fulfillment

Source: Zielinska B.: GP, No. 4, 1967, p. 3; Golebiowski M. and B. Zielinska: GP, No. 4, 1968, p. 4; Golebiowski M. and B. Zielinska: GP, No. 4, 1969, p. 5; Golebiowski M. and B. Zielinska: GP, No. 4, 1970, p. 10; Golebiowski M. and B. Zielinska: GP, No. 4, 1971, p. 232; Golebiowski M. and B. Zielinska: GP, No. 5, 1972, p. 261; Golebiowski M. and B. Zielinska: GP, No. 4, 1973, p. 238; Golebiowski M. and B. Zielinska: GP, No. 4, 1974, p. 225; Golebiowski M. and B. Zielinska: GP, No. 4, 1975, p. 263.

Note: In all cases initial plan fulfillment report figures are used, which may differ from the data in the preceding tables, derived from statistical yearbooks.

TABLE B.78

Annual Plan and Fulfillment of Growth Rates of Industrial Output in CMEA Countries, 1966–74

Years	Bulgaria		Czechoslovakia		GDR		Hungary		Poland		Rumania		USSR	
	P	F	P	F	P	F	P	F	P	F	P	F	P	F
1966	10.6	12.2	5.5	7.4	6.1	6.5	4.6	6-6.5	6.5	7.4	10.5	11.7	6.7	8.6
1967	11.4	13.4	n.a.	7.1	6.0	6.8	4-6	9.0	6.2	7.7	11.3	13.5	7.3	10.0
1968	10.6	11.8	7.1	5.2	6.4	6.1	6-7	5.0	7.1	9.4	10.5	11.6	8.1	8.1
1969	11.6	11.0	5.7	5.2	7.0	8.0	6.0	3.0	8.2	8.9	10.8	10.7	7.3	7.0
1970	8.2	9.3	6.7	7.7	8.5	6.4	6.0	7.0	7.3	8.3	11.3	12.0	6.3	8.3
1971	10.0	9.5	5.9	6.9	5.6	5.4	5-6	5.5	6.8	8.1	9.0	11.5	6.9	7.8
1972	7.7	8.3	5.8	6.4	5.5	6.3	5-6	5.6	7.0	10.8	11-12	11.7	6.9	6.5
1973	9.9	10.6	5.8	6.8	6.5	6.8	5.5-6	7.2	9.8	12.0	16.2	14.7	5.8	7.4
1974	11.0	8.5	5.8	6.2	6.7	7.4	5.5-6	8.2	11.1	11.5	16.7	15.0	6.8	8.0

P – Plan
F – Fulfillment

Source: Zielinska B.: GP, No. 4, 1967, p. 1; Golebiowski M. and B. Zielinska: GP, No. 4, 1968, p. 1; Golebiowski M. and B. Zielinska: GP, No. 4, 1969, p. 1; Golebiowski M. and B. Zielinska: GP, No. 4, 1970, p. 6; Golebiowski M. and B. Zielinska: GP, No. 4, 1971, p. 232; Golebiowski M. and B. Zielinska: GP, No. 5, 1972, p. 261; Golebiowski M. and B. Zielinska: GP, No. 4, 1973, p. 238; Golebiowski M. and B. Zielinska: GP, No. 4, 1974, p. 226; Golebiowski M. and B. Zielinska: GP, No. 4, 1975, p. 268.

Note: In all cases initial plan fulfillment report figures are used, which may differ from the data in the preceding tables, derived from statistical yearbooks.

TABLE B.79

Annual Plan and Fulfillment of Growth Rates of Agricultural Output in CMEA Countries, 1966-74

Years	Bulgaria		Czechoslovakia		GDR		Hungary		Poland		Rumania		USSR	
	P	F	P	F	P	F	P	F	P	F	P	F	P	F
1966	10.7	15.0	7.9	10.0	3.3	3.0	5.0	5-6	-3.0	5.5	5.0	11.2	8-10	10.0
1967	5.0	0	3.0	3.6	2.0	2.0	2-3	0	-2.3	2.3	5.8	6.9	4.0	1.0
1968	12.6	-8.7	2.0	3.6	2.4	3.0	2-4	2	-1.5	4.0	n.a.	-3.6	7.4	3.5
1969	16	2.0	n.a.	0.9	5.0	-7.0	2-3	5.6	-1.7	-4.7	4.0	4.8	6.1	-3.0
1970	10.7	4.0	3.1	1.3	2.6	3.5	1.0	-4.5	2.8	1.9	16.0	5.2	8.5	8.7
1971	7.1	3.1	4.4	2.8	3.2	-0.8	7-8	9-10	2.4	3.7	n.a.	18.2	5.5	0
1972	6.0	4.8	3.1	3.6	4.5	10.7	2-3	4.0	4.6	8.1	21.0	9.0	n.a.	-4.6
1973	7.8	3.0	4.3	3.4	4.9	1.0	2.0	5.0	2.1	7.8	19.0	0.2	12.6	14.0
1974	5.7	0	3.8	3.0	n.a.	0	2-2.5	3.7	4.3	2.0	21.5	0	6.4	-3.7

P - Plan
F - Fulfillment

Source: Zielinska B.: GP, No. 4, 1967, p. 2; Golebiowski M. and B. Zielinska: GP, No. 4, 1968, p. 3; Golebiowski M. and B. Zielinska: GP, No. 4, 1969, p. 3; Golebiowski M. and B. Zielinska: GP, No. 4, 1970, p. 8; Golebiowski M. and B. Zielinska: GP, No. 4, 1971, p. 234; Golebiowski M. and B. Zielinska: GP, No. 5, 1972, p. 263; Golebiowski M. and B. Zielinska: GP, No. 4, 1973, p. 239; Golebiowski M. and B. Zielinska: GP, No. 4, 1974, p. 229; Golebiowski M. and B. Zielinska: GP, No. 4, 1975, p. 271.

Note: In all cases initial plan fulfillment report figures are used, which may differ from the data in the preceding tables, derived from statistical yearbooks.

TABLE B.80

Annual Plan and Fulfillment of Growth Rates of Investment Expenditures in CMEA Countries, 1966-74

Years	Bulgaria		Czechoslovakia		GDR		Hungary		Poland		Rumania		USSR	
	P	F	P	F	P	F	P	F	P	F	P	F	P	F
1966	16.2	24.0	8.0	8.9	10.0	7.0	7.0	6.7	5.8	7.6	14.0	10.2	6.4	6.0
1967	5.6	15.6	n.a.	2.6	9.0	9.0	4.5	15.0	8.1	11.3	16.0	17.0	7.9	8.0
1968	10.0	11.0	6.0	8.6	10.0	10.0	n.a.	4.0	5.8	8.5	11.0	11.1	5.7	8.0
1969	0	0.8	6.5	14.0	10.0	13.0	7.4	9.0	8.8	8.2	7.0	5.1	8.4	4.0
1970	2.5	10.6	-4.0	6.1	11.4	7.0	6.7	12.0	2.5	5.2	8.0	9.3	7.8	9.0
1971	4.9	1.7	4.3	4.8	-1.5	0	8.4	20.0	7.2	7.3	n.a.	10.8	7.2	7.0
1972	6.4	10.0	7.4	8.3	2.0	3.0	0	-2.0	9.6	22.6	5.0	10.5	5.7	7.0
1973	8.0	6.9	6.8	8.4	9.0	8.5	3.0		12.9	23.0	7.6	9.1	3.5	4.4
1974	13.0	9.0	9.6	8.7	5.3	4.0	11.0	11.0	12.4	22.6	18.1	17.0	6.5	7.0

P - plan
F - fulfilment

Source: Zielinska B.: GP, No. 4, 1967, p. 4; Golebiowski M. and B. Zielinska: GP, No. 4, 1968, p. 6; Golebiowski M. and B. Zielinska: GP, No. 4, 1969, p. 6; Golebiowski M. and B. Zielinska: GP, No. 4, 1970, p. 11; Golebiowski M. and B. Zielinska: GP, No. 4, 1971, p. 236; Golebiowski M. and B. Zielinska: GP, No. 5, 1972, p. 265; Golebiowski M. and B. Zielinska: GP, No. 4, 1973, p. 242; Golebiowski M. and B. Zielinska: GP, No. 4, 1974, p. 231; Golebiowski M. and B. Zielinska: GP, No. 4, 1975, p. 267.

Note: In all cases initial plan fulfillment report figures are used, which may differ from the data in the preceding tables, derived from statistical yearbooks.

TABLE B.81

Annual Plan and Fulfillment of Growth Rates of Labor Productivity in CMEA Countries, 1966-74

Years	Bulgaria		Czechoslovakia		GDR		Hungary		Poland		Rumania		USSR	
	P	F	P	F	P	F	P	F	P	F	P	F	P	F
1966	n.a.	3.6	n.a.	4.2	n.a.	5.3	n.a.	4.7	n.a.	3.6	n.a.	8.3	n.a.	5.6
1967	n.a.	7.6	n.a.	6.4	n.a.	5.1	n.a.	5.3	n.a.	3.6	n.a.	8.9	n.a.	6.8
1968	n.a.	9.0	n.a.	3.8	n.a.	6.2	n.a.	1.1	n.a.	5.8	n.a.	7.0	n.a.	5.0
1969	9.8	6.3	n.a.	4.6	9.0	8.0	n.a.	-1.0	4.9	5.2	6.8	5.5	n.a.	4.8
1970	9.0	7.2	6.8	7.9	9.4	5.0	n.a.	6.8	5.0	6.5	8.9	9.0	5.2	7.0
1971	7.7	6.0	5.7	6.5	5.4	4.5	5.0	5.3	6.3	4.9	n.a.	5.9	n.a.	6.3
1972	n.a.	5.9	4.5	5.8	5.0	5.0	n.a.	6.5	5.7	6.0	8.0	7.0	6.1	5.2
1973	7.0	8.0	5.4	5.6	5.7	5.8	n.a.	5.8	6.3	8.5	9.1	9.1	5.0	6.0
1974	9.0	6.0	4.8	5.4	6.0	6.3	5.5	7.2	8.1	9.6	9.6	7.0	6.0	6.5

P - plan
F - fulfilment

Source: Golebiowski M. and B. Zielinska: GP, No. 4, 1970, p. 7; Golebiowski M. and B. Zielinska: GP, No. 4, 1971, p. 233; Golebiowski M. and B. Zielinska: GP, No. 5, 1972, p. 262; Golebiowski M. and B. Zielinska: GP, No. 4, 1973, p. 239; Golebiowski M. and B. Zielinska: GP, No. 4, 1974, p. 228; Golebiowski M. and B. Zielinska: GP, No. 4, 1975, p. 271.

Note: In all cases initial plan fulfillment report figures are used, which may differ from the data in the preceding tables, derived from statistical yearbooks.

SELECTED BIBLIOGRAPHY

Alton, T. P., ed. Statistics on East European Economic Structure and Growth. OP-48. New York, 1975.*

Angelov, A. S., and M. N. Kostov. Finansov plan i finanse-pravna norma (The Financial Plan and Fiscal-Legal Regulation). Sofia, 1972.

Arrow, K. J. The Limits to Organization. New York, 1974.

———, and F. H. Hahn. General Competitive Analysis. San Francisco, 1971.

Atanasov, B. Pazarniat mekhanizum pri novata sistema (Market Mechanism in the New System). Sofia, 1969.

Bergson, A. Real National Income of Soviet Russia since 1928. Cambridge, Mass., 1961.

———. The Economics of Soviet Planning. New Haven, Conn., 1964.

———. Essays in Normative Economics. Cambridge, Mass., 1966.

———. Soviet Post-War Economic Development. Stockholm, 1974.

Berliner, J. S. Factory and Manager in the USSR. Cambridge, Mass., 1957.

Bornstein, M., ed. Plan and Market. New Haven, Conn., 1973.

Brown, A., and E. Neuberger, eds. International Trade and Central Planning. Berkeley, Calif., 1968.

Brown, J. F. Bulgaria under Communist Rule. New York, 1970.

Csikos-Nagy, B. Socialist Economic Planning. Budapest, 1973.

---

*References to other occasional papers edited by T. P. Alton can be found in Chapters 1, 2, and 3 and Appendix B.

Dellin, L. A. D., ed. Bulgaria. New York, 1957.

Denison, E. Why Growth Rates Differ. Washington, D.C., 1967.

Dobb, M. H. Papers on Capitalism, Development and Planning. New York, 1967.

―――. Welfare Economics and the Economics of Socialism. Cambridge, Mass., 1969.

Domar, E. Essays in the Theory of Economic Growth. New York, 1957.

Donnithorne, A. China's Economic System. New York, 1967.

Eckstein, A., ed. Comparison of Economic Systems. Berkeley, Calif., 1971.

―――, W. Galenson, and T. C. Liu, eds. Economic Trends in Communist China. Chicago, 1968.

Ellman, M. Soviet Planning Today. Cambridge, Mass., 1971.

Erlich, A. Soviet Industrialization Debate 1924-1928. Cambridge, Mass., 1960.

Feiwel, G. R. The Economics of a Socialist Enterprise. New York, 1965.

―――. The Soviet Quest for Economic Efficiency. New York, 1967, 1972.

―――. New Economic Patterns in Czechoslovakia. New York, 1968.

―――. Poland's Industrialization Policy. Vol. 1 of Industrialization and Planning under Polish Socialism. New York, 1971.

―――. Problems in Polish Economic Planning. Vol. II of Industrialization and Planning under Polish Socialism. New York, 1971.

―――. The Intellectual Capital of Michal Kalecki. Knoxville, 1975.

―――, ed. New Currents in Soviet-Type Economies. Scranton, Pa., 1968.

Friss, I., ed. *Reform of the Economic Mechanism in Hungary.* Budapest, 1969.

Gado, O., ed. *Reform of the Economic Mechanism in Hungary.* Budapest, 1972.

Gerschenkren, A. *Economic Backwardness in Historical Perspective.* Cambridge, Mass., 1962.

Gishev, V., et al., eds. *Petiletkite v NR Bulgaria* (*FYPs in the National Republic of Bulgaria*). Sofia, 1971.

Godina purvu na Shesta Petiletka (*The First Year of the Sixth FYP*). Sofia, 1970.

Goldman, M. *Controlling Pollution.* Englewood Cliffs, N.J., 1967.

Goldmann, J., and K. Kouba. *Economic Growth in Czechoslovakia.* White Plains, N.Y., 1969.

Gomulka, S. *Inventive Activity, Diffusion and the Stages of Economic Growth.* Arhus, 1971.

Grossman, G., ed. *Essays in Socialism and Planning in Honor of Carl Landauer.* Englewood Cliffs, N.J., 1970.

Handel zagraniczny a wzrost krajow RWPG (*Foreign Trade and Growth of CMEA Countries*). Warsaw, 1969.

Hardt, J. *Tariff, Legal and Credit Constraints on East-West Trade Commercial Relations.* Ottawa, 1975.

Hirshman, A. *The Strategy of Economic Development.* New Haven, Conn., 1958.

Holzman, F. D. *Foreign Trade under Central Planning.* Cambridge, Mass., 1974.

Horvat, B. *Business Cycles in Yugoslavia.* White Plains, N.Y., 1971.

Ikonomicheskata politika na Bulgarskata komunisticheska partiia (*Economic Policy of the Bulgarian Communist Party*). Sofia, 1974.

Kalecki, M. Z zagadnien gospodarczo-spolecznych Polski Ludowej (Socioeconomic Questions of People's Poland). Warsaw, 1964.

_____. Introduction to the Theory of Growth in a Socialist Economy. Oxford, 1969.

Kanterovich, L. V. The Best Use of Economic Resources. Oxford, 1965.

Kaser, M. Comecon. London, 1967.

Khadzhinikolov, V., et al. Nova i nainova stopanski istoriia (New and Newest Economic History). Sofia, 1966.

Kiriakov, K., ed. Sotsialno-ikonomicheski problemi na izgrazhdaneto na razvito sotsialistichesko obshtestvo v Bulgaria (Socioeconomic Problems of Building a Developed Socialist Society in Bulgaria). Sofia, 1974.

Koopmans, T. C. Three Essays on the State of Economic Science. New York, 1957.

Kornai, J. Anti-Equilibrium. Amsterdam, 1971.

_____. Rush Versus Harmonic Growth. Amsterdam, 1972.

Kotsev, G., P. Skachkova, and I. Radylova. Vliyane na podgotviyanata nova sistema na tseni na edro vrkv razkhadite za proizvedstvoto na promishlenosta (Influence of the Prepared New Wholesale Price System on the Production Costs of Industry). Sofia, 1969.

Kudrova, E. Statistika natsionalnogo dokhoda evropeyskhikh sotsialisticheskikh stran (National Income Statistics of European Socialist Countries). Moscow, 1969.

Kuznets, S. Modern Economic Growth. New Haven, Conn., 1966.

_____. Population, Capital and Growth. New York, 1973.

Lange, O. The Working Principles of the Soviet Economy. New York, 1944.

_____. Political Economy. Vol. I. New York, 1963.

Lazarov, K. Ikonomichesko razvitie na Narodna Republika Bulgaria (Economic Development of the National Republic of Bulgaria). Sofia, 1961.

Lippincott, B. E., ed. On the Economic Theory of Socialism. Minneapolis, 1938.

Maddison, A. Economic Growth in Japan and USSR. New York, 1969.

Marer, P. Postwar Pricing and Price Patterns in Socialist Foreign Trade (1946-1971). Bloomington, Ind., 1972.

Marinov, K. Niakoi ikonomogeografski problemi v razvitieto na bulgarskoto narodno stopanstvo vuv vruzka z mezhdunarotnoto sotsialistichesko razdelenie na truda (Some Problems of Economic Geography in the Development of Bulgarian National Economy in Connection with International Division of Labor). Sofia, 1961.

Mateev, E. Balans na narodnoto stopanstvo (The Balance of the National Economy). Sofia, 1972.

Meade, J. E. The Theory of Indicative Planning. New York, 1970.

Mishan, E. J. The Costs of Economic Growth. New York, 1967.

Montias, J. M. Economic Development in Communist Rumania. Cambridge, Mass., 1967.

Moss, M., ed. The Measurement of Economic and Social Performance. New York, 1973.

Musgrave, R. A. Financial Systems. New Haven, Conn., 1969.

Natan, Z. Istoriia ekonomicheskogo razvitiia Bulgarii (History of Economic Development of Bulgaria). Moscow, 1961.

National Bureau of Economic Research. Economic Growth. New York, 1972.

Normativni aktove i metodicheski ukazania po prilaganeto na ikonomicheskia mekhanizum (Regulations and Methodological Directives in Application to the Economic Mechanism). Sofia, 1974.

Nove, A. The Soviet Economy. New York, 1968.

SELECTED BIBLIOGRAPHY

Ohkawa, K., and H. Rosovsky. Japanese Economic Growth. Stanford, 1973.

Peneva, B. Proportsiata sredstva za proizvodstvo-predmeti za potreblenie v NR Bulgaria (Proportion of Means of Production-Objects of Consumption in the National Republic of Bulgaria). Sofia, 1975.

Perkins, D. H. Market Control and Planning in Communist China. Cambridge, Mass., 1966.

Popov, N., ed. Ikonomika na Bulgaria v shest toma (The Bulgarian Economy in Six Volumes). Sofia, 1969.

Problemy teorii gospodarki socjalistycznej (Problems of the Theory of Socialist Economy). Warsaw, 1970.

Prybyla, J. S. The Political Economy of Communist China. Scranton, Pa., 1970.

Pryor, F. T. The Communist Foreign Trade System. Cambridge, Mass., 1963.

──────. Property and Industrial Organization in Communist and Capitalist Nations. Bloomington, Ind., 1973.

Przemysl w Polsce i wybranych krajach 1950-1968 (Industry in Poland and Selected Countries). Warsaw, 1970.

Razshireno sotsialistichesko vuzproizvodstvo v NRB (Expanded Socialist Reproduction in the National Republic of Bulgaria). Sofia, 1973.

Robinson, J. The Accumulation of Capital. New York, 1956.

──────. Economic Management China 1972. London, 1973.

Rozwoj gospodarczy krajow RWPG 1950-1968 (Economic Development of CMEA Countries 1950-1968). Warsaw, 1969.

Rusinov, S. Economic Development of Bulgaria after the Second World War. Sofia, 1970.

Salter, W. E. G. Productivity and Technical Change. Cambridge, Mass., 1969.

Schumpeter, J. A. Capitalism, Socialism and Democracy. New York, 1947.

Shamliev, B. Spravochnik na agitatora; tsifri i fakti (Manual for the Party Activist; Figures and Facts). Sofia, 1961.

Spulber, N. The Economics of Communist Eastern Europe. New York, 1957.

Tinbergen, J. Economic Policy. Chicago, 1967.

_____, et al. Optimum Social Welfare and Productivity. New York, 1972.

U.S., Congress. Joint Economic Committee. Economic Developments in Countries of Eastern Europe. Washington, D.C., 1970.

_____. Reorientation and Commercial Relations of the Economies of Eastern Europe. Washington, D.C., 1974.

Vneshnay torgovlya SSSR za 1971 god (Foreign Trade of the USSR in 1971). Moscow, 1972.

Vodenicharov, A., et al., eds. Spravochnik na aktivista (Manual for the Activist). Sofia, 1971.

Ward, B. The Socialist Economy. New York, 1967.

Weintraub, A., et al. The Economic Growth Controversy. White Plains, N.Y., 1973.

Wilczynski, J. The Economics and Politics of East-West Trade. New York, 1969.

Wiles, P. J. D. Communist International Economics. New York, 1969.

Zauberman, A. Industrial Progress in Poland, Czechoslovakia and East Germany. London, 1964.

_____. Aspects of Planometrics. New Haven, Conn., 1967.

Zhivkov, T. The New System of Economic Management. Sofia, 1966.

_____. Fundamental Trends in the Further Development of the System of Public Administration in Our Society. Sofia, 1968.

_____. Problems of the Construction of an Advanced Socialist Society in Bulgaria. Sofia, 1969.

_____. Bulgaria along the Road to an Advanced Socialist Society. Sofia, 1971.

ABOUT THE AUTHOR

GEORGE R. FEIWEL, a 1969 Guggenheim Fellow, received his Ph.D. from McGill University. He has lectured at leading European universities and on several occasions was Senior Visitor to the Faculty of Economics, University of Cambridge and Associate of Harvard University, Russian Research Center. In 1973 he was Visiting Professor at the Institute for International Economic Studies, University of Stockholm. In 1975 he was Honorary Research Associate, Department of Economics, Harvard University. He is now Distinguished Professor of Economics, University of Tennessee and the Chancellor's Research Scholar. In addition to a number of contributions to scholarly journals, his works include Economics of a Socialist Enterprise (1965), Soviet Quest for Economic Efficiency (1967, 1972), New Economic Patterns in Czechoslovakia (1968), Industrialization and Planning under Polish Socialism, 2 vols. (1971), selected by Choice as one of the outstanding academic books of the year, and the widely acclaimed The Intellectual Capital of Michal Kalecki (1975).

RELATED TITLES
Published by
Praeger Special Studies

CHANGE AND ADAPTATION IN SOVIET AND EAST
EUROPEAN POLITICS
      edited by Jane P. Shapiro
      Peter J. Potichnyj

EAST EUROPEAN COOPERATION: The Role of
Money and Finance
      Jozef M. van Brabant

ECONOMIC DEVELOPMENT IN THE SOVIET UNION
AND EASTERN EUROPE, Vol. 1: Reforms, Technology, and Income Distribution
      edited by
      Zbigniew M. Fallenbuchl

ECONOMIC DEVELOPMENT IN THE SOVIET UNION
AND EASTERN EUROPE, Vol. 2: Sectoral Analysis
      edited by
      Zbigniew M. Fallenbuchl

MODERNIZATION IN ROMANIA SINCE WORLD WAR II

      Trond Gilberg

PERSONAL AND SOCIAL CONSUMPTION IN EASTERN
EUROPE: Poland, Czechoslovakia, Hungary, and
East Germany
      Bogdan Mieczkowski

TECHNOLOGY TRANSFER TO EAST EUROPE: U.S.
Corporate Experience
      Eric W. Hayden

THE POLITICAL ECONOMY OF EAST-WEST TRADE
      Connie M. Friesen

YUGOSLAV ECONOMIC DEVELOPMENT AND
POLITICAL CHANGE: The Relationship Between
Economic Managers and Policy-Making Elites
      Richard P. Farkas